Archimedes

New Studies in the History and Philosophy of Science and Technology

Volume 61

Archimedes has three fundamental goals: to further the integration of the histories of science and technology with one another; to investigate the technical, social and practical histories of specific developments in science and technology; and finally, where possible and desirable, to bring the histories of science and technology into closer contact with the philosophy of science.

The series is interested in receiving book proposals that treat the history of any of the sciences, ranging from biology through physics, all aspects of the history of technology, broadly construed, as well as historically-engaged philosophy of science or technology. Taken as a whole, Archimedes will be of interest to historians, philosophers, and scientists, as well as to those in business and industry who seek to understand how science and industry have come to be so strongly linked.

Submission / Instructions for Authors and Editors: The series editors aim to make a first decision within one month of submission. In case of a positive first decision the work will be provisionally contracted: the final decision about publication will depend upon the result of the anonymous peer-review of the complete manuscript. The series editors aim to have the work peer-reviewed within 3 months after submission of the complete manuscript.

The series editors discourage the submission of manuscripts that contain reprints of previously published material and of manuscripts that are below 150 printed pages (75,000 words). For inquiries and submission of proposals prospective authors can contact one of the editors:

Please find on the top right side of our webpage a link to our *Book Proposal Form.*

More information about this series at https://link.springer.com/bookseries/5644

Susanna Berger • Daniel Garber
Editors

Teaching Philosophy in Early Modern Europe

Text and Image

Springer

Editors
Susanna Berger
USC Dornsife, Department of Art History
and School of Philosophy
University of Southern California
Los Angeles, CA, USA

Daniel Garber
Department Philosophy
Princeton University
Princeton, NJ, USA

ISSN 1385-0180 ISSN 2215-0064 (electronic)
Archimedes
ISBN 978-3-030-84620-6 ISBN 978-3-030-84621-3 (eBook)
https://doi.org/10.1007/978-3-030-84621-3

This Springer imprint is published by the registered company Springer Nature Switzerland AG
The registered company address is: Gewerbestrasse 11, 6330 Cham, Switzerland

Acknowledgments

This book grew out of a conference held at Princeton University in 2015. We thank the speakers and our co-organizers Anthony Grafton, Jennifer Rampling, and Roger Ariew for their help in putting together this event. We also thank the Princeton Society of Fellows and the David A. Gardner '69 Magic Project Grant for their generous support. Finally, we are grateful to Jed Buchwald for his enthusiasm for this edited volume.

Contents

About the Authors

Roger Ariew, Distinguished University Professor, Department of Philosophy, University of South Florida, is author of *Descartes and the First Cartesians* (Oxford University Press, 2014), *Descartes and the Last Scholastics* (Cornell University Press, 1999), second revised and expanded edition as *Descartes among the Scholastics* (Brill, 2011), coauthor of *Historical Dictionary of Descartes and Cartesian Philosophy* (2nd ed., Rowman & Littlefield, 2015), and editor and translator of such works as *Descartes, Philosophical Essays* (Hackett, 2000) and *Pascal, Pensées* (Hackett, 2005). He is currently working on *Descartes' Correspondence: A Historical-Critical Edition and English Translation*, 8 vols., with Theo Verbeek, Erik-Jan Bos, et al.

Susanna Berger is Associate Professor of Art History and Philosophy at the University of Southern California. In her research and teaching, she explores diverse facets of visual art and intellectual history in early modern Europe (ca. 1500–1800), from overlooked printed and drawn images of a philosophical nature to well-known works in the history of European painting and architecture by Caravaggio, Borromini, and Rembrandt, among others. Her first book, *The Art of Philosophy: Visual Thinking in Europe from the Late Renaissance to the Early Enlightenment* (Princeton University Press, 2017), is a transnational study of the relations between images and philosophical knowledge in early modern France, Italy, England, Germany, and the Netherlands.

Ann Blair teaches Book History and Early Modern European History in the Department of History at Harvard University. Her research focuses on methods of intellectual work among scholars and authors in Europe ca. 1500–1700, including methods of reading and note-taking, of working with amanuenses, and the many purposes of paratexts. Her publications include *Too Much To Know: Managing Scholarly Information Before the Modern Age* (Yale University Press, 2010) and *Information: A Historical Companion* (Princeton University Press, 2021), co-edited with Paul Duguid, Anja-Silvia Goeing, and Anthony Grafton.

Domenico Collacciani is a postdoctoral fellow at Université Paul Valery Montpellier 3. He holds a PhD in philosophy from the universities of Rome and Caen. His research interests are modern philosophy and the history of science, in particular Spinoza and the reception of Descartes. He is currently working on a project on Gassendi's philosophy in Europe. His recent publications include: *Lectures du Traité Théologico-Politique* (L'Harmattan, 2012), co-edited with Francesco Toto and Blanche Gramusset-Piquois; "Devenir Cartesien? La méthode de l'ontologie de Gerhard de Neufville à Johann *Clauberg*," in *Le Études philosophiques*, 3 (2020); "The Reception of L'Homme among the Leuven Physicians," in Delphine Antoine-Mahut and Stephen Gaukroger, eds., *Descartes' Treatise on Man and its Reception* (Springer, 2016).

Daniel Garber is the A. Watson Armour III University Professor of Philosophy at Princeton University. Garber's principal interests are the relations between philosophy, science, religion, and society in the period of the Scientific Revolution. In addition to numerous articles, Garber is the author of *Descartes' Metaphysical Physics* (University of Chicago Press, 1992), *Descartes Embodied* (Cambridge University Press, 2001), and *Leibniz: Body, Substance, Monad* (Oxford University Press, 2009) and is co-editor with Michael Ayers of the *Cambridge History of Seventeenth-Century Philosophy* (Cambridge University Press, 1998). He is currently working on a project to understand the emergence of a self-consciously "new" philosophy in the early modern period.

Raphaële Garrod is Associate Professor of Early Modern French and Tutorial Fellow in French at Magdalen College, University of Oxford.

Anja-Silvia Goeing is Professor of Education at the University of Zurich and an associate in history at Harvard University. She is a specialist of early modern higher education and information history. Her recent books include the monograph *Storing, Archiving, Organizing: The Changing Dynamics of Scholarly Information Management in Post-Reformation Zurich* (Brill, 2017); and two co-edited volumes, the first with Ann Blair, Paul Duguid, and Antony Grafton, *Information: A Historical Companion* (Princeton University Press, 2021), the second with Glyn Parry and Mordechai Feingold, *Early Modern Universities: Networks of Higher Learning* (Brill, 2021). Twitter @debatesovert

Alexander Marr is Professor of the History of Early Modern Art at the University of Cambridge and a Fellow of Trinity Hall. His most recent books are *Rubens's Spirit: From Ingenuity to Genius* (Reaktion Books, 2021) and the co-authored *Logodaedalus: Word Histories of Ingenuity in Early Modern Europe* (University of Pittsburgh Press, 2018). He is currently writing a book about Hans Holbein and ingenuity, called *Holbein's Wit*.

Richard J. Oosterhoff is Lecturer in Early Modern History at the University of Edinburgh. His recent publications include *Making Mathematical Culture: University and Print in the Circle of Lefèvre d'Étaples* (Oxford University Press, 2018), the co-authored monograph *Logodaedalus: Word Histories of Ingenuity in*

Early Modern Europe (Pittsburgh University Press, 2018), and the collection *Ingenuity in the Making: Materials and Technique in Early Modern Art and Science*, edited with José Ramón Marcaida and Alexander Marr (Pittsburgh University Press, 2021).

Martine Pécharman is CNRS Research Director (CRAL, EHESS, Paris). Her research focuses primarily on early modern theories of language and logic, with special interests in Hobbes, Port-Royal, Locke, and Condillac. Recently, she has co-edited with Philippe Hamou *Locke and Cartesian Philosophy* (Oxford University Press, 2018). She is currently working on a volume for The Clarendon Edition of the Works of Thomas Hobbes.

Renée Raphael is Associate Professor of History at the University of California, Irvine. She has published on Galileo, the history of the book, and early modern universities; her current research focuses on mining and bureaucracy in the early modern Iberian world.

Sophie Roux is Professor of History and Philosophy of Science at École Normale Supérieure, Paris, where she is currently director of the research team *La République des savoirs* (CNRS-Collège de France-ENS). Her main research interest is in the history of early modern science and early modern philosophy, and more specifically in the history of Cartesianism and anti-Cartesianism in France. She has also written papers concerned with philosophy of science (e.g., on thought experiments or on mathematization) and history of philosophy of science (e.g., on Pierre Duhem, on Louis Couturat, or on French historical epistemology).

Abbreviations

AT	René Descartes. 1964–1974. *Oeuvres...*, ed. C. Adam and P. Tannery, 2nd. ed. Paris: Vrin.
BGE	Bibliothèque de Genève
BL	British Library
BM	British Museum
BnF	Bibliothèque nationale de France
BRB	Bibliothèque royale de Belgique
BSG	Bibliothèque Sainte–Geneviève
CB	Pierre Bayle. 1999–2017. *Correspondance*, ed. Elisabeth Labrousse, Antony McKenna, et alia, vols. 1–15. Oxford: The Voltaire Foundation.
CSM	René Descartes. 1984–85. *The Philosophical Writings...*, vols. 1–2, trans. John Cottingham, Robert Stoothoff, and Dugald Murdoch. Cambridge, UK: Cambridge University Press.
CSMK	René Descartes. 1991. *The Philosophical Writings...*, vol. 3, trans. John Cottingham, Robert Stoothoff, Dugald Murdoch, and Anthony Kenny. Cambridge, UK: Cambridge University Press.
HAB	Herzog August Bibliothek
OD	Pierre Bayle. 1964–1968. *Oeuvres diverses*, vols. 1–4, ed. Elisabeth Labrousse. Hildesheim: Georg Olms Verlag.
PE	Eugene F. Rice Jr., ed., 1972. *The Prefatory Epistles of Jacques Lefèvre d'Étaples and Related Texts*. New York: Columbia University Press.

Chapter 1
Introduction

Susanna Berger and Daniel Garber

Philosophy was a central discipline in the early modern period. In the philosophy classroom in European universities, students learned how to reason and argue, how to think about morality and the greatest good, as well as physics and metaphysics, cosmology, biology, and the ultimate metaphysical categories of reality. Since virtually every educated European (at least the men) in the period went through this curriculum in logic, moral philosophy, natural philosophy, and metaphysics, no matter what they did afterward, understanding what was taught and how it was taught illuminates nearly every corner of literate culture in the period.

In particular, we believe that understanding how philosophy was taught is especially important for understanding the history of philosophy and the history of science in the period. In the sixteenth and seventeenth centuries, philosophy and what we now call science were part and parcel of the same cluster of knowledge, and were taught together in the universities. All of the thinkers of the period who are now standard figures taught in histories of philosophy, histories of science, and general intellectual histories, figures who make up the cultural legacy of early modern Europe, figures like Bacon (1617–1621), Descartes (1596–1650), Hobbes (1588–1679), Galileo (1564–1642), Leibniz (1646–1716) and Newton (1642–1727), all studied philosophy at school. How they learned and what they learned will certainly illuminate their thought for modern readers. Furthermore, one of the standard tropes among many of the figures whom we still read is their opposition to philosophy as taught at the universities. Descartes begins his famous *Discourse on the Method* with a satiric retelling of his days at school, and why he escaped as soon as

S. Berger (✉)
University of Southern California, Los Angeles, CA, USA
e-mail: scberger@usc.edu

D. Garber (✉)
Princeton University, Princeton, NJ, USA
e-mail: dgarber@princeton.edu

S. Berger, D. Garber (eds.), *Teaching Philosophy in Early Modern Europe*, Archimedes 61, https://doi.org/10.1007/978-3-030-84621-3_1

he could; Hobbes's *Leviathan* is filled with diatribe against what students are taught at university, with a recommendation that they should be studying Hobbes's own writings instead. To understand what these figures were reacting against, to understand the way in which they thought that they were modern, we need to understand the experience of students in the early modern classroom. Moreover, some of the chapters that follow undermine and complicate the standard conception of a strict opposition to Aristotelian scholastic philosophy among the so-called new philosophers by showing how a number of anti-Aristotelian thinkers made use of Aristotelian argumentative tools and texts and by revealing how innovative, new ideas emerged within university settings typically dismissed by scholars then and now for their conservatism.

A number of disciplines have turned to the close examination of the teaching of philosophy in the period. Historians of art, visual culture, philosophy, and science have been researching how philosophy was taught and how students within the university were exposed to the new intellectual currents outside of the university.[1] *Teaching Philosophy* aims to build on emerging areas of scholarship, by exploring the roles of texts and images in teaching practices in philosophy in early modern Europe, with particular focus on France and Italy.

While the essays in this volume focus on the transmission of philosophy from a variety of different perspectives, there are some themes that we would like to emphasize. If Laurence Brockliss's *French Higher Education in the Seventeenth and Eighteenth Centuries. A Cultural History* (1987) can still be cited today as by far the most valuable synthetic account of French academic culture in the early modern period, the book's description of the role of visual representations in the pedagogical experience of French University students leaves something to be desired.[2] Brockliss writes,

> Teaching-aids were few and far between. From the beginning of the period it was customary to hang a map of the world on the door of the rhetoric classroom, but the practice did not become commonplace in the lower classes until the eighteenth century. It was only then too that a blackboard and chalk began to appear. For most of the period, the stone walls were left bare and covered with whitewash.[3]

Research in the histories of science, philosophy, and visual and material culture has started to call into question such perfunctory dismissals of visual aids in early modern University culture. In particular, scholars have explored images—from frescoes to oil paintings to prints and drawings—pertaining to the ideas of the so-called new philosophers, such as Galileo or Newton.[4] Yet the functions of images in the transmission and generation of ideas pertaining to Aristotelian philosophy is only just

[1] See, for instance, Blair 1993; eadem 1997; eadem 2010a; Ariew and Grene 1997; Brockliss 1981a; idem 1981b; idem 1987; idem 1990; idem 2006; Garber 1988; and Schmutz 2008.

[2] For such an assessment of Brockliss's book, see Schmutz 2008, 366.

[3] Brockliss 1987, 56.

[4] See, for instance, Reeves 1997; Bredekamp 1999; Freedberg 2002; Smith 2004; and Feingold 2004.

beginning to be examined.[5] More work remains to be done on both Aristotelian and anti-Aristotelian documents related to university culture in such forms as illustrated broadsides and thesis prints, manuscript lecture notebooks, *alba amicorum* (friendship albums), and printed books.[6] We would like to emphasize that the observation and generation of visual representations were important tools of philosophical thinking and instruction. Some of the chapters in the book uncover the relative merits of texts and images in the communication of philosophical ideas and show how both often worked in tandem to convey knowledge to students or to generate novel interpretations.

Furthermore, a number of the chapters that follow can be situated within a new area of cultural history that has developed over the last twenty to thirty years that investigates knowledge institutions and aims to understand how information has been structured and handled in earlier times.[7] Scholars have explored how collections and pedagogical aids functioned to organize the "information explosion" that historians have argued that Europeans were subject to from around 1550 to 1750 and whose sources have been tied to the growing number of printed books, world exploration and the finding of territories previously unknown to Europeans, the rediscovery of ancient texts, and an intense passion for the assembling of information.[8] This volume introduces visual analogs to textual modes of managing information that were used by Aristotelian and anti-Aristotelian scholars and students in the period. These could take the forms of, for instance, tables, dichotomies, or maps of philosophical knowledge. Moreover, just as cultural historians today are interested in how information and knowledge has been organized in the past, the chapters that follow reveal that early modern thinkers and educators themselves grew increasingly self-aware about the ways in which they were transmitting, gathering, and generating observations and insights. This heightened self-criticality and concern with didactic method is manifest from, for example, early modern reflections on the benefits and dangers of ingenuity as well as from an increased focus on the form of such pedagogical materials as philosophical textbooks and broadsides.

Chapters 2 and 3 center on the university textbook, a genre that Charles B. Schmitt helped to call to scholars' attention in an important essay published in 1988 in *The Cambridge History of Renaissance Philosophy*.[9] The term "textbook" is modern and may be understood as a book designed for employment in classrooms; a

[5] Berger 2017; and Vanpaemel et al. 2012.

[6] For sources on illustrated thesis prints, see Gieben 1993, 273–74 n. 2; and Rice 1999, 165 n. 1. On frontispieces, see Corbett and Lightbown 1979; and Remmert 2005. On images in student notebooks, see Vanpaemel et al. 2012. For important studies of images, texts, and instruments as vehicles through which knowledge was disseminated in early modern Europe, see Kusukawa and Maclean 2006.

[7] On information history, see Grafton 2010, 95–101; Burke 2000; and Blair 2003, 11–28. See the "Cultures of Knowledge" project at Oxford University: http://www.culturesofknowledge.org.

[8] Blair 2010b, 11.

[9] Schmitt 1988.

distinguishing characteristic of the textbook is its tendency to summarize.[10] In his essay, Richard J. Oosterhoff observes that Jacques Lefèvre d'Étaples's deepest impact on the history of thought in early modern Europe was through his textbooks that thousands of students used during their undergraduate education. The textbooks produced by d'Étaples (c. 1455–1536) and his circle were an early example of this genre to appear in print and, surprisingly, even such university critics as Thomas More (1478–1535) praised them; in accentuating the popularity of university textbooks in humanist circles, Oosterhoff counters standard oppositions between university and humanist cultures. Visual thinking infuses the textbooks of d'Étaples and his circle; the books feature inventive typography and a range of visual representations, from diagrams to tables. In his analysis Oosterhoff challenges Walter Ong's seminal assessment of the textbooks as works that hindered free thought and led to "the decay of dialogue"; to the contrary, Oosterhoff demonstrates how the textbooks' visual and material forms aimed to generate active thought.[11] Moreover, he reveals that their employment of dialogue elevated students as knowledge sources. He specifies that the textbooks promoted promising students as "ingenuous" individuals; that is to say such students were characterized by open natures and a readiness for serious studies, without relying on the privilege of noble birth.

Roger Ariew's chapter examines the popular philosophy textbook by Eustachius a Sancto Paulo (1573–1640) that drew admiration from both Descartes and Leibniz. In addition to discussing Eustachius's work, Ariew traces the rise and decline of synoptic tables in scholastic and Cartesian textbooks and posits that the inclination and subsequent disinclination for these schemas mirrored transformations in scholastic textbooks over the course of the seventeenth century more generally. Ariew ties the popularity of tables in the early 1600s to the abandonment of the commentary tradition; he argues that tables, which enable observers to see "at a glance" how topics pertain to one another, filled the void left by commentaries that had cited where particular subjects are discussed in longer texts. Ariew contends that the decline of the synoptic table towards the end of the seventeenth century advertised the rise of a new type of Cartesian textbook.

Whereas the chapters of Ariew and Oosterhoff explore summations of knowledge offered via textbooks, Susanna Berger's contribution considers summations of knowledge presented in philosophical broadsides. In particular, Berger examines broadsides designed by Martin Meurisse (1584–1644) and Jean Chéron (1596–1673) in collaboration with the engraver Léonard Gaultier (1560/61–1635) and the publisher Jean Messager (1572–1649). The prints of Meurisse, Chéron, Gaultier, and Messager feature an innovative combination of word and image that encourages visual thinking among their observers: Not only do the broadsides use the space of the page to chart theoretical relationships, but they also offer visual commentaries that enrich philosophical concepts.[12] These broadsides were employed as thesis

[10] Schmitt 1988, 792–804 (792).

[11] Ong 1983.

[12] Concerning visual thinking in philosophical broadsides, see Berger 2017.

prints for use in public examinations called disputations. In her chapter, Berger reflects on the ways in which these engravings relate to printed maps produced in this period. She argues that like maps, these philosophical engravings fabricate understandings of a field through the selective presentations of topics and a range of descriptive mechanisms that enhance the observer's comprehension, while advertising and advancing particular political interests. Her chapter includes a close analysis of a previously undiscussed section of a broadside summarizing natural philosophy and concludes with a consideration of the ways in which philosophical broadsides visualize the dangers of traveling beyond the known boundaries of learning.

In the chapter by Dominco Collaciani and Sophie Roux, we turn to another kind of pedagogical text, the thesis booklet, which like the thesis print was employed in disputations. Their chapter offers a close analysis of the mathematical thesis booklets defended at the Jesuit college in Paris, the Collège de Clermont between 1637 and 1682. Because the Jesuit *Ratio Studiorum* placed much more severe restrictions on the philosophy curriculum than on the mathematical, the mathematical theses ranged over a much wider variety of material than the philosophical theses did, and ventured into the consideration of issues then under discussion in the wider intellectual community, including Copernicanism and Cartesianism. Collaciani and Roux focus on the treatment of optics in these thesis booklets. They show the way in which the thesis booklets put themselves into the dialogue between Aristotelian ideas of sense perception and some of the newer theories then under debate, including Cartesian theories. They also emphasize the way in which optical experience and optical devices are reflected in the many engravings found in the thesis booklets, bringing these visual tools and experiments into the theses and into the classroom. And finally, they show the evolving attitudes toward Descartes, as he moves from one among many *novatores* in the beginning of the period studied to a central but somewhat dangerous figure by the end.

Raphaële Garrod's chapter offers a close analysis of an emblem on ingenuity from Antoine de Bourgogne's emblem book *Mundi lapis lydius...* (*Compass of the World*, 1639), which was owned by the library of the Jesuit College of La Flèche, Descartes's *alma mater*. Emblems consisted of three components: a motto (*inscriptio*), an image (*pictura*), and an epigram in prose or verse offering an explanation (*subscriptio*).[13] Insofar as they juxtapose word and image, emblems exemplify the early modern fascination with combining visual and textual modes of thinking. In de Bourgogne's emblem on ingenuity, text and image work in tandem to encapsulate an ambivalent attitude towards this notion that, Garrod argues, was prevalent in Jesuit culture at large. On the one hand, the emblem presents a condemnation of extreme ingenuity as excessive subtlety and on the other, it celebrates ingenuity as a stylistic, witty performance.

---

[13] For a bibliography of emblem books, see Praz 1975. The literature on emblems is vast and growing; see, for instance, Saunders 2000; and eadem 1999–2002.

The next two chapters center on the teaching practices of particular philosophy professors: the focus of Chap. 7 is Calvinist professor Jean-Robert Chouet (1642–1731) and that of Chap. 8 is Pierre Bayle (1647–1706), a French Calvinist philosopher and the author of a seminal biographical dictionary. Ann Blair and Anja-Silvia Goeing's contribution studies eight student manuscripts produced between 1667 and 1685 from physics classes taught by Chouet, whose instruction the authors speculate was representative of philosophy pedagogy at the time more generally. Blair and Goeing consider what the manuscripts show concerning shifts and continuities over the course of Chouet's teaching career. Although the authors observe that explanatory images are not central to Chouet's physics teaching, they argue that the manuscripts evince a concern with visual modes of ordering information: Not only do they contain such forms identified with printed books as title pages, tables of contents, and alphabetical indices, but the students also emphasized distinct parts of the lectures by varying the size and thickness of letterings and through the inclusion of blank spaces.

Martine Pécharman turns to the instruction of Bayle at the protestant Academy of Sedan from 1675 through 1681. Although visual representation played an important part in much philosophical teaching of the early modern period, it certainly was not central to all philosophical instruction—it plays, for instance, no role in Pécharman's discussion of Bayle's pedagogical practice. In her chapter, Pécharman examines Bayle's philosophy course in terms of its relationship to Bayle's later philosophical development, but also and especially in its own right. As she reveals, Bayle manipulates the argumentative instruments of scholastic dialectic in order to defend anti-Aristotelian ideas in physics. Bayle is not alone in deploying this stratagem: Hobbes, although not an Aristotelian, made use of Aristotelian concepts to promote his anti-Aristotelian ideas.[14] Perhaps both Bayle and Hobbes adopted this rhetorical move from Quintilian, who notes: "It is sometimes also possible to take some remark or action of ... your opponent or your opponent's advocate in order to strengthen your point."[15] For Bayle, moreover, a knowledge of Aristotelian logic was a precondition to championing the ideas of the new philosophers.

Chapters 9 and 10 move to philosophical instruction in Italy. In her essay, Renée Raphael studies teaching notes and published philosophy courses to expose the ways in which experiments were presented to students during the second half of the seventeenth century at the Collegio Romano. Here, as well, images play a relatively minor role; Raphael observes that it was common to find compendia pertaining to natural philosophy instruction at the Collegio without any illustrations, although some textbooks from the late seventeenth century featured visual representations of the world systems and Raphael discusses a number of sketches. In her chapter, Raphael presents a slew of different descriptions of experiments from pedagogical texts that feature a ranging degree of textual and visual details. Because most professors did not present sufficient information for students to recreate the

[14] Berger 2017, chapter 5.

[15] Quintillian 2014, vol. 2, Book 5, chapter 11, 455.

experiments, Raphael concludes that it is improbable that the descriptions of experiments in the courses functioned either as accounts of experiments professors had themselves witnessed or as stimuli for the students' own experimental practice. She establishes that insofar as professors did include some textual particulars and visual representations of experimental practices and tools, they did so in order to demonstrate their own expertise as readers of the most up-to-date literature.

Although visual representations play a relatively minor role in the discussions of teaching practices in the chapters by Pécharman and Raphael, they are at the core of Alexander Marr's contribution. Here, Marr presents a drawing made in Rome by the French artist Simon Vouet (1590–1649) as well as an engraving made after the drawing by Johannes Tröschel (1585–1628), a printmaker from Nuremberg; the images show eight satyrs debating around a table that displays an anamorphosis of an elephant. Marr speculates that the design was originally intended to function as the upper half of a thesis print. He maintains that it should be interpreted as a Christian allegory of the "virtue of knowledge." Marr's chapter builds on Garrod's considerations of ingenuity in relation to early modern pedagogical culture, insofar as he posits that the image celebrates a "collective ingenuity" that was typical of philosophical endeavors and creativity in the visual arts in the courtly culture of early modern Italy.

Teaching Philosophy breaks new ground in a number of ways. Firstly, it seeks to bring text-based scholars in the history of philosophy together with social and cultural historians to examine the interaction between tradition and innovation in the early modern classroom, the site where traditional views of the world were transmitted to the generation that was to give birth to modern philosophy and science. Secondly, it draws together scholars who are centered on ideas and words with other scholars who focus on the role of images in the classroom and the intellectual world in this central period of history. We hope that it will advance our comprehension of how philosophy was understood and transmitted in this rich and crucial era. In this way we hope to enrich our grasp of an important aspect of the background to the emergence of modern thought in one of the most fertile periods of European intellectual history.

Bibliography

Ariew, Roger, and Marjorie Grene. 1997. The Cartesian Destiny of Form and Matter. *Early Science and Medicine* 2 (3): 300–325.

Berger, Susanna. 2017. *The Art of Philosophy: Visual Thinking in Europe from the Late Renaissance to the Early Enlightenment*. Princeton: Princeton University Press.

Blair, Ann. 1993. The Teaching of Natural Philosophy in Early Seventeenth-Century Paris: The Case of Jean Cécile Frey. *History of Universities* 12: 95–158.

———. 1997. *The Theater of Nature: Jean Bodin and Renaissance Science*. Princeton: Princeton University Press.

———. 2003. Reading Strategies for Coping with Information Overload Ca. 1550–1700. *Journal of the History of Ideas* 64 (1): 11–28.

———. 2010a. The Rise of Note-Taking in Early Modern Europe. *Intellectual History Review* 20 (3): 303–316.
———. 2010b. *Too Much to Know: Managing Scholarly Information Before the Modern Age*. New Haven: Yale University Press.
Bredekamp, Horst. 1999. *Thomas Hobbes Visuelle Strategien. Der Leviathan: Urbild Des Modernen Staates*. Berlin: Akademie.
Brockliss, Laurence. 1981a. Aristotle, Descartes and the New Science: Natural Philosophy at the University of Paris, 1600–1740. *Annals of Science* 38 (1): 33–69.
———. 1981b. Philosophy Teaching in France, 1600–1740. *History of Universities* 1: 131–168.
———. 1987. *French Higher Education in the Seventeenth and Eighteenth Centuries: A Cultural History*. Oxford/New York: Clarendon Press.
———. 1990. Copernicus in the University: The French Experience. In *New Perspectives on Renaissance Thought*, ed. John Henry and Sarah Hutton. London: Duckworth.
———. 2006. The Moment of No Return: The University of Paris and the Death of Aristotelianism. *Science & Education* 15 (15): 259–278.
Burke, Peter. 2000. *A Social History of Knowledge: From Gutenberg to Diderot*. Cambridge: Polity Press.
Corbett, Margery, and Ronald Lightbown. 1979. *The Comely Frontispiece: The Emblematic Title-Page in England, 1550–1660*. Boston: Routledge & Kegan Paul.
Feingold, Mordechai. 2004. *The Newtonian Moment: Isaac Newton and the Making of Modern Culture*. New York/Oxford: The New York Public Library/Oxford University Press.
Freedberg, David. 2002. *The Eye of the Lynx: Galileo, His Friends, and the Beginnings of Modern Natural History*. Chicago: University of Chicago Press.
Garber, Daniel. 1988. Descartes, the Aristotelians and the Revolution That Did Not Happen in 1637. *The Monist* 71: 471–486.
Gieben, Servus. 1993. Iconografia antoniana in due fogli di tesi del Museo Francescano di Roma. *Il Santo* 2 (3): 273–298.
Grafton, Anthony. 2010. Jumping Through the Computer Screen. *The New York Review of Books* 57 (20): 95–101.
Hotson, Howard. 2007. *Commonplace Learning: Ramism and Its German Ramifications, 1543–1630*. Oxford: Oxford University Press.
Kusukawa, Sachiko, and Ian Maclean, eds. 2006. *Transmitting Knowledge: Words, Images, and Instruments in Early Modern Europe*. Oxford: Oxford University Press.
Ong, Walter J. 1983. *Ramus, Method, and the Decay of Dialogue: From the Art of Discourse to the Art of Reason*. Chicago: University of Chicago Press.
Praz, Mario. 1975. *Studies in Seventeenth-Century Imagery*. Second expanded ed. Rome: Edizioni di storia e letteratura.
Quintillian. 2014. *The Orator's Education*. Translated by Donald A. Russell. Cambridge, MA: Harvard University Press.
Reeves, Eileen. 1997. *Painting the Heavens: Art and Science in the Age of Galileo*. Princeton: Princeton University Press.
Remmert, Volker. 2005. *Widmung, Welterklärung und Wissenschaftslegitimierung: Titelbilder und ihre Funktionen in der Wissenschaftlichen Revolution*. Wiesbaden: Harrassowitz.
Rice, Louise. 1999. Jesuit Thesis Prints and the Festive Academic Defense at the Collegio Romano. In *The Jesuits: Cultures, Sciences, and the Arts, 1540–1773*. Toronto: University of Toronto Press.
Saunders, Alison. 1999–2002. *A Bibliography of French Emblem Books of the Sixteenth and Seventeenth Centuries*. Geneva: Droz.
———. 2000. *The Seventeenth-Century French Emblem: A Study in Diversity*. Geneva: Librairie Droz.

Schmitt, Charles B. 1988. The Rise of the Philosophical Textbook. In *The Cambridge History of Renaissance Philosophy*, ed. Quentin Skinner Schmitt and Eckhard Kessler. Cambridge: Cambridge University Press.

Schmutz, Jacob. 2008. Le petit scotisme du Grand Siècle: Étude doctrinale et documentaire sur la philosophie au Grand Couvent des Cordeliers de Paris, 1517–1771. *Quaestio* 8: 365–472.

Smith, Pamela H. 2004. *The Body of the Artisan: Art and Experience in the Scientific Revolution*. London: University of Chicago Press.

Vanpaemel, Geert, Katharina Smeyers, An Smets, and Diewer van der Meijden, eds. 2012. *Ex cathedra: Leuvense collegedictaten van de 16de tot de 18de eeuw*. Leuven: Peeters.

Chapter 2
The Dialogue of Ingenuous Students: Early Printed Textbooks at Paris

Richard J. Oosterhoff

Abstract Dialogues and the "dialogic" often hold a privileged place in accounts of the Renaissance and early modern learning. University textbooks therefore are presented as a site of social control, imposing fixed formulas rather than stimulating independent judgment. This essay shows how the program of printed textbooks by the Renaissance Paris arts master Jacques Lefèvre d'Étaples (c. 1455–1536) and his circle were intended to support dynamic mental habits, in which students were expected to embody real knowledge. These university textbooks respond not only to a humanist fashion for dialogue, but also to a medieval tradition of "outsider" knowledge which sought to ensure that university habits answered broader social needs. This matters for understanding early modern universities, because Lefèvre's deepest influence on early modern intellectual culture was less on high humanist polemic and more on the thousands of students who used his textbooks as their path through the BA cursus, and on those writers who modelled their textbooks on his.

Keywords Dialogue · Lefèvre d'Étaples · Textbooks · Typography · Print · Virtue · Character difficulty · Labour · Genius · Ingenuous

In 1512 an anonymous scholar updated the "who's who" of European scholarship, the *De scriptoribus ecclesiasticis* (*On Writers of the Church*) of Trithemius, adding the name Jacques Lefèvre d'Étaples (c. 1455–1536).[1] The editor was probably French, since the expanded edition was printed in Paris and his additions mostly were of French authors. Jacques Lefèvre d'Étaples headed the list. He was the Paris humanist whose *Quincuplex Psalterium* (*Five-Fold Psalter*) (1509) and *Epistolae Sancti Pauli* (*Letters of Saint Paul*) (1512) were setting new standards in biblical scholarship. In 1512 Lefèvre was already a gray eminence, and younger humanists

[1] Trithemius 1512.

R. J. Oosterhoff (✉)
University of Edinburgh, Edinburgh, Scotland
e-mail: richard.oosterhoff@ed.ac.uk

S. Berger, D. Garber (eds.), *Teaching Philosophy in Early Modern Europe*, Archimedes 61, https://doi.org/10.1007/978-3-030-84621-3_2

such as Erasmus of Rotterdam challenged Lefèvre's authority in order to establish their own.[2] Most famously, in 1517 Erasmus would contest Lefèvre's reading of the New Testament letter to the Hebrews, suggesting that Lefèvre's scholarship was impaired by an imperfect mastery of Greek.[3] A measure of Lefèvre's stature in biblical scholarship was Francis I's slip in conversation with the great Greek scholar Guillaume Budé, mistaking Lefèvre for Erasmus.[4]

This is the image of Lefèvre, the "prince of French humanism," that Renaissance historians usually repeat. In *De scriptoribus ecclesiasticis*, however, Lefèvre was primarily praised as the "single glory of all France" (*totius Galliae unicum decus*) for his accomplishments as a writer of textbooks for the arts curriculum. Pseudo-Trithemius listed various introductions to Aristotelian logic, ethics, natural philosophy, and metaphysics. Another third of the list included books on arithmetic, music, and astronomy, Lefèvre's contribution to the mathematical arts of the quadrivium.[5] It was to these books that Thomas More urged students to go, instead of miring themselves in the sophisms of older textbooks.[6] Close study of book purchases in sixteenth-century Oxford and Cambridge show that these textbooks represented the new curriculum introduced by educational reformers.[7] These were republished repeatedly in Paris, and then excerpted and imitated by figures such as Symphorien Champier, Hieronymus Gebwiler, Matthias Ringmann, Pedro Ciruelo and by publishers in cities as different as Cologne, Alcalá, and Venice.[8] I would suggest that Lefèvre's deepest influence on early modern intellectual culture was less on high humanist polemic and more on the thousands of students who used his textbooks as their path through the BA cursus.

Lefèvre and his circle's textbook project was influential partly due to precedence—their textbook project, beginning in the 1490s, was among the first to reconfigure the whole arts curriculum using the new technology of print.[9] It was also distinctive in character. Despite being inextricable from the university, these books captured the admiration of university critics. As More's letter to Dorp makes clear, these textbooks were seen as an alternative to sterile formalism. Lefèvre was seen as a more direct route to real knowledge, one that would not ensnare students in intellectual folly—"supersophistical trifles," as More called them.

---

[2] In 1501 Lefèvre noted that his student Charles de Bovelles was twenty, and that he was twice that age. On dating, see the comments and bibliography in Rice 1972, XIn1. Basic bibliography includes Renaudet 1953; Bedouelle 1976.

[3] This debate is recounted in Mann 1934 and most recently Schönau 2020.

[4] Bedouelle 1995, 23–42.

[5] Trithemius 1512, 215^{v}–16^{r}.

[6] More 1986, vol. 15, 22–23.

[7] Hannam 2007. On the curricular pressures of the sixteenth century: Curtis 1959; McConica 1979, 291–317.

[8] A near-complete bibliography of textbooks and editions by Lefèvre is in Rice 1972, 535–568. See also Lohr 1988, 138–142.

[9] I pursue this argument in more detail in Oosterhoff 2018, chap. 4. The delayed uptake of print for core textbooks is described by Corsten 1987, 83–123.

In what follows I will focus on dialogues in this body of texts. Dialogues and the "dialogic" often hold a privileged place in accounts of the Renaissance and early modern learning.[10] Even those who find the specifics of Mikhail Bakhtin's "dialogic imagination" overstated, nevertheless privilege dialogue as a space for dissent or democratic openness.[11] Early universities are usually associated with the hierarchies of scholasticism. University textbooks therefore are presented as a site of social control, imposing fixed formulas rather than stimulating independent judgment. As one instance, Walter Ong saw early modern print especially as a deathly force against dialogue, rendering philosophical textbooks as stolid structures of static reason.[12] My argument shall push in another direction. First I will outline Lefèvre and his circle's program of printed textbooks, and suggest that their visual and material structures were intended to support dynamic mental habits. Then I will show that their use of dialogue was motivated quite deliberately by a specific set of social ideals, in which students were expected to be a source of real knowledge. Finally, in the last section I will suggest how these university textbooks respond not only to a humanist fashion for dialogue, but to a medieval tradition of "outsider" knowledge which sought to ensure that university habits answered broader social needs.

2.1 Movement of Mind and Page

By the time the Parisian redactor of Trithemius wrote in 1512, an entire generation of students at the Collège du Cardinal Lemoine would have been able to survey the whole arts cursus from textbooks written by Lefèvre, Clichtove, and other regents in the Paris college. These books included epitomes, introductions, paraphrases, commentaries and dialogues that seem roughly indexed according to the three years of the BA *cursus*, capped with a study of moral philosophy for licentiates preparing to take the MA. It is not entirely clear in what order these books were written, and they probably first circulated in manuscript within the college community. We can see this in Lefèvre's first printed publication of 1492, his *Paraphrases philosophiae naturalis* (*Paraphrases on Natural Philosophy*). But later letters show that Lefèvre considered his dialogues on metaphysics, though printed first in 1494, to have already been published in 1490, evidently in manuscript.[13] Indeed, many of Lefèvre's shorter epitomes, introductions, and dialogues may have originated as manuscript helps.[14] Their uncertain status in the 1490s suggest how experimental print was within university education of that decade.

[10] The topic has exploded since Hirzel 1895. Most relevant are Marsh 1980; Cox 1992; Hösle 2012; Traninger 2012.

[11] E.g. Goldhill 2008. The classic work is Bakhtin 1981.

[12] Ong 1958.

[13] Lefèvre mentions the date of composition in his edition *Aristotelis castigatissime recognitum opus metaphysicum*, 1515, 125^{r} (Rice 1972, 22).

[14] On this evidence, see Oosterhoff 2018, 87–95.

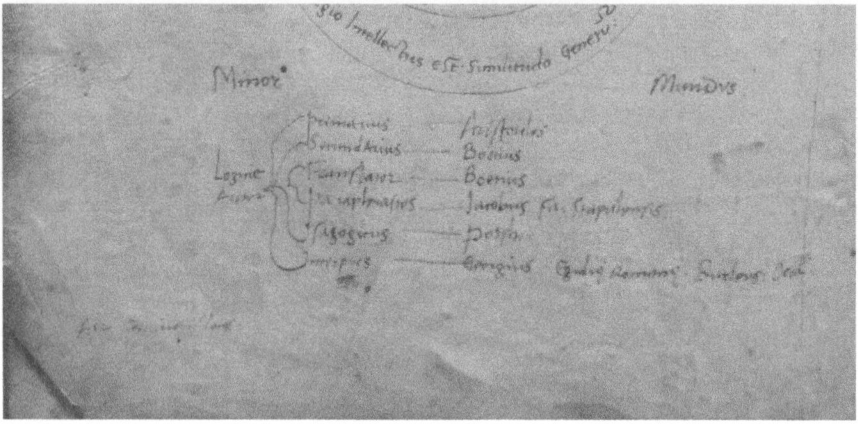

Fig. 2.1 Notes on the hierarchy of authors, by Beatus Rhenanus at the Collège du Cardinal Lemoine sometime between 1503 and 1507. Lefèvre, *Libri logicorum* (*Books of Logic*) (Paris: Hopyl and Estienne, 1503), Bibliothèque humaniste de Sélestat, K 1047, guard page, detail. (By kind permission of the Maire de Sélestat)

A student-driven manuscript circulation helps explain the goal of the most popular of Lefèvre's books, his brief "arts" or "introductions." They aimed not to replace Aristotle's own texts, but instead to facilitate access to them. This is especially clear for logic. Lefèvre's printed *introductio* to logic (1496) offered these as a kind of primer, which would help students avoid entanglement in the abstruse technicalities of the highly developed school logic that had become the staple of first-year university training in the fourteenth and fifteenth centuries. He advised tarrying only long enough to gain a sense of the terrain, "for it is enough to have touched on these things in passing (like scouts passing through enemy lines)."[15] The point of an introduction was to give just enough familiarity that one might successfully pass through the *cursus*, to give the more reliable authorities themselves the attention they deserved. Beatus Rhenanus, writing in 1503, annotated his copy of Lefèvre's logical textbook with a hierarchy of various authorities: Lefèvre was the paraphraste, positioned above mere commentators (*interpretes*) and below Aristotle himself (Fig. 2.1).

Such an ideal of knowledge shaped the structure and use of textbooks. To be sure, this was an intellectualist account of knowledge that privileged mental abstractions. Aristotle was the most prized authority because he offered the most resources for the sapiential vision of learning that Lefèvre taught. In his dialogues introducing the *Metaphysics* Lefèvre presented an account of wisdom that was encyclopaedic in scope: the student could become wise by gathering within himself the universal

[15]Lefèvre 1496b, sig. a1^{v} (Rice 1972, 39): "Satis enim est ea vel in transcursu (velut qui exploratores hostile agmen transcurrunt) attigisse."

statements of each knowledge domain.[16] Knowledge was properly a knowledge of essences or universals. At one point, Lefèvre described a kind of "hidden analogy" which underlay the whole of Aristotle's natural philosophy.[17] By this Lefèvre suggested that all the many modes of learning—whether lone propositions, broader paraphrases of arguments, or worried and developed in dialogues—could be distilled to essential propositions. Learning aimed to develop intellectual vision above all else.

It would be a mistake, however, to dismiss this intellectualist account of learning as therefore uninterested in the humbler everyday elements of teaching and learning, its practices, materiality, and visuality. Those historians who have studied these books have remarked upon the innovative typography and array of diagrams, tables, and visual aids that distinguish them.[18] The self-conscious attention to genre—paraphrases, dialogues, and so on—also reveals a sustained attention to the role of the senses and habits in helping student to learn. An Aristotelian account of the virtues, in which knowledge was a habitus, acquired through long, embodied habituation, reinforced the value of the entire structure of practice.[19] Thus the hierarchy of knowledge made it necessary to attend to details of textbooks, genre, and student habits. The goal of an *ars* or *introductio* was to make accessible the structure of more basic disciplines, so that students could rapidly acquire the habitus needed for ever higher learning.

For these reasons, Lefèvre and his students set special emphasis on mathematics. It was not that mathematics was the highest domain of study; in at least one place, Lefèvre set mathematical philosophy in itself on the lowest conceptual rung.[20] But precisely for that reason he set mathematics in the formative first year of the curriculum—unusually, at Paris—complaining that it had been a subject long overlooked within the university. Whereas medieval and Renaissance universities had always first set a student to logic in their first year, under Lefèvre's influence students began with mathematical learning.[21] This new profile of mathematics can be seen in the

[16] I offer a reading of this in Oosterhoff 2019, 73–95 (85–88). For more on metaphysics this context, see e.g. Faye and Hirstein 2002.

[17] Lefèvre 1492, sig. b2r: "Id insuper te latere non debet per totam Aristotelis philosophiam abditam latentemque esse quandam secretam Analogiam perinde atque per totum corpus sparsus fususque tactus est [cf. Cicero, *De natura deorum* II.56]."

[18] E.g. Klinger-Dollé 2016.

[19] This is made explicit in Lefèvre and Clichtove 1502a. See Oosterhoff 2018, 47–55, and Kraye 1995, 96–117.

[20] Lefèvre 1515, 1v (PE 356). "Infimum mathematicum, ut quod de accidente sit; post quod ascendendo naturale ac physicum, ut quod de substantia est, sed mobili ac media. Post physicum vero sursum vergendo solum restat divinum philosophiae genus, ut quod de substantia, sed non mobili ac media verum…" (The lowest is mathematical philosophy, for its object is accident. Next is physical philosophy, whose object is substance, but mobile and median substance. We rise then to metaphysical philosophy, whose subjectis also substance, but immobile and wholly divine substance).

[21] I argue this from a study of Beatus Rhenanus' book purchases and annotations: see Oosterhoff 2018, 56–85 and Appendix.

suite of mathematical textbooks Lefèvre, Clichtove, and Charles de Bovelles published at Cardinal Lemoine. Mathematics here was not valuable as an end in itself, but as a foundation that served to correct the usual logical shaping of student intuitions and reasoning habits.

Lefèvre and those around him therefore found mathematics a good model for mastering the disciplines in general. Learning should begin with internalizing first principles or axioms.[22] This is clearest in his early mathematical textbooks of the 1490s, which follow the usual convention of using page layout to set off axioms or "elements" as the main text. Proofs and demonstrations were, in medieval mathematics, often treated as commentaries—easily modified or replaced by later writers.[23] Lefèvre, for instance, produced an edition of Jordanus' *Elementa Arithmetica* (*Arithmetical Elements*) (1496) that replaced most of the original proofs with his own. In more his basic mathematical textbooks, he pushed this convention to an extreme, entirely omitting proofs. His widely popular *Epitome in libros arithmeticos Boetii* (*Summary of the Arithmetical Books of Boethius*) was simply a collection of propositions that flowed from a set of first principles or "elements."[24] This could be condensed even further. The entire handbook on arithmetic could fit on one page, in a table or *formula* that reduced the discipline to a hierarchy of terms (Fig. 2.2). The figure would allow readers, Lefèvre pointed out, to "fix" the entirety of the discipline in the mind, before then moving to definitions, properties, and lastly to demonstrations.[25]

Such cognitive tools for arranging the elements of an entire discipline on a page, whether as lists of terms or propositions, features in most of Lefèvre's introductions, extending the idea of elements well beyond mathematics.[26] In his first printed book, the *Paraphrases philosophiae naturalis* of 1492, Lefèvre already had prefaced the book with a similar *formula* (Fig. 2.3). Then he offered lists of theses and *conclusiones*, numbered and indexed to their positions within the discursive paraphrases and dialogues of natural philosophy to follow. Therefore, the narrative priority of axioms and enunciations was closely linked with the brevity of these introductory textbooks: they enabled a student to move back and forth between condensed proposition and expanded discourse, from the flash of an idea to the process of thinking it through.

---

[22] Aristotle made a similar point several times: *Analytica priora* 46a3–27; *Analytica posteriora* I.7 (71b19–25); *Metaphysica* I.2 (982b11–28); see also 1000a6–10, 1000b21–1001a3, 1060a27–36, 1075b13–14. On the broader commitment to this approach, see Oosterhoff 2014, 1–19; idem 2018, 106–11.

[23] Goulding 2010, 151–54.

[24] Lefèvre 1496a. Later introductions to geometry and perspective published with Lefèvre's *Epitome* follow the same format: Lefèvre, Clichtove, and Bovelles 1503.

[25] Lefèvre 1496a, sig. b9^r: "primum sibi vendicat arithmetice locum que ut rite cognoscatur quemadmodum et cetere certis eget adminiculis inter que primo menti figenda est universorum circa que versatur subiecta formula, mox singulorum diffinitiones, post quas numerorum affectiones proprietatesque, post proprietates quo ex loco universe sunt monstrande."

[26] On the language of "elementating", see Oosterhoff 2018, 106–11, 117; Zepeda 2015, 48–76.

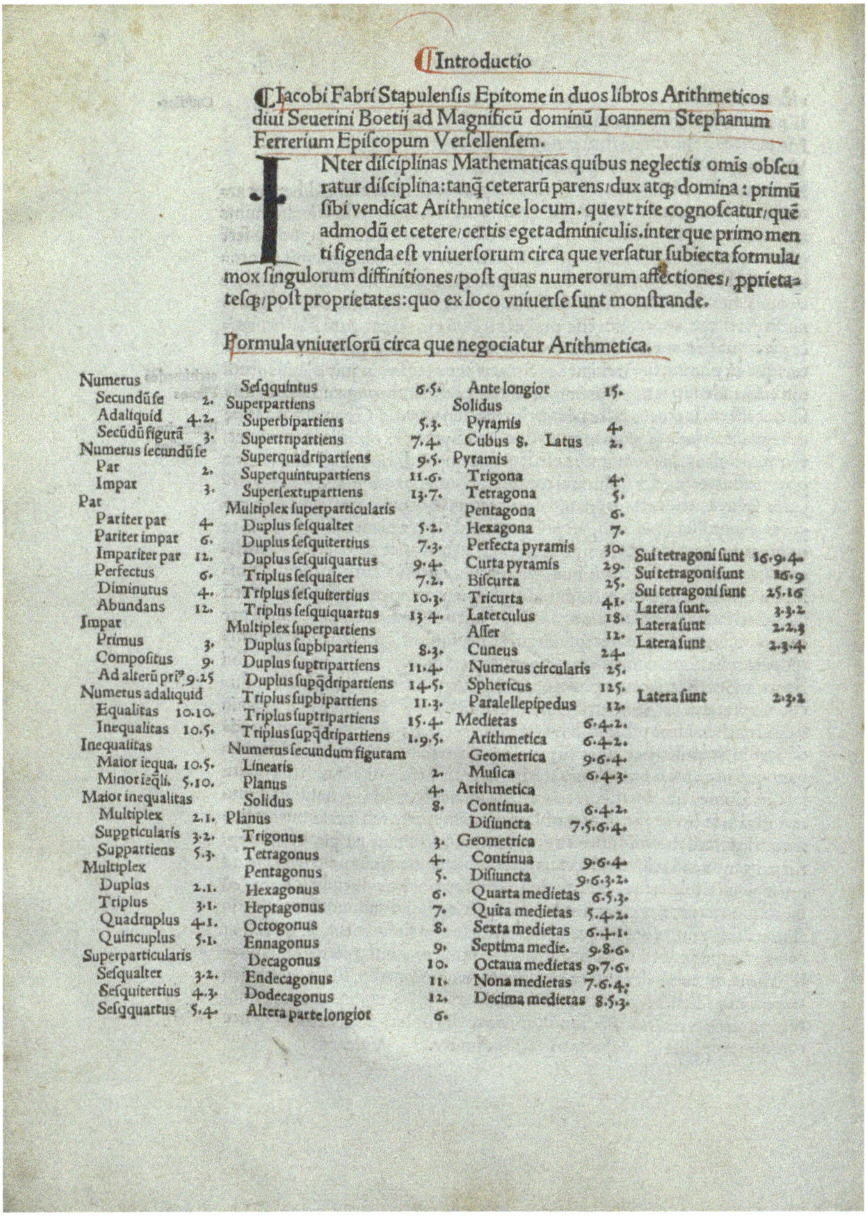

Fig. 2.2 The table (*formula*) introducing an epitome of Boethian arithmetic, first published with Lefèvre, *Elementa arithmetica, etc.* (1496), h8r; here from Lefèvre, Clichtove, and Bovelles, *Epitome compendiosaque introductio in libros arithmeticos* (1503). ETH-Bibliothek Zürich, Rar 230, https://doi.org/10.3931/e-rara-30983 / Public Domain Mark

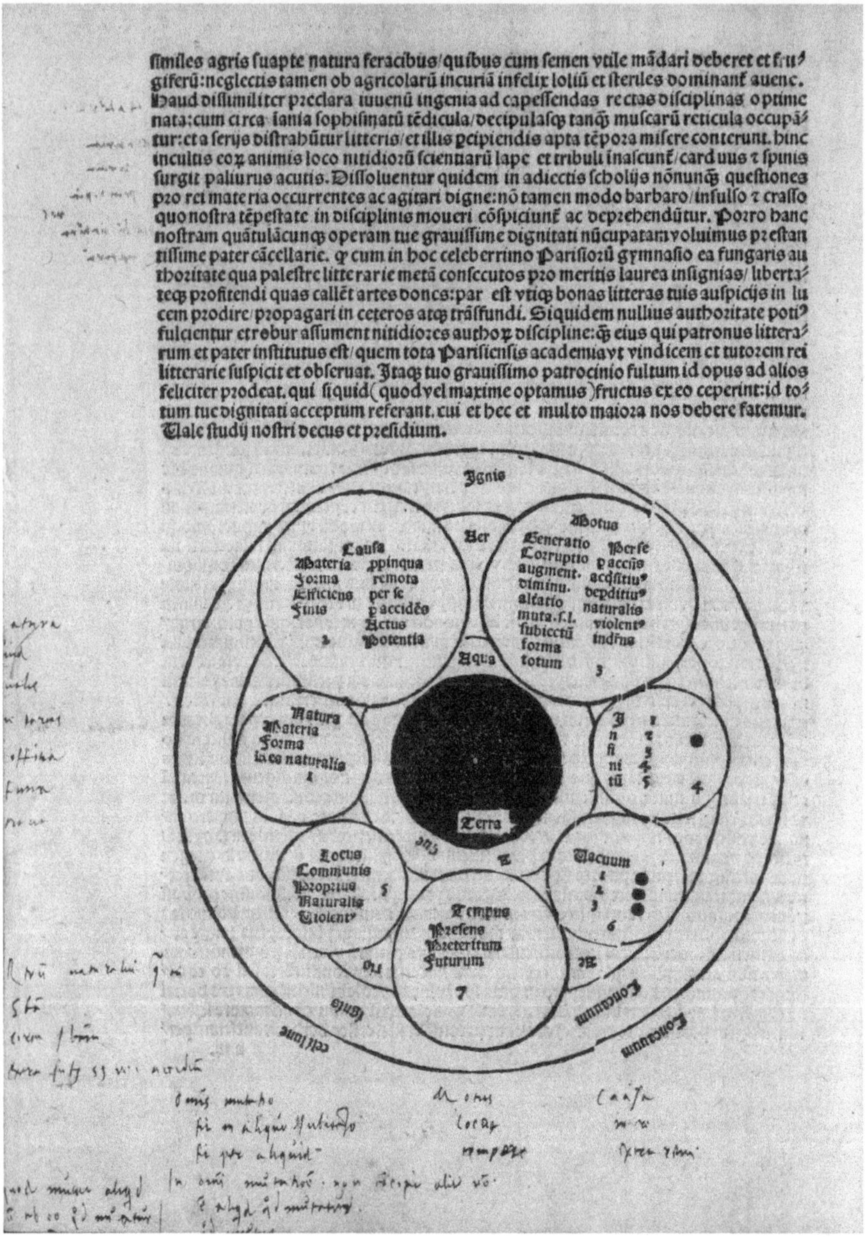
similes agris suapte natura feracibus/ quibus cum semen utile mādari deberet et fru-
giferū: neglectis tamen ob agricolarū incuriā infelix loliū et steriles dominant auene.
Haud dissimiliter preclara iuuenū ingenia ad capessendas rectas disciplinas optime
nata: cum circa inania sophismatū tēdicula/ decipulasq tanq muscarū reticula occupā-
tur: et a serijs distrahūtur litteris/ et illis pcipiendis apta tēpora misere conterunt. hinc
incultis eorum animis loco nitidiorū scientiarū lape et tribuli inascunt/ carduus et spinis
surgit paliurus acutis. Dissoluentur quidem in adiectis scholijs nōnunq questiones
pro rei materia occurrentes ac agitari digne: nō tamen modo barbaro/ insulso et crasso
quo nostra tēpestate in disciplinis moueri cōspiciunt ac deprehendūtur. Porro hanc
nostram quātulācunq operam tue grauissime dignitati nūcupatam voluimus prestan-
tissime pater cācellarie. qui cum in hoc celeberrimo Parisiorū gymnasio ea fungaris au-
thoritate qua palestre litterarie metā consecutos pro meritis laurea insignias/ liberta-
temq profitendi quas callēt artes dones: par est utiq bonas litteras tuis auspicijs in lu-
cem prodire/ propagari in ceteros atq trāsfundi. Siquidem nullius authoritate potius
fulcientur et robur assument nitidiores authorum discipline: q eius qui patronus littera-
rum et pater institutus est/ quem tota Parisiensis academia vt vindicem et tutorem rei
litterarie suspicit et obseruat. Itaq tuo grauissimo patrocinio fultum id opus ad alios
feliciter prodeat. qui siquid (quod vel maxime optamus) fructus ex eo ceperint: id to-
tum tue dignitati acceptum referant. cui et hec et multo maiora nos debere fatemur.
Vale studij nostri decus et presidium.

Fig. 2.3 The diagram introducing Lefèvre's *Paraphrases philosophiae naturalis*, first published 1492, b2r; here from Lefèvre and Clichtove, *Totius philosophie naturalis paraphrases* (*Paraphases on the Whole of Natural Philosophy*), *commentariis adiectis* (Paris: Henri Estienne [I], 1510). ETH-Bibliothek Zürich, Rar 6491, https://doi.org/10.3931/e-rara-25809 / Public Domain Mark

In use, these formal arrangements of ideas in space are not static but kinetic. In the same way that a mathematical proof and diagram coax their reader through a persuasive motion, towards the habitus of knowledge, these textbooks endeavour to help students perform the mental motions that belong to each discipline.[27] Visually, the Fabrist diagrams belong at the front edge of the sixteenth-century explosion of schematic representations of the disciplines.[28] Walter Ong famously associated such visual schemes with fixed knowledge and the decay of dialogue.[29] But the Fabrist textbooks invite the opposite response, aiming to stimulate the to-and-fro that belongs to dialogue.

2.2 Dialogues and the Ingenuous Student

The dialogues of Lefèvre also set in motion the sociability these textbooks were intended to support. The *Easy Dialogue on Physics* (1492) opens with a businessman putting his young son into the care of two teachers. The boy's first task is to read a book of "introductory physics" while his new masters take a short walk. After the boy has had a chance to examine the basic concepts, they review by pointing to a diagram in a book. In fact, the diagram turns out to be the very figure that introduces the volume in which the dialogue appears: the masters ask the boy to explain the concepts he has reviewed—nature, cause, motion, place, vacuum, and so on—and the dialogue then uses his answers and questions to direct the course of the lesson in Aristotelian physics. The mise-en-abîme suggests the mnemonic and dialogic function Lefèvre intended his *introductiones* to fulfill. I shall also suggest that, Lefèvre used dialogues to frame the character of the good student as open, ingenuous.

Lefèvre often used dialogue. Besides the two on physics, I have already mentioned the two on metaphysics. There are more. Very occasionally, he incorporated short dialogic sections into his commentaries, as he did in a page on Porphyry's *Isagoge*, entitled "Disceptatio de universali."[30] In his *De magia naturali*, published in manuscript during the mid-1490s, he included two books framed as dialogues between himself, his patron Jean de Ganay, and his student Josse Clichtove.[31] Admirers later incorporated Lefèvre as a voice in their own dialogues, such as those by Alain Varènes and Giulio Landi.[32] Pedro Ciruelo's brief dialogue on astronomy may have been modelled on Lefèvre's example—certainly he borrowed liberally

[27] Three arguments that this is the correct way to understand mathematical proofs: Mancosu 1996, Netz 1999, Wagner 2017.

[28] See especially Siegel 2009 and Berger 2017, references at 17–18.

[29] Ong 1958.

[30] Lefèvre 1503, 23^{v}.

[31] On these, see now Mandosio 2018.

[32] de Varennes 1512, Landi 1564.

from Lefèvre's textbooks in order to publish a new curriculum of mathematics in Spain.[33]

Lefèvre explained his choice of dialogue by recounting a conversation with his student and ammanuensis, Guillaume Gontier. Lefèvre should write in the genre of dialogues, Gontier had said, for "if you use it, you will advise those who are learning how they should ask questions and how they should answer them; simultaneously, you will usefully counsel both student and teacher."[34] What kind of model is intended here? There are many medieval and renaissance examples of dialogues that might be called catechetical, in which the back and forth is merely a device to facilitate transmission of text. We could compare the widely republished textbook of Gregor Reisch, the *Margarita philosophica* (*The Philosophical Pearl*, first printed in 1503). The bulk of the work is dialogue, but the student mostly plays dummy to the master's ventriloquism. But Lefèvre's use of the genre is keenly aware of itself. With his explanation of his own choice, Lefèvre draws attention to two contexts of dialogue. First, he addresses his patron Germain de Ganay, saying "you asked me why, most learned Germain, why I wrote these as dialogue and not through flowing prose."[35] Then, his response to Ganay's question is not even Lefèvre's own—he claims his student Gontier had the idea. To the reader, Lefèvre offers the conceit of intellectual responsibility shared between himself, patron, and student. His dialogues emerge out of real world dialogue.

Not all dialogue was good. Lefèvre's textbooks eschew most of the trappings of *quaestiones* which took some of their structure from medieval university disputation, a form of dialogue closely related to oral performance.[36] Concern about disputation practices may have been a main reason that Lefèvre introduced some of the most difficult theories of the late medieval *calculatores* in his *Introductory Dialogues on Difficult Physics*. Lefèvre joined university critics from Leonardo Bruni and Jean Gerson to the university reformers in his own day in worrying that the practices of disputation and exercises such as *sophismata* were turning students into small-minded logical experts, trading broad learning for narrow technical points. Lefèvre took on the topics most blamed for such bad habits—such as the latitude of forms—and used dialogue to show how students might tackle them without getting trapped in unhealthy, disputatious behavior.[37] While sophista could be a neutral term for a student at a certain stage in their required course of disputations, pedagogues pushing for change such as Lefèvre overwhelmingly drew on the term's negative connotations.[38]

---

[33] "Dialogus disputatorius in additiones opusculi de sphæra mundi," in Ciruelo 1498 n4^{v}-n7^{r}.

[34] Lefèvre 1494, b1^{v} (Rice 1972, 22): "Si ita feceris, [Gontier] inquit, admonebis qui docturi erunt quo pacto interrogare debeant, interrogataque docere, et simul utiliter discipulo consules et docenti."

[35] Lefèvre 1494, b1^{v} (Rice 1972, 22): "Interrogasti me, doctissime Germane, cur per dialogos et non potius oratione perpetuo discurrente commentariolos tibi dicatos composuerim."

[36] Novikoff 2013.

[37] I argue this in "Decline of the Calculators in Paris c. 1500: Humanism and Print" (forthcoming).

[38] The term's humanist meanings are, however, complex, as shown by MacPhail 2011.

In contrast, good dialogue helped to reveal and develop the souls of students. An influential framework for this task was Aristotle's moral philosophy, particularly the *Nicomachean Ethics*. Whereas Aristotle's work is largely aimed at the exceptional citizen, Lefèvre's *Ars moralis* (first printed 1494) introduced a vision of the *Ethics* particularly refocused for students. The terms generally come from Aristotle: virtues are the mean between overabundance or defect of a quality, so that bravery (*fortitudo*) lies between foolhardy carelessness (*audacia*) and timidity (*timiditas*). The virtues are either moral or intellectual in kind. The distinctive focus of Lefèvre and his student is a matter of emphasis, not specific concepts. Unlike the renditions of Aristotle such as those by Thomas Aquinas, which distilled the virtues to cardinal virtues such as prudence, temperance, courage, and justice, Lefèvre retained the longer lists of Aristotle and set special emphasis on what we might call "conversational" moral virtues: affability, truth, companionability, justice, and friendship (*affabilitas*, *veritas*, *comitas*, *iustitia*, *amicitia*). The sections on these virtues are the longest in the book.[39]

Clichtove's commentary of 1502 helps unfold the way such virtues were taught at the Collège du Cardinal Lemoine. In each case, the related vices damage the quality of intellectual conversation, leading to sophistry. Affability can become an unhealthy desire to please, making one an "Egyptian sophist" who promises to be anything at all.[40] Truthfulness disintegrates into arrogance or dissimulation by claiming either more or less than one knows; here the example is those "sophists who claim themselves to be skilled and ready to teach in all knowledge, so as to acquire wealth."[41] The student who practices the virtue of companionability (*comitas*) when treated to "jokes and witticisms" (*ioci salesque*) finds the mean between raucous hilarity and stolid silence.[42] This point struck a chord with Clichtove, seasoned teacher of adolescents: he noted the importance of recreation in the midst of scholarly labour, commending the ball games played by Roman exemplars such as Scipio, Lelius, Valerius, and Scaevola, and warning off the dangers of obscene laughter, "unchaste" jokes, and dice.[43]

Friendship thus emerges as a virtue deserving particular care and attention for the sake of knowledge itself. Like Aristotle, Lefèvre and Clichtove presented friendship as a civic virtue too: friendships for mutual gain bind merchants together; and friendships between rulers and ruled, husbands and wives, parents and children bind together the commonwealth. But the highest form of friendship is the shared

---

[39] Within the *Artificialis introductio*, *amicitia* is the virtue allotted the most pages (Lefèvre and Clichtove 1502a, 33^{v}-41^{r}); the next is *iusticia* (27^{r}-33^{r}).

[40] Lefèvre and Clichtove 1502a, 23^{v}: "Et is nichilo differre videtur ab Egyptio sophista, qui et arbor fiebat, et aqua, et fera et quicquid liberet, ut acutor est Basilius in libro de liberis educandis." See Basil of Caesarea, *Address to Young Men on the Right Use of Greek Literature*, §9.

[41] Lefèvre and Clichtove 1502a, 25^{r}: "sophistas qui scientiam omnem se callere et tradituros profitentur, ut divitias ea ratione sibi comparent. Quales describit Plato Protagoram, Euthidemum, Dionysodorum et ceteros."

[42] Lefèvre and Clichtove 1502a, 25^{v}-26^{r}.

[43] Lefèvre and Clichtove 1502a, 26^{v}.

goodwill of equals, directed to common ends. In the standard translations, Aristotle had called this *amicitia honesta*. Lefèvre and Clichtove offered a telling new phrase: *amicitia studiosa*. The adjective *studiosa* alludes to learning, and especially to desire. Lefèvre points out that "studious friends" are rare; Clichtove argues that this is "because it is hard to love deeply many people who have been shown by long familiarity, on account of their integrity."[44] The *studiosus* desires good for the sake of the other, so that he desires for his friend that which he wants for himself: for this reason "a friend is another self."[45] Finding another self is rare.

Not anyone can become a *studiosus*, then. Like most manuals reflecting on intellectual formation, these pedagogues wrestled with a language of inborn nature to consider each student's raw potential. One popular term was found in the rhetorical language of Cicero and Quintilian: the term *ingenium*, referring to those intellectual qualities "inborn" by nature.[46] Clichtove did not dwell on this term, however, and instead preferred the term *ingenuous*. Discussing the intellectual virtues, he considered what qualities might predispose one to various forms of study. He reprised the older language of "servile" arts, for those "mechanical" crafts that deal with mere technique and matter. "For this reason," he said, "among the ancients long ago, they were suited by ingenuous mind to letters and virtue, while by hard *ingenium* they were only suited to the crafts."[47] The intellectual life of the *studiosi* belongs to the ingenuous.

The word ingenuous had ambivalent meanings. Calepino, the age's most widely used dictionary, chiefly linked the word with being freeborn, *in libertate natus est*, but also noted then that *rursus ingenuum dicitur naturale*, "again, the ingenuous thing is said to be 'natural.'"[48] Pedagogical manuals were often written by impecunious scholars looking to tutor the sons of aristocratic patrons, and they regularly alluded to the well born nature of their prospective students. Such manuals used the language of ingenuity to identify these students: freeborn, well born, or nobly bred.[49] Clichtove and Lefèvre never, so far as I have found, used the word *ingenuous* in this sense. Their readership in such textbooks was not patrons, but youth from a range of social standings. For them, the ingenuous youth will have an open nature,

[44] Lefèvre and Clichtove 1502a, 36ʳ: "Nam difficile est multos longa consuetudine probatos propter honestatem vehementer amare."

[45] Lefèvre and Clichtove 1502a, 40ʳ: "Nam studiosus sese habet ad amicum perinde atque ad seipsum. Est enim amicus alter ipse." This helps contextualise a line often associated with Erasmus' later use of it in Adage 2. See further Eden 2001.

[46] Marr, Garrod, Marcaida, and Oosterhoff 2018, 19–20, 36–46.

[47] Lefèvre and Clichtove 1502a, 43ʳ: "Quare apud antiquos ingenuo perdiu animo ad litteras et virtutes, duro autem ingenio ad artes solum accommodabantur."

[48] Calepino 1522, s.v. Dictionaries are of course conservative sources, obliquely reflecting the range of use: cf. Robert Estienne's *Dictionarium latinogallicum*, which also focuses on connections to *honnesteté*, though hinting at openness too: "Ingenuè, adverbium, *franchement et sans crainte*." Estienne 1543, 371.

[49] E.g. the classic examples collected in Kallendorf 2002.

of the kind who might, through shared studies, college sociability, and conversation, become a true friend.

The dialogues demonstrate how this kind of engagement might lead to better knowledge. In part, they often give a constructive role to the student, who becomes an active participant in the pursuit of new knowledge. In the *Introductory Dialogue on Difficult Physics*, the youth Neanias often reflects insights that prompt his teachers to marvel. At one point a teacher considers the very phenomenon: "It seems wrong to me for an old man to marvel at a boy, since wonder arises from ignorance, even of those things boys seem to understand. But it seems to me worse to preserve one's ignorance to the grave, so I profess that I marvel at this boy!"[50] In the give and take of dialogue, for a moment the boy becomes teacher.

The dialogue therefore shows the fertility of those who strive together in friendly wrestling for truth. At the head of the *Difficult Physics*, Lefèvre presents an enigmatic set of notes on the difficulty of ascertaining knowledge:

> ¶ Analogy is a definite proportion of one thing to another, by which uncertain things are proved by certain things—this method was wonderfully begun by Plato, completed by Aristotle, unknown to many, and hidden from most philosophers. ¶ Proteus was the son of Ocean and Tethyos, a very great prophet who was unwilling to release truthful sayings unless forced and overcome. ¶ Milo of Croton was the strongest of athletes who killed a bull in the Olympic games with the blow of his bare hand, then carried it the length of the field holding his breath, and ate the whole bull on the same day.[51]

The point about analogy reflects Lefèvre's preoccupation with applying the language of mathematical proportion to a wide range of phenomena.[52] It also echoes ideas of the hidden concord between dissimilar things—and people, as Lefèvre had put it in his introduction to the *Ethics*.[53] Here the obscurity of analogy is stressed; it is a method for difficult knowledge. The theme of difficulty is compounded by the examples of Proteus and Milo of Croton, who become parables of the extreme labour needed to uncover truth.[54] For Bacon, significantly, Proteus especially would

---

[50] Lefèvre and Clichtove 1502b, 127^{r}: "Turpe michi videtur senem virum, puerum admirari. Siquidem admiratio ex ignorantia nascitur, eorum etiam que pueri michi cognoscere videntur. Sed turpius suam ignorantiam ad sua sepulcra servare michi visum est. Puerum tamen fateor sum admiratus."

[51] Lefèvre and Clichtove 1502b, 127^{r}. "¶Analogia certa rei ad rem proportio, que incerta certis probat—a Platone mirifice inchoata, ab Aristotele completa, multis ignota, et a plerisque philosophorum occultata. ¶ Proteus filius Oceani et Tethyos, vates maximus, nonnisi coactus victusque volens ora veridica solvere. ¶ Milo Crotonensis athleta fortissimus, qui nude manus ictu in certamine olympico taurum interfecit, stadio uno spiritu retento portavit, quem totum die illo comedit."

[52] This is a major theme of Oosterhoff 2018, *passim* and esp. 77–85.

[53] Concord binds friends together, just like goodwill: Lefèvre and Clichtove 1502a, 40^{r-v}.

[54] Proteus appeared in the *Odyssey* 4.382–569; Plato *Euthydemus* 288b; Virgil *Georgics* 4.387–529; Ovid *Metamorphoses* 8.731, 11.221–56; Diodorus Siculus *Antiquities of Egypt* 1.62. The identification of Proteus as a *vates maximus* suggests that Lefèvre had Diodorus at least partly in mind, while a later reference to Aristaeus reveals Virgil as another source.

stand in for nature's reluctance to divulge her secrets.[55] But here Proteus and Milo represent the sociable labour of conversation itself. This becomes clear at the dialogue's end, when one interlocutor nearly seems to have exhausted his knowledge—but the others are sure he knows more: "Just as Aristeus once bound up the varied and multiformed truthtelling Proteus with arms and chains, and so compelled him to tell the truth, so Homophron and I bind you in our arms, and we compel you to teach us what we asked."[56] The compulsion draws on the bonds of friendship, notes the put-upon speaker: "you're a friend (*comes*), and beautifully you compel me—fearing the worthy arms of Milo—to assent to your requests."[57] Friendship does not evade difficulty, but because friends share desires, they can push difficult conversations to the point of bearing good fruit.

2.3 Sources of Ingenuous Affect

The ideal of the ingenuous student saw intellectual fertility not as the gift of noble birth but as deep moral character prepared for laborious study. The ideal had several sources. Socially, an argument from nobility would not have been useful to the majority of students at Paris. Lefèvre and Clichtove, like many of their students, came from comfortable but not particularly aristocratic families in the villages of Picardy. Indeed, a prosopographical study would set them firmly within the rising classes of humanists and bureaucrats who used the universities—which doubled in number between 1400 and 1500—to move upward socially, claiming a nobility of virtue rather than birth.[58]

The specific shape of this virtue, however, was concretely Christian. (Even Lefèvre's commentary on the *Nicomachean Ethics* framed the virtue of love as modelled on Christ's love.[59]) Therefore the models for reforming knowledge were not just the rhetoricians and generals of ancient Rome, or the seven wise men of Greece (though Lefèvre and Clichtove regularly mentioned them). To position the ingenuous student, it will be helpful to consider the medieval idea of *reformatio*. This ideal had always included the early Church too, in which the simplicity of the apostolic life was held out as the model for Christian character. Saint Paul had

[55] For the early modern period see, inter alia, Wind 1958, 158–75; Burns 2001, 969–80; Pesic 2010, 57–82.

[56] Lefèvre and Clichtove 1502b, 148v-49r: "ut Aristaeus olim varium, multiformemque veridicum tamen Protea brachiis et vinculis implicuit, compulitque verum fateri, ita quoque ego et Homophron brachiis te implicabimus, cogemusque quod quaerimus te nos docere. One. Comis es o Enanti, et pulchre me cogis emerita Milonis brachia timentem, vestris petitionibus acquiescere."

[57] Lefèvre and Clichtove 1502b, 149r: "Comes es o Enanti, et pulchre me cogis emerita Milonis brachia temendem vestris petitionibus acquiescere."

[58] On the numbers of universities, see Verger 1992, 35–74. An outline of this intellectual class might be seen in Verger 2000; Copeland 2001.

[59] Oosterhoff 2016, 549–583, at 563.

hinted at the paradox this might mean for philosophy, pointing out that the Christian gospel was "folly" to the Greeks (1 Cor. 1.23). The fool of faith, therefore, might have wisdom to teach those learned in the schools. Wrestling with this dynamic infused some of the largest dissenting movements of the later middle ages, including Lollards, Hussites, and the Rhineland beguines and beghards. One target of this dissenting rhetoric was the universities, seen to foster philosophical pride rather than submitting to the spiritual needs of the people.

Lefèvre's most prized models were some of the most influential voices in this critique of university philosophy: the Catalan mystic Ramon Llull (1232–316), the Paris university chancellor Jean Gerson (1363–1429), and the German Cardinal Nicholas of Cusa (1401–1464). Each of these are very different thinkers, and simply doing justice to the scholarship on their places in late medieval cultural and intellectual history would take us far afield.[60] What is relevant is the fact that they were taken, by Lefèvre and his circle, as an antidote to university problems. They were perceived, in particular, to share a concern for balancing affect and emotion in intellectual pursuits.

One example will help to make the point. Ramon Llull was the Catalan merchant who, after a dramatic conversion, taught himself Latin and Arabic and travelled from Paris to Tunis to share his "great art" as a tool of evangelism. On the one hand, Lullian arts—wheels and diagrams that allowed various attributes to be combined into revealing new combinations of logic and attributes of nature—are an intellectual's dream. The arts compute, necessarily and universally.[61] Yet Llull's Latin was clumsy and his stance firmly outside the university. This, as much as the logical aspirations, attracted Lefèvre's circle, who were responsible for a revival of the Catalan's works.[62] In 1499 Lefèvre first edited several of them, enjoining readers to overlook their autodidactic awkwardness: "do not hold back because this man was an idiota and illiterate, an ordinary inhabitant of the wild wastes and vast solitude; indeed, it is believed he was given a certain infusion from above, by which he far excels the wise men of our age."[63] Despite their rude exterior, Lefèvre thought the works were especially *useful* because of their practical teachings. There are plenty of books that deal with the *univeralium cognitio*, "the understanding of universals"; what was needed remained the *modus operandi*, "method of doing."

---

[60] Starting points: Renaudet 1953; Rice 1972; [Mazour-]Matusevich 2002, 237–263.

[61] An assumption is that computations dealing with words necessarily reflects possibilities for things. See introduction to Hösle and Büchel 1985.

[62] Victor 1975, 504–534; idem 1976, 313–345.

[63] Llull 1499, [1]^{v}: "Neque vos quicquam deterreat quod vir ille idiota fuerit et illiteratus horride rupis et vaste solitudinis assiduus accola, nam et creditur quadam superna infusione dignatus, qua sapientes huius seculi longe precelleret." In fact, Lefèvre himself decided to raise the literary standard of the work: "Quod minus humile, depressum, deiectumque propter stillum visum est, aliquantulum erexi, nichil de sententia mutans, quod Remundus passim in libris suis benignissime fieri postulavit."

Doing, for Llull, required loving speech to be the master virtue.[64] In 1505 Lefèvre edited the *Contemplationes* and *De amico et amato* (*On the Lover and the Beloved*).[65] The first work structures the Lullian art around prayers on God's attributes, creation, and humanity. God's *misericordia* provides the possibility of a *concordia* binding all things together. In Lefèvre's edition, the book is followed by the *De amico et amato*, which sets out 365 aphorisms as a dialogue between a devout Christian (the Friend) and God (the Beloved). Again, the possibility of the believer's intellectual ascent to God only happens as a process of love—first God's merciful self-giving, then the believer's answering desire.

Lefèvre's preface to this edition frame scholarly friendship in this context of divine desire. It begins with a memory soaked in affect: Lefèvre's first meeting with Llull's work in 1491, in the south of France, where the merest sight of the book had overwhelmed him with desire to read it.[66] The man who originally intended to sell him the book, seeing his response, insisted he take it for free: "such was his gentleness that he wanted me to take the book, and he likewise was in tears."[67] Lefèvre suggests that he treasured the book as an enduring "consolation," during a time when he wanted to become a hermit—but his circle of friends held him back. Lefèvre mentions a range of monastic and university reformers of the 1490s, including Jean Mombaer, Jean Rolin, Jan Standonck, and Philippe Bourgoing, who advised that, even though Lefèvre "burned with desire to take flight of the world," he was too ill and tired from overwork to pursue the rigors of the eremetic life.[68] He therefore "returned to the earlier arts … since I could not do it in solitude."[69] This remarkable account therefore describes the process by which Lefèvre came to devote himself to writing university *artes* and introductory textbooks in the first place. The textbook emerges from a sense of vocation imbued with the affective priorities of the Christian eremite.

~

Scholars have long tried to place a false dichotomy between universities and humanism.[70] I have argued that this textbook program of Lefèvre's circle (the first comprehensive such project to have been tried in print) successfully aimed at students, authorities, and topics belonging to the university context. Its original impetus includes devotional and intellectual forces from outside the university, but it was not—as is often said of the Italian cases—the pure revivification of ancient

[64] On love as the "master virtue," see Johnston 1996, 151–155.

[65] Llull 1505.

[66] Llull 1505, [1]^{v}. Edited in Rice 1972, 277.

[67] Llull 1505, [1]^{v}: "ille vero (que sua erat modestia) voluit ut librum caperem, et sublachrimabatur pariter."

[68] Llull 1505, [1]^{v} [cf. Rice 1972, 277]: "dum mundi fugam protelo desiderio tamen estuans."

[69] Llull 1505, [1]^{v}: "ad priores artes revolutus …cum id in solitudine non possim."

[70] The commonplace is still repeated by leading scholars, despite the weight of a generation of scholarship such as Lines 2006.

texts as a kind of a "Second Latin culture" that stimulated an alternative to the universities.[71]

Dialogues particularly reveal how our modern assumptions about literary authorship limit our attention to the range of early modern philosophical production. In the Bakhtinian model, the dialogic is aligned with the vatic, even what we might call the disruptive power of genius, understood in a Romantic sense of unfettered openness or freedom from rules. The genre of dialogue thus provides the freedom needed for genius to flourish—typical examples might be Galileo's *Dialogi* or Descartes' *La recherche de la vérité par la lumière naturelle*. Such examples offer a dichotomy too rigid for understanding the range of early modern dialogue, especially within the context of pedagogy. Infatuation with such genius has blinded us to the shallower, but much broader, intellectual production of early modern textbooks. These texts do wrestle with the vatic, including innate ability and inspiration through prayer, but they shift focus from the dialogic interlocutor as *genius* to his *ingenuous* nature. The ideal of the ingenuous student does have implications for social status, yet moves us away from the judgment that early modern education was simply a matter of supplying states with ambitious bureaucrats.[72] For the ingenuous student was not judged by birth or by a freedom from rules. Both nobility of birth and genius operate under the tyranny of fortune. Instead, the ingenuous student was an ideal who could be judged by his character and efforts to speak in ways ruled by desire for the good of the other—in a word, this ideal student was to be governed by studious friendship, namely, a shared desire for truth. The fact that ideals surely outstripped reality, then as now, is no reason to ignore them and their power.

Bibliography

Bakhtin, Mikhail. 1981. *The Dialogic Imagination: Four Essays*. Austin: University of Texas Press.

Bedouelle, Guy. 1976. *Lefèvre d'Étaples at l'intelligence des Écritures*. Geneva: Droz.

———. 1995. Lefèvre d'Etaples et Erasme: une amitié critique. In *Jacques Lefèvre d'Etaples (1450?-1536)*, Actes du colloque d'Etaples les 7 et 8 novembre 1992, ed. François Pernot, 23–42. Paris: Honoré Champion Éditeur.

Berger, Susanna. 2017. *The Art of Philosophy: Visual Thinking in Europe from the Late Renaissance to the Early Enlightenment*. Princeton: Princeton University Press.

Burns, William E. 2001. 'A Proverb of Versatile Mutability': Proteus and Natural Knowledge in Early Modern Britain. *Sixteenth Century Journal* 32 (4): 969–980.

Calepino, Ambrogio. 1522. *Vocabularium, thesaurus copiosissimus*. Toscolano.

Ciruelo, Pedro. 1498. *Uberrimum sphere mundi commentum intersertis etiam questionibus domini Petri de Aliaco*. Paris: Jehan Petit.

Copeland, Rita. 2001. *Pedagogy, Intellectuals, and Dissent in the Later Middle Ages: Lollardy and Ideas of Learning*. Cambridge: Cambridge University Press.

Corsten, Severin. 1987. Universities and Early Printing. In *Bibliography and the Study of 15th-Century Civilization*, vol. 5, British Library Occasional Papers. London: British Library.

[71] Witt 2012.

[72] Grafton and Jardine 1986.

Cox, Virginia. 1992. *The Renaissance Dialogue: Literary Dialogue in Its Social and Political Contexts, Castiglione to Galileo*. Cambridge: Cambridge University Press.
Curtis, M.H. 1959. *Oxford and Cambridge in Transition, 1558–1642*. Oxford: Clarendon Press.
de Varennes, Alain. 1512. *De amore dialogus, de luce dialogi, etc.* Paris: Henri Estienne.
Eden, Kathy. 2001. *Friends Hold All Things in Common: Tradition, Intellectual Property, and the Adages of Erasmus*. New Haven: Yale University Press.
Estienne, Robert. 1543. *Dictionarium latinogallicum*. Paris: R. Estienne.
Faye, Emmanuel, and James Hirstein. 2002. "Metaphysica collecta." Un cours de métaphysique fabriste pris en note par Beatus Rhenanus. Présentation et édition partielle. In *Chemins de la pensée médiévale*, ed. Paul J.J.M. Bakker, Emmanuel Faye, and Christophe Grellard, 169–191. Turnhout: Brepols.
Goldhill, Simon, ed. 2008. *The End of Dialogue in Antiquity*. Cambridge: Cambridge University Press.
Goulding, Robert. 2010. *Defending Hypatia: Ramus, Savile, and the Renaissance Rediscovery of Mathematical History*. New York: Springer.
Grafton, Anthony, and Lisa Jardine. 1986. *From Humanism to the Humanities: Education and the Liberal Arts in Fifteenth- and Sixteenth-Century Europe*. Cambridge, MA: Harvard University Press.
Hannam, James. 2007. *Teaching Natural Philosophy and Mathematics at Oxford and Cambridge 1500–1570*. PhD Dissertation, University of Cambridge.
Hirzel, Rudolf. 1895. *Der Dialog: Ein literarhistorischer Versuch*. Leipzig: S. Hirzel.
Hösle, Vittorio. 2012. *The Philosophical Dialogue: A Poetics and a Hermeneutics*. Translated by Steven Rendall. Notre Dame: University of Notre Dame Press.
Hösle, Vittorio, and W. Büchel, eds. 1985. *Raimundus Lullus, Die neue Logik. Logica Nova*, Philosophische Bibliothek 379. Hamburg: Felix Meiner Verlag.
Johnston, Mark D. 1996. *The Evangelical Rhetoric of Ramon Llull: Lay Learning and Piety in the Christian West Around 1300*. Oxford: Oxford University Press.
Kallendorf, Craig W., ed. 2002. *Humanist Educational Treatises*. Cambridge, MA: Harvard University Press.
Klinger-Dollé, Anne-Hélène. 2016. *Le De sensu de Charles de Bovelles (1511): conception philosophique des sens et figuration de la pensée*. Geneva: Droz.
Kraye, Jill. 1995. Renaissance Commentaries on the Nichomachean Ethics. In *The Vocabulary of Teaching and Research between Middle Ages and Renaissance*, ed. Olga Weijers. Turnhout: Brepols.
Landi, Giulio. 1564. *Le attioni morali*. Venice: Gabriel Giolitto de' Ferrari.
Lefèvre d'Étaples, Jacques, ed. 1492. *Paraphrases philosophiae naturalis*. Paris: Henri Estienne.
———, ed. 1494. *Introductio in metaphysicorum libros Aristotelis*, ed. Josse Clichtove. Paris: Johannes Higman.
———, ed. 1496a. *Arithmetica elementa; Musica elementa; Epitome in libros arithmeticos divi Severini Boetii; Rithmimachie ludus que et pugna numerorum appellatur*. Paris: Johannes Higman and Wolfgang Hopyl.
———, ed. 1496b. *Introductiones logicales in suppositiones, in predicabilia, in divisiones, in predicamenta, in librum de enunciatione, in primum priorum, in secundum priorum, in libros posteriorum, in locos dialecticos, in fallacias, in obligationes, in insolubilia*. Paris: Guy Marchant.
———, ed. 1503. *Libri logicorum ad archteypos recogniti cum novis ad litteram commentariis ad felices primum Parhisiorum et communiter aliorum studiorum successus in lucem prodeant ferantque litteris opem*. Paris: Wolfgang Hopyl and Henri Estienne.
———, ed. 1515. *Aristotelis castigatissime recognitum opus metaphysicum*. Translated by Cardinal Bessarion and Johannes Argyropoulos. Paris: Henri Estienne.
Lefèvre d'Étaples, Jacques, and Josse Clichtove. 1502a. Artificialis introductio per modum Epitomatis in decem libros Ethicorum Aristotelis adiectis elucidata [Clichtovei] commentariis. Paris: Wolfgang Hopyl and Henri Estienne.
———. 1502b. *Totius philosophiae naturalis paraphrases, adiecto commentario*. Paris: Henri Estienne.

Lefèvre d'Étaples, Jacques, Josse Clichtove, and Charles Bovelles. 1503. *Epitome compendiosaque introductio in libros arithmeticos divi Severini Boetii, adiecto familiari [Clichtovei] commentario dilucidata. Praxis numerandi certis quibusdam regulis (auctore Clichtoveo). Introductio in geometriam Caroli Bovilli. Astronomicon Stapulensis.* Paris: Wolfgang Hopyl and Henri Estienne.

Lines, David A. 2006. Humanism and the Italian Universities. In *Humanism and Creativity in the Renaissance: Essays in Honor of Ronald G. Witt*, ed. Christopher S. Celenza and Kenneth Gouwens. Leiden: Brill.

Llull, Ramon. 1499. *Hic continentur libri Remundi pij eremite*, ed. Jacques Lefèvre d'Étaples. Paris: Jean Petit.

———. 1505. *Contenta. Primum volumen Contemplationum Remundi duos libros continens. Libellus Blaquerne de amico et amato*, ed. Jacques Lefèvre d'Étaples. Paris: Guy Marchant for Jean Petit.

Lohr, Charles H. 1988. *Latin Aristotle Commentaries: II Renaissance Authors*. Florence: Leo S. Olschki.

MacPhail, Eric. 2011. *The Sophistic Renaissance*. Geneva: Droz.

Mancosu, Paolo. 1996. *Philosophy of Mathematics and Mathematical Practice in the Seventeenth Century*. Oxford: Oxford University Press.

Mandosio, Jean-Marc. 2018. Introduction. In *La Magie naturelle / De Magia naturali: I. L'Influence des astres*. Authored by Jacques Lefèvre d'Étaples, ed. and trans. Jean-Marc Mandosio. Paris: Les Belles Lettres.

Mann, Margaret. 1934. *Érasme et les débuts de la réforme française (1517–1536)*. Paris: H. Champion.

Marr, Alexander, Raphaële Garrod, José Ramon Marcaida, and Richard Oosterhoff. 2018. *Logodaedalus: Word Histories of Ingenuity in Early Modern Europe*. Pittsburgh: Pittsburgh University Press.

Marsh, David. 1980. *The Quattrocento Dialogue: Classical Tradition and Humanist Innovation*. Cambridge, MA: Harvard University Press.

[Mazour-] Matusevich, Yelena. 2002. Jean Gerson, Nicholas of Cusa and Lefèvre d'Étaples: The Continuity of Ideas. In *Nicholas of Cusa and His Age: Intellect and Spirituality: Essays Dedicated to the Memory of F. Edward Cranz*, ed. Thomas P. McTiche and Charles Trinkaus. Leiden: Brill.

McConica, James. 1979. Humanism and Aristotle in Tudor Oxford. *The English Historical Review* 94 (371): 291–317.

More, Thomas. 1986. Letter to Martin Dorp. In *The Complete Works of St. Thomas More*, ed. D. Kinney, vol. 15. New Haven/London.

Netz, Reviel. 1999. *The Shaping of Deduction in Greek Mathematics: A Study in Cognitive History*. Cambridge: Cambridge University Press.

Novikoff, Alex J. 2013. *The Medieval Culture of Disputation: Pedagogy, Practice, and Performance*. Philadelphia: University of Pennsylvania Press.

Ong, Walter J. 1958. *Ramus, Method, and the Decay of Dialogue: From the Art of Discourse to the Art of Reason*. Cambridge, MA: Harvard University Press.

Oosterhoff, Richard. 2014. *Idiotae*, Mathematics, and Artisans: The Untutored Mind and the Discovery of Nature in the Fabrist Circle. *Intellectual History Review* 24: 1–19.

———. 2016. Lovers in Paratexts: Oronce Fine's Republic of Mathematics. *Nuncius* 31 (3): 549–583.

———. 2018. *Making Mathematical Culture: University and Print in the Circle of Lefèvre d'Étaples, Oxford-Warburg Studies*. Oxford: Oxford University Press.

———. 2019. Jacques Lefèvre d'Étaples and Charles de Bovelles on Platonism, Theurgy, and Intellectual Difficulty. In *Plotinus' Legacy: The Transformation of Platonism from the Renaissance to the Modern Era*, ed. Stephen H. Gersh. Cambridge, UK: Cambridge University Press.

Pesic, Peter. 2010. Shapes of Proteus in Renaissance Art. *Huntington Library Quarterly* 73 (1): 57–82.

Renaudet, Augustin. 1953. *Préréforme et humanisme à Paris pendant les premières guerres d'Italie, 1494–1517*. 2nd ed. Paris: Édouard Champion.
Rice, Eugene F., ed. 1972. *The Prefatory Epistles of Jacques Lefèvre d'Étaples and Related Texts*. New York: Columbia University Press.
Schönau, Christoph. 2020. Jacques Lefèvre d'Etaples und die Bibel: Die Auseinandersetzung mit Erasmus um die Auslegung von Hebr 2,7. In *Renaissance und Bibelhumanismus*, ed. J. Marius, J. Lange van Ravenswaay, and Herman J. Selderhuis, 121–136. Göttingen: Vandenhoeck & Ruprecht.
Siegel, Steffen. 2009. *Tabula: Figuren der Ordnung um 1600*. Berlin: Akademie Verlag.
Traninger, Anita. 2012. *Disputation, Deklamation, Dialog: Medien und Gattungen europäischer Wissensverhandlungen zwischen Scholastik und Humanismus*. Stuttgart: Franz Steiner Verlag.
Trithemius, Johannes. 1512. *De scriptoribus ecclesiasticis*. Paris: Bertold Rembolt and Jean Petit.
Verger, Jacques. 1992. Patterns. In *A History of the University in Europe, Vol. 1: Universities in the Middle Ages*, ed. H. de Ridder-Symoens and Jacques Verger. New York: Cambridge University Press.
———. 2000. *Men of Learning in Europe at the End of the Middle Ages*. Translated by Steven Rendall. Notre Dame: University of Notre Dame Press.
Victor, Joseph M. 1975. The Revival of Lullism at Paris, 1499–1516. *Renaissance Quarterly* 28 (4): 504–534.
———. 1976. Charles de Bovelles and Nicholas de Pax: Two Sixteenth-Century Biographies of Ramon Lull. *Traditio* 32: 313–345.
Wagner, Roi. 2017. *Making and Breaking Mathematical Sense*. Princeton: Princeton University Press.
Wind, Edgar. 1958. *Pagan Mysteries in the Renaissance*. London: Faber and Faber.
Witt, Ronald G. 2012. *The Two Latin Cultures and the Foundation of Renaissance Humanism in Medieval Italy*. Cambridge: Cambridge University Press.
Zepeda, Henry. 2015. Euclidization in the Almagestum Parvum. *Early Science and Medicine* 20 (1): 48–76.

Chapter 3
"*Le meilleur livre qui ait jamais été fait en cette matière*": Eustachius a Sancto Paulo and the Teaching of Philosophy in the Seventeenth Century

Roger Ariew

Abstract Following the praise by René Descartes of Eustachius a Sancto Paulo's *Summa Philosophiae Quadripartita* as "the best book ever made on this subject," I attempt to explain the virtues of Eustachius' *Summa*. I contrast Eustachius' compendium with other textbooks from the time, such as those of the Jesuits of Coimbra and Franciscus Toletus (referred to by Descartes from his schooldays at La Flèche), and describe some changes already taking place in textbooks, conceived mostly as commentaries on Aristotle's works. As the requirement for direct commentary wanes, textbooks become more creative with their arrangement and distribution of materials, with what they cover and what they emphasize. Ultimately, works such as Eustachius' *Summa* uniformly omit all the Aristotelian texts; unfettered by the need to stick closely to the tradition of formal commentary, they break new grounds. Although the radical reconceptualization of the curriculum is fairly exciting, there is something left behind in that process. Commentaries on the works of Aristotle have the virtue that everybody who went to school could remember and quickly find where a particular topic is discussed, across all commentaries. Eustachius, understanding that students new and old would need something to replace the familiar ordering, introduced analytic tables, a pedagogical device reminiscent of Porphyrian trees, which could be "perceived at a single glance" and allow one to capture the relations among the topics of concern. I discuss Eustachius' tables and their influence: tables are emblematic of the great changes occurring to school texts during the first half of the seventeenth century.

Keywords Analytic Tables · Cartesian Logic · Late Scholastic Ethics · Late Scholastic Logic · Late Scholastic Textbooks

---

R. Ariew (✉)
Department of Philosophy, University of South Florida, Tampa, FL, USA
e-mail: rariew@usf.edu

S. Berger, D. Garber (eds.), *Teaching Philosophy in Early Modern Europe*, Archimedes 61, https://doi.org/10.1007/978-3-030-84621-3_3

When, in 1640, Marin Mersenne was sending out the manuscript of the *Meditations*, Descartes thought himself at war with the Jesuits. He wrote to Mersenne that he will not travel that winter, since he is "expecting the objections of the Jesuits in 4 or 5 months," and believed that he "must put himself in the proper posture to await them."[1] For that reason, he told Mersenne that he felt like "reading some of their philosophy—something I have not done in twenty years—in order to see whether it now seems to me better than I once thought." Thus, he requested that Mersenne send him "the names of authors who wrote philosophy textbooks and who have the most following among the Jesuits, and whether there are any new ones in the last twenty years." Descartes reminiscing about the Jesuit textbook authors he had known at La Flèche, then said he remembered "only the Coimbrans, Toletus, and Rubius," but also requested from Mersenne "to know whether there is someone who has written an abridgment of all of Scholastic philosophy and who has a following, for this would spare me the time to read their heavy tomes." Descartes recalled: "It seems to me that there was a Carthusian or a Feuillant who had accomplished it, but I do not remember his name"[2] We do not have Mersenne's reply, about the "Carthusian or Feuillant," but, presumably, he identified Eustachius a Sancto Paulo as the Feuillant Descartes remembered having written a summary of all of Scholastic philosophy in one volume, since, in his next letter to Mersenne Descartes wrote: "I bought the *Philosophy* of Brother Eustachius a Sancto Paulo," and added that Eustachius' *Summa* seems to him to be "the best book ever made on this subject."[3] Descartes continued to look for other Scholastic textbooks, seeking one as excellent as Eustachius', but written by a Jesuit; in that process he looked over the *Philosophy* of Abra de Raconis (who, unfortunately for Descartes, was not a Jesuit). He wrote to Mersenne, "I saw the *Philosophy* of Mr. Raconis, but it is far less suited for my purpose than that of Father Eustachius. And as for the Coimbrans, they are too long; I would wholeheartedly wish that they wrote as succinctly as the other [Eustachius], and I would much prefer to deal with the great Society [that is, the Jesuits] than a single individual."[4] Descartes also initiated another project, the precursor to the *Principles*: "My plan is to write, in sequence, a complete course of my philosophy in the form of theses, in which, without superfluous discourse, I will simply put all my conclusions, along with the true reasons from which I draw them, something I believe I can do in very few words. In the same book, I intend to have printed a textbook course in ordinary philosophy, such as perhaps the one by Brother Eustachius, with my notes at the end of each question, where I will add the various opinions of others, what we should believe concerning all of them, and perhaps at the end I will make a comparison between these two philosophies."[5] A month later, he informed Mersenne that he had begun the project; but unfortunately, it was soon

---

[1] Descartes 1996, III, 184–5.

[2] Descartes 1996, III, 185.

[3] Descartes 1996, III, 232.

[4] Descartes 1996, III, 251.

[5] Descartes 1996, III, 233.

aborted: "I am sorry about the death of Father Eustachius; for although it gives me more freedom to write my Notes on his Philosophy, I would nevertheless have preferred to do it with his permission and while he was alive."[6]

Descartes' praise for Eustachius' *Summa* seems to have been genuine. Eustachius did manage something of a feat, having succeeded in publishing the contents of the whole collegiate curriculum in one handy volume. The contrast between "the heavy tomes" of the Jesuits that Descartes remembered and Eustachius' *Summa* must have been remarkable. Still, I believe this by itself is not sufficient to explain the great popularity of Eustachius' *Summa*. The work was issued in different editions almost every other year from 1609 to 1648 and one its parts, Ethica, continued to be published a century later, well into the 1730s;[7] there were many other philosophical compendia written during the period, but none of them matched Eustachius' in popularity or longevity. In 1663, Leipzig University Professor Johannes Adam Schertzer published *Breviarium Eustachianum,* a 759 page "abbreviation" of Eustachius' *Summa*. Schertzer's work found its way into G. W. Leibniz' thesis, *Disputatio metaphysica de principio individui,* defended at Leipzig that same year.[8] More than six decades after Descartes' remark, when Leibniz was writing about pedagogy to the Jesuit Des Bosses, he brought up Eustachius' *Summa* as the exemplar of the philosophical textbook. He said to his correspondent Bartholomew Des Bosses: "As for you, worthy Sir, I see that you are embarking on the correct path of reforming and improving philosophy for use in the Schools, so that the youth are not infected with objectionable principles. … A teacher of theology encounters philosophy everywhere. Rather than a comprehensive philosophy course, which your time now does not allow, I would suggest you compose a philosophical summary, such as Eustachius a Sancto Paulo once prepared: it will serve as a textbook for those attending your theology lectures, and when published will someday reach the rest of the world."[9] So, taking a cue from Descartes and Leibniz, I wish to explain the virtues of Eustachius' *Summa,* and not just by referring to its (relatively) small size. I

[6] Descartes 1996, III, 286. Descartes had previously indicated that he does "not want to do anything with the writings of a living person without his permission, something I think should be easily obtained when my intention is known, which is to consider the one I will choose as the best of all the ones who wrote a philosophy" (Descartes 1996, III, 234). But he added: "I entirely lost the intent to refute this philosophy. For I see that it is so absolutely and so clearly destroyed by the sole establishment of mine that there is no need of any other refutation" (Descartes 1996, III, 470).

[7] Lohr 1988, II, 136 counts 16 editions from 1609 to 1648, but he misses the editions of the *Ethica*.

[8] Leibniz 1923– , VI.1, Section 13: "Atque ita intelligo Excell. Scherzerum, Praeceptorem meum summo studio observandum q. 43. Breviar. Metaph. Eustachii de S. Paulo." Section 19: "Defendit Scotum ex recentioribus Petrus Fonseca, quamvis a Murcia pro nostra sententia citetur, v. Met. c. 6. q. 5. et Eustachius a St. Paulo l. c."

[9] Leibniz 1875–90, II, 294–95: "Te vero, Vir Eximie, cum rectum iter ingredi videam emendandae atque exponendae Philosophiam ad usum scholae, ut juventus non poenitendis principis imbuatur… Et Theologiam docendi ubique sese offeret philosophia. Pro cursu philosophiae ampliore, quem nunc non patitur tempus tuum, Breviariam philosophicum condi suaderem, quale olim Eustachius a S. Paulo confecit; id auditoribus Theologia tuis *procheiron* loco foret, aliquando et in caeteros iturum."

first need to describe some changes already taking place in Scholastic textbooks that will culminate in Eustachius' *Summa,* namely those clearly observable in the textbooks of the Jesuits from Coimbra and Franciscus Toletus, referred to by Descartes from his time at La Flèche.

Reflecting some widespread pedagogical reforms, the writing of textbooks multiplied in earnest during the second half of the sixteenth century. At the time the Jesuits, following the model of the University of Paris, were in the process of reorganizing and standardizing their collegiate curriculum. Textbooks, both Jesuit and non-Jesuit, reflected those changes. For example, the Jesuits of the University of Coimbra, the Conimbricenses, wrote volumes by committee, presenting the works of Aristotle that were taught in their curriculum; they patterned themselves on the model of the great commentaries, each volume treating a specific text (*Physica, de Anima de Caelo*, etc.), but with an elaborate scholarly apparatus, giving both Aristotle's Greek text and a new Latin translation, as well as paraphrases or commentaries (*explanationes*) and *quaestiones*—the latter being the analysis of standard problems relevant to the text under discussion.[10] This pattern was generally accepted by other early textbook writers, although some of the Coimbran editions and other commentaries, such as those of Toletus, omitted the Greek versions of Aristotle. Still, unlike their commentary on the *Physica* or *de Caelo*, the Coimbrans' *Dialectica* was not a treatment of a single Aristotelian work, but a series of commentaries strung together according to the traditional order of Aristotle's *Organon*. But even this got modified in several ways, with some additions to the discussion and a fair amount of picking and choosing among the texts of the *Organon***.**

For example, after the preliminary questions, the Coimbran *Dialectica* begins in earnest with a commentary on Porphyry's *Isagoge*. It then gives the complete Latin texts of the *Categories* and *On Interpretation* with commentary and related questions, but it excerpts the *Prior Analytics,*[11] and continues with various chapters from Part I of *Posterior Analytics*[12] but omits Part II completely. Finally, only the first three chapters of the *Topics* are tackled and none of the text of the *Sophistical Refutations* is provided, though a couple of related questions are discussed. In their preface "To the Reader," the Coimbrans assert that they did not provide the full texts of the *Topics* and *Sophistical Refutations* in order to save their readers from an "unprofitable labyrinth"; they refer them to Petrus Fonseca's *Dialectial Institutions* instead.[13]

---

[10] These volumes came out between 1592 and 1606. They were originally led by Petrus de Fonseca who published his own *Dialectical Institutions* (1564) and *Commentary on Aristotle's Metaphysics* (1577–1589). Ultimately, they were directed by Manuel de Gois, who oversaw the publication of the full curriculum of commentaries: *Physics* (1592), *On the Heavens, Meteorology, Short Treatises on Nature*, and *Ethics* (1593), *On generation and Corruption* (1597), *On the Soul* (1598), and *Dialectics* (1606).

[11] The *Dialectics* provide only the first fourteen chapters and chapter 28 of *Prior Analytics*.

[12] The *Dialectics* provide only chapters 1–13, 27–29, and 33 from *Posterior Analytics* Part I.

[13] "Ne in Topicorum, et Elenchorum labyrinthos Lectorem induceremus, summam compendiosè texere constituimus, et curiosos earum fabricarum aucupes mittere ad introductionem D. Petri

Toletus' *Commentary on Aristotle's Logic* also begins with Porphyry's *Isagoge* (following the preliminary questions). And, after his commentary and questions on the *Categories*, Toletus inserts Gilbertus Porretanus' *Six Principles,* the latter being a rearrangement of Aristotle's categories.[14] He continues with *On Interpretation*, skips the *Prior Analytics* entirely, and gives a complete discussion of the *Posterior Analytics*, including its Book II, which is neglected by the Coimbrans. Although Toletus' *Logic* is a hefty book, there is no exposition of the *Topics* or the *Sophistical Refutations*. Perhaps because of such apparent deficiencies, the later editions of Toletus' Logic are supplemented by the folios of Ludovicus Carbone's *Additamenta.* These consist of a Preface with questions about the *Prior Analytics* and a Treatise on Syllogism, followed by a Treatise on the Instruments of Science, namely, definition, demonstration, topical syllogism, enthymeme, and example. The *Additamenta* end with a treatise on Precognition, that is, the conditions required for demonstration.[15] None of the *Additions* are properly commentaries on Aristotle's texts, nor do they provide the relevant original Aristotelian texts, but, clearly, they deal with questions discussed in connection with Aristotle's *Organon*—here primarily from the *Prior* and *Posterior Analytics*—that are missing in Toletus' Logic.[16]

There is also a major change in the Coimbrans' treatment of *Ethics*. With Logic the problem faced by the Coimbrans is that of making sense of different works as a single treatise, unified according to a relatively new schema. With Ethics the task is somewhat different, given that Aristotle's *Nicomachean Ethics* is the single work at the center of their discussions.[17] However, producing a treatise on the ten books of

---

Fonsecae è nostra Societate, cuius doctrinam, seu primam lac Dialecticae suis studiosis haec instillat Academia," Conimbricenses 1606, n.p. See also the end of the commentary on the *Topics,* that is, Conimbricences 1606, II, 749, where the authors justify the omission of the remainder of Aristotle's text and its interpretation again by referring to Fonseca's *Dialectical Institutions*.

[14] In Gilbertus Porretanus' *Six Principles* the ten Aristotelian categories are divided into two classes, namely, forms inhering in the subject: substance, quantity, quality, and relation (see Toletus 1572, chap 1. De forma, fols. 98–99) and the subordinate remaining six: action, passion, when, where, position, and habit (see chap. 2–7, De actione, De passione, De quando, De ubi, De positione, De habitu, fols. 99–102).

[15] *Additamenta ad commentaria D. Francisci Toleti in Logicam Aristotelis. Praeludia in libros Priores Analyticos; Tractatus de Syllogismo; de Instrumenti sciendi; et de Praecognitionibus, atque Praecognitis* (1597). Also published in Toletus 1597.

[16] The liberty to discuss whatever question a professor might wish and in whatever order is not something one should take for granted. In fact, it was specifically prohibited by Claudio Acquaviva (5th General of the Jesuits: 1581–1615) in his directives preliminary to the future Ratio Studiorum, *De ratione studiorum vitandisque novis opinionibus* (1586): "When his text [Saint Thomas] is read in classes, his order has to be followed in his questions and articles, except those omitted because of their difficulty; and in this latter case, the professor will indicate briefly the subject of the articles and questions and will invite his listeners to read them as they are presented. [...] Let no one adopt new opinions in the questions already treated by other authors; similarly, let no one introduce new questions in the matters related in some way to religion or having some importance, without first consulting the Prefect of studies or the Superior." See Rochemonteix 1899, IV, 11n–12n.

[17] Textbook authors know of other Aristotelian ethical works: both Dupleix and Bouju cite the *Magna Moralia*; Bouju also frequently cites the *Eudemian Ethics*.

the *Nicomachean Ethics* results in almost the same problem as before, in that its ten books do not themselves seem sufficiently unified—at the very least, one can wonder about the two apparently disparate topics collected in book III of the *Ethics* (on voluntariness and involuntariness and on the virtues of fortitude and temperance) or about the doubling of the discussion of happiness in books I and X. Thus, the task of producing an ethics becomes the one of re-conceptualizing the materials of the *Nicomachean Ethics* into a single unified treatise. Although there were commentaries on the *Nicomachean Ethics* written during the sixteenth and seventeenth centuries,[18] even the Coimbrans do not attempt to write their *Ethics* in that form; instead, they collect *Disputations* on the main issues associated with Aristotle's books of the *Nicomachean Ethics,* which they arrange according to a new schema, mostly following suggestions made by Thomas Aquinas. The Coimbrans' *Ethics,* consists of nine books of disputations (that is, sets of *quaestiones*) on: (1) the good; (2) the end; (3) happiness; (4) will, intellect, and appetite; (5) the good and evil of human actions; (6) the passions; (7) virtues in general; (8) prudence; and finally, (9) justice, fortitude, and temperance.[19]

Once the requirement for direct commentary is dispensed with, textbooks become even more creative with their arrangement and distribution of materials, with what they cover and what they emphasize. Ultimately, works such as Eustachius a Sancto Paulo's *Summa* uniformly omitted all the Aristotelian texts. Like Carbone and the Coimbran *Ethica* before him, Eustachius arranged the *quaestiones* more or less in the way in which the philosophy curriculum would have presented them—that is, Logic and Ethics, Physics and Metaphysics—in a single volume. If we look a bit more closely at the individual parts of the curriculum, however, Eustachius' *Summa*, unfettered by the need to stick closely to the tradition of formal commentary, breaks other new ground. For example, Part I of the *Summa*, that is, Dialectics or Logic, is structured into a tripartite treatise about the operations of the mind. After a Preface consisting of the usual questions on the subject and divisions of dialectics, and whether it is art or science, theoretical or practical, Eustachius discusses the first operation of the mind, which concerns "things presented to it by a kind of simple vision, without affirmation or denial," that is, "simple apprehension." This operation is the simplest of all, since it can be expounded by using only one word or by several words without the copula. Simple apprehension or cognition of the mind is divided into two kinds, the first confused, the second distinct: "The first involves only the intellection of what a word signifies; the second involves not merely what the word signifies but also a clear and distinct conception of the thing that the word signifies. The first may be said to be the apprehension of a word or a term, the second of a nature and essence. The former is shared by everyone who is familiar with language, including common people and peasants; the latter is found

[18] See Jill Kray in Garber and Ayers 1997, 1281–82.

[19] See Index Superiorum Disputationum, et Quaestionum in Ethica Aristotelis ad Nichomachum, in Conimbricenses 1593 (unpaginated front matter).

in the wise, who have explored the natures of things."[20] Simple apprehension is to be contrasted with the second operation, in which the mind "compares things and assents or dissents to them by denying," and the third operation where "from the many things thus collected together the mind infers something distinct from them by a process of reasoning or argument"; the third operation is then called "discourse or argument."[21]

Thus, instead of giving a six-part commentary on Aristotle's six books, Eustachius rearranges the materials into a tripartite schema, with the first part, simple apprehension, corresponding to the materials treated by the *Isagoge* and *Categories* plus the *Six Questions*, and the second, on judgment, the subject treated by *On Interpretation,* while the third, on argument, the materials of the *Prior* and *Posterior Analytics*, *Topics,* and *Sophistical Refutations*. However, although Eustachius' three parts still take on much of the standard materials, they are no longer merely about terms or even about things conceived in the mind, but about the actual operations of the understanding and their perfectibility. This becomes the predominant way of thinking about logic;[22] in fact, the major issue in Logic becomes whether it is properly divisible into three parts or whether one needs a fourth part. (I will return briefly to this topic toward the end.)

[20] Eustachius 1629 [1609], I, 14: "illa dicitur, cum quia vocabulum significet duntaxat intelligitur; haec vero, cum non tantum quid vox significet, sed etiam quid sit res significata clare et distincte concipitur: illa termini seu vocis, haec naturae et essentiae apprehensio dicitur: illa omnibus qui voces norunt, ut plebeiis et rusticis, communis est; haec in sapitentibus duntaxat, qui exploratas rerum naturas habent, reperitur." The Cartesian overtones are unmistakable; Le Grand draws them out: "We are said to perceive a thing, when we clearly and distinctly conceive the Object offer'd to us ... and stop there without forming any Judgment concerning them. And the Form, by the immediate perception whereof we are conscious of knowing any Object, is called an *Idea*. ... So that when we frame the *Species* of any thing by Thinking, that first view of our Mind, by which it represents and conceives the Thing as present, is called *Perception*, or in other words, *The first operation of the Mind*, or *Simple Apprehension*," Le Grand 1972 [1694], 1 col. a-2 col. b. In Le Grand's Latin edition, the passage is given as: "**Percipere** aliquis dicitur, dum rem propositam clarè et distinctè concipit. ... ibi subsistendo, et nullum de illis judicium formando. Atque forma, per cujus immediatam perceptionem, ipsius ejusdem cognitionis, conscii sumus, *Idea* appellantur. ... Sic ut cùm rei alicujus speciem Cogitatione efformamus, primus ille mentis nostrae intuitus, quo rem sibi presentem exhibet, aut concipit, Perceptio, seu aliis nominibus, *prima mentis operatio, aut simplex Apprehesio nuncupatur*," Le Grand 1680, 2.

[21] Eustachius 1629 [1609], Prima Pars Dialecticae, De iis quae ad primam mentis operationem spectant, I, 12 *et seq.*

[22] The claim is not that Eustachius invented the view, but that he was likely the first to use it to organize his Logic. The doctrine can be found in Toletus' *Commentary on Aristotle's Logic,* which, as we have said, is organized in a mostly traditional way. In his 6th preliminary question, Toletus rehearses several possible divisions of Logic, including one using the "operations of the intellect": "Triplex autem est intellectus operatio: Prima, simplicium apprenhensio, qua res ipsas singulas per se intelligimus: Secunda, compositio, seu Divisio, qua res intellectas unam alteri copulamus: Tertia, ratiocinatio est, qua ex noto ignotum discurretes decidimus: quae operationes similes sunt sermoni," (1572, fol. 9). Petrescu 2018 traces the doctrine back to some debates among Scotists, who are using aspects of Aristotle's *De anima* (book 3) to organize the logical works.

Although the radical reconceptualization of the curriculum seems rather exciting, there is something left behind in the process. Commentaries on the works of Aristotle have the virtue that everybody who went to school could remember and quickly find where a particular topic is discussed— for example, time, space, and void, in Book 4 of the *Physics*. This would be true across all commentaries. Something was needed that could take the place of such an arrangement. So Eustachius, understanding that students new and old would need a pedagogical device to replace the familiar ordering, introduced analytic tables, not unlike Porphyrian trees, that could be "perceived at a single glance" and allow one to capture the relations among his topics. (See Fig. 3.1 for Eustachius' "General Schema" for Logic.) Eustachius provides myriads of subordinate tables as well. He gives a

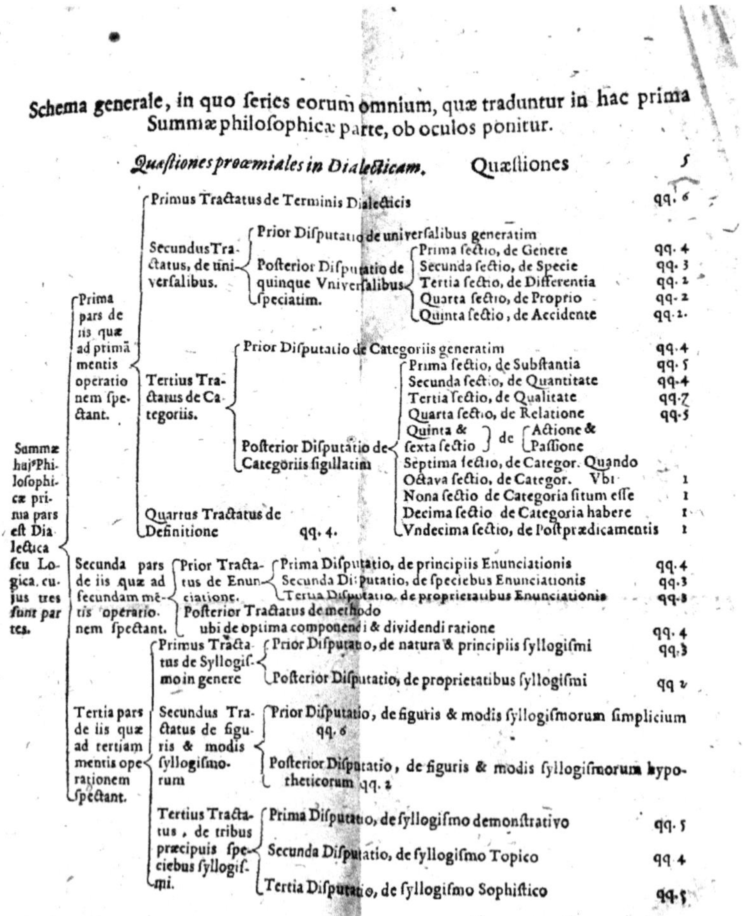

Fig. 3.1 Eustachius' "General Schema" for Logic

General Schema for each of the parts of Philosophy.[23] The schemas or tables go hand-in-hand with the reconceptualization of the materials.

For Ethics, Eustachius follows the Coimbran pattern, but conceives it in a tripartite fashion (see Fig. 3.2). After the preliminary questions, the first part of Eustachius' *Morals*, entitled On Happiness, is bifurcated into the good and the end and happiness itself (corresponding to the Coimbran parts 1–3). Eustachius' part two concerns the Principles of Human Actions and discusses in succession: internal principles of human action, such as will and appetite; acquired principles, such as habit; and external principles, including God and Angels (corresponding to the

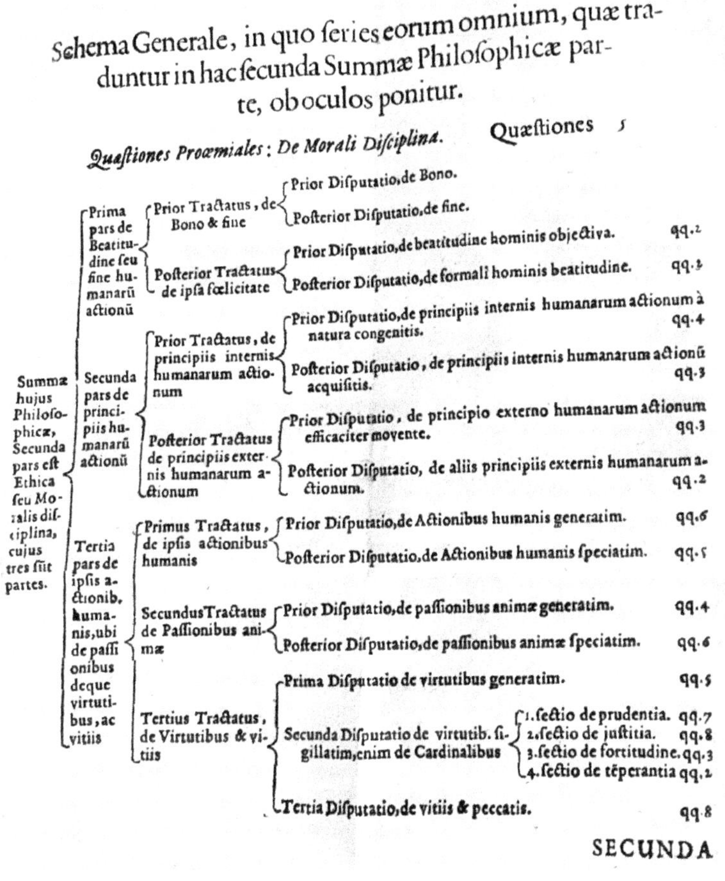

Fig. 3.2 Eustachius' "General Schema" for Ethics

[23] I don't know whether to call Eustachius' *Totius Corporeae Machinae ex 12 Coelis et 4 Elementis Compactae Descriptio* a table, but it is a fold-out like his other tables. And, like his other tables, it certainly contains an amazing amount of information in one figure.

Coimbran parts 4–5). Eustachius' third part is about Human Actions, that is, Passions, Virtues, and Vices. It is further divided into several disputations and questions: concerning passions, such as love and hate; the virtues in general; prudence; justice; fortitude; and temperance; and ending with a short disputation on vice and sin (corresponding to the Coimbran parts 6–9). Again, all this is given in a single "General Schema."

Others tried different devices. Let me briefly mention two of them: one by Théophraste Bouju and another by Abra de Raconis. A structural aspect of Bouju's *Corps de Toute la Philosophie* is that, like the Coimbrans, he provides Aristotle's text in both Greek and Latin, though unlike them, he furnishes only relevant quotations taken from all over the Aristotelian corpus. Bouju seems somewhat embarrassed by his quotations in Greek and Latin, given that he is writing a textbook in French (that is, for those who were not educated in School Latin or Greek). The verso to the title page indicates that "the Greek and Latin texts of Aristotle at the head of some chapters and articles in this work do not interrupt and are not an impediment to the French text that follows immediately after, having been put there only for the reasons brought forth on page 23." His defense of this procedure on pages 23–24 is spirited; here are two points from the general discussion: "I only wanted to relate some of Aristotle's texts at the head of the chapters of this work, taken from various places of his writings on each of the topics I treat, so that, in the order I relate them from various places, his own words can render his doctrine much clearer than when these passages are separated from one another, as they are in his books.... And if I did not produce his text in the language in which he wrote them, someone could have doubted whether the Latin version was any good."[24] As for Abra de Raconis, an unusual feature of his *Summa* can be seen in Part III (*Physica*). The Preface talks about the Tree of Physics "Arbor Physicae." De Raconis compares the whole of physics to a tree whose roots are the first principles and causes of natural bodies, whose bark is the accidents of natural bodies, whose trunk is the world, and whose branches are the heavens, the elements, and mixed bodies. The rest of the work is organized along those lines, with the various parts of the work being labeled roots, trunk, bark, branches, and twig, denoting the order of dependency or subalternation of the issues.[25] Bouju and de Raconis' devices are interesting for a variety of

[24] Verso to the title page of Bouju 1614: "Les textes Grecs-Latins d'Aristote, qui sont à la teste de quelques Chapitres et articles de cet œuvre, n'interrompent point le François, car il va tout de suitte, sans y estre astraint, ayans esté mis seulement pour les raisons portées en la page 23." Bouju 1614, 23-24: "Mais neantmoins j'ay voulu rapporter grand nombre de textes d'Aristote en teste des chapitres de cet œuvre, pris de divers endroicts de ses escrits sur chaque matière que je traicte, pour luy faire dire par ses propres paroles, en l'ordre que je les rapporte de divers lieux, sa doctrine plus clairement beaucoup, que quand ces passages sont separez les uns des autres, comme ils se trouvent en ses livres. ... Et dautant que si je ne rapportois son texte en la langue qu'il l'a escrit, quelqu'un pourroit doubter si la version Latine seroit bonne."

[25] "Huius arboris anatomia, nobis in toto hoc opere proponitur, in qua quidem arbore, radices primum scilicet Principia, et causas corporis naturalis; corticem, accidentia corporis naturalis; truncum, mundum: et ramos, coelos nempe, elementa, mixta spectabimus.... " de Raconis 1651, pars 3, p. 1.

reasons, but neither seems to have made the kind of splash that Eustachius' Tables did. Bouju's work had only one edition; French-language readers clearly preferred Scipion Dupleix's *Corps de philosophie*, which was reprinted many times in the first half of the seventeenth century.[26] De Raconis' *Summa* was more successful, with six editions in a dozen printings from 1617 to 1651.[27]

I return to Eustachius' Logic. As I said, most textbook authors followed Eustachius in his tripartite arrangement, which became the predominant way of thinking about logic. For example, De Cerizier's *Logic* (part I of *Le Philosophe français*) was arranged according to the three operations of the understanding, that is, a first part on *apprehension simple*, a second on *enunciation*, and a third on *discours ou raisonnement*.[28] That is also the structure given by the Scotist Claudius Frassen to his *Logic* (part I of his *Philosophica Academica*): the standard preliminary questions are followed by a discussion on the three operations of the mind, *apprehensio simplex*, *judicium*, and *discursus*; the structure of the book is therefore tripartite, following these operations. In contrast, Pierre Gautruche, in his *Philosophiae Institutio*, agreeing with the general line about there being three operations of the understanding,[29] added a fourth part to his Logic. The Preliminary Disputation on the operations of the mind (*de operationibus logicis, seu intellectualibus*) that would seem to give rise to a tripartite treatise is followed instead by four more disputations, on terms, then on enunciation, then argument or discourse, and finally on knowledge or (*scientia*), which is "the principal effect of the operations of logic."[30] Gautruche thus separates argumentation, about various kinds of syllogisms, from demonstration, or how we acquire knowledge that is certain. His bifurcation of reason into syllogistic or formal argument and demonstration, the latter being discourse that meets material conditions for certainty and knowledge, reflects some division about how to deal with the structure of Logic: should one treat syllogistic reasoning separately from scientific reasoning? Moreover, should one treat these two independently from method or order?

As one can see in his "General Schema," Eustachius inserts a brief discussion of method as part of his analysis of the second operation of the mind, enunciation, before his discussion of the third operation, discourse. In contrast with Eustachius' exposition, others do not treat method in conjunction with judgment, but with argument and discourse. Authors as diverse as Frassen and the Thomist Antoine Goudin claim that method should not be distinguished from the third part of logic as a fourth

[26] Another device is evident in Dupleix's *La curiosité naturelle redigée en questions selon l'ordre Alphabetique* (1631); it is a popular work that was translated into English and contained in his *Corps de Philosophie* (Dupleix 1620). Dupleix arranges his questions in alphabetical order because he cannot give universal causes for his curiosities; there are just no analytical tables that can be constructed for the materials.

[27] Lohr 1988, II, 3.

[28] Ceriziers 1643.

[29] Gautruche 1655, 46.

[30] Gautruche 1665, 137.

part unto itself.[31] Louis de Lesclache, however, although following the tripartite division of logic as the three operations of the mind, decided that method or order needed to be discussed in a fourth part; for Lesclache, Logic deals with "the three actions of the understanding," namely, "conception, judgment, and consequence," and the fourth concerning "method," that is, "resolution and composition": "This science called Logic is not content to regulate the three actions of our understanding, but also gives us precepts to invent and rightly dispose of many items of knowledge; that is why it is divided into four parts. The first three will give us the rules for rightly conducting the three actions of our reason and the fourth will establish the method we must follow in all sciences and in all discourses."[32]

We should note that the four-part structure of Lesclache's Logic resembles greatly that of the *Port-Royal Logic,* published some ten years later.[33] We should also add that Lesclache's textbook *La philosophie expliquée en tables* was the direct heir of Eustachius' *Summa* in an important and unusual way: the whole of Lesclache's textbook consisted of tables, beginning with a table on the usefulness of tables (see Fig. 3.3). Apparently, Lesclache had "prodigious success" in teaching Philosophy because, it is said, "the synoptic form of his tables rendered its study extremely simple."[34] Still, Lesclache's tables were criticized as unoriginal in a pamphlet published in 1650; the anonymous author declares: "If I did not find the origin of his tables in Abraham de Guise, Kekkermann, Eustachius a Sancto Paulo, Hallier, Hoeckelshoven, Timpler, Zuingger, and Crassot, I would have admitted that he is their author."[35] The author thus provides us with a sketch of the origin of

[31] Frassen 1668, 91-93; Goudin, 1726 [1668], I, 96-100 or 1864, I, 207-213.

[32] Lesclache 1651 [1650], pp. 4-6: "Cette science, qui reçoit le nom de Logique, ne se contente pas de régler les trois actions de notre entendement, elle nous donne encore des preceptes pour inventer & pour bien disposer plusieurs connoissances, c'est pourquoi elle sera divisée en quatre parties. Les trois premieres nous donneront les règles pour bien conduire les trois actions de notre raison, & la quatriéme établira la methode qu'il faut suivre dans toutes les sciences & dans tous les discours." Lesclache also published the sections of his textbook separate volumes, often without the author's name, date, publisher's name, or place of publication. We can note that there are copies of *La Philosophie expliquée par tables, Abregé de la Philosophie en Tables [La Logique], La Philosophie divisée en cinq parties. 2 La Seconde Partie De La Philosophie Ou* La Science Generale Divisee en quatre Parties, and *Seconde partie de la philosophie ou la Science Générale* (the latter two Paris: Chastellain), all from 1650.

[33] See Ariew 2015.

[34] Michaud and Michaud 1819, 282: "**Lesclache (Louis de)**, instituteur, né vers 1620, dans un village près de Clermont en Auvergne, après avoir fait d'assez bonnes études, vint à Paris, où il ouvrit une école de grammaire et de philosophie, qui eut d'abord un succès prodigieux, parce que la forme synoptique de ses tableaux en rendait l'étude extrêmement facile. Il avait eu le malheur de contracter un mariage mal assorti; et sa femme dissipa en peu de temps les économies qu'il avait pu faire. Les progrès de la philosophie de Descartes firent déserter son école; et il se vit forcé de quitter Paris, pour aller enseigner dans les provinces. Il s'établit d'abord à Lyon, et ensuite à Grenoble; mais sa méthode ne réussit ni dans l'une ni dans l'autre de ces villes. Il revint à Lyon; et il y mourut de chagrin, le 17 août 1671, dans un age peu avancé."

[35] "Quoy que je luy dispute la verité de beaucoups de ses discours populaires, je suis content neanmoins qu'il s'en arribuë l'invention. Bien-davantage, pour satisfaire entièrement à son dezir, si je ne treuvois l'original de ses tables dans Abraham de Guise, Keckerman, Eustache de S. Pol,

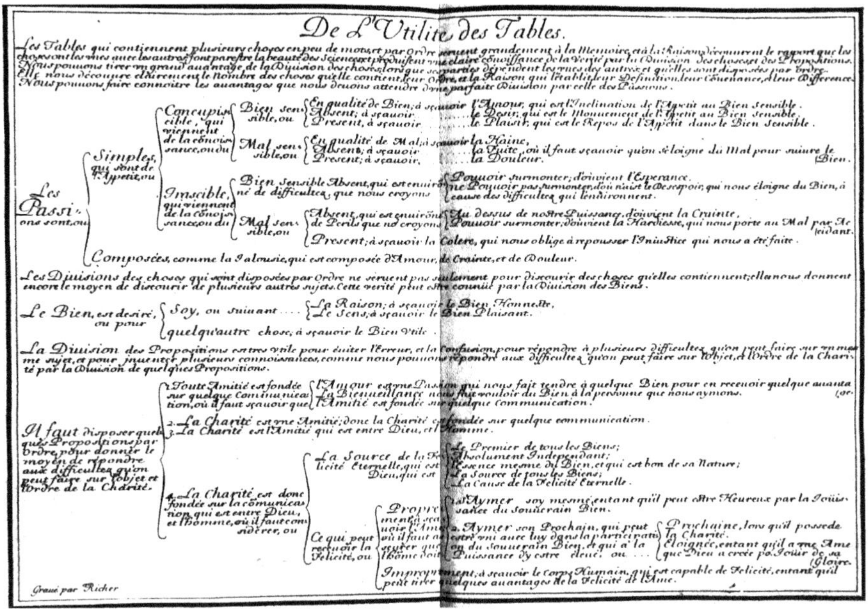

Fig. 3.3 Lesclache's Table about the Usefulness of Tables

philosophical tables from Eustachius and his contemporaries, but also referring to Theodorum Zuingger who published a commentary of Galenic medicine with Tables as early as 1561.[36]

I should conclude with one last example, representing the first Cartesian textbook to depict the whole Philosophy curriculum, namely, Jacques Du Roure's *La philosophie divisee en toutes ses parties, établie sur des principes évidents et expliquée en Tables et par Discours, ou particuliers, ou tirez des anciens ou des nouveaux auteurs; et principalement des Peripateticiens et de Descartes.* As his title indicates, Du Roure offers tables for every subject he discusses. (See Fig. 3.4 for Du Roure's Table of Descartes' Logic.) In his "General Discourse," Du Roure provides a brief history of philosophical tables, one similar to the one given by Lesclache's anonymous critic: "We consider here the main authors of tables and then reflect on

Hallier, Pacius, Hoeckelshoven, Timplerus, Zuingger, et Crassot, j'avouërois qu'il en est l'autheur." Anonymous 1650, 7-8. The authors being referred to are: Nicholas-Abraham de la Framboisière, *Oeuvres* (1613)—I owe this reference to Jacob Schmutz; Bartholomäus Keckermann, *Systema Logicae* (1611); Eustachius a Sancto Paulo; François Hallier, *Analysis Logicae* (1630); Johannis ab Hoeckelshoven, *Practicorum, sive Philosophiae Practicae Libri III* (1604); Clemens Timpler, *Tabulae Totius philosophiae practicae methodum adumbrantes* in *Philosophiae practicae systema methodicum* (1611); Theodurum Zuingger, *In Arte Medicinalem Galeni, Tabulae et Commentarii* (1561); and Jean Crassot, *Corporis Philosophiae* (1619).

36 While Eustachius is acknowledged a forerunner of Tables by his near contemporaries, it is clear that he did not innovate them; he simply adapted them for his purposes from another tradition, perhaps Lullian or Ramist.

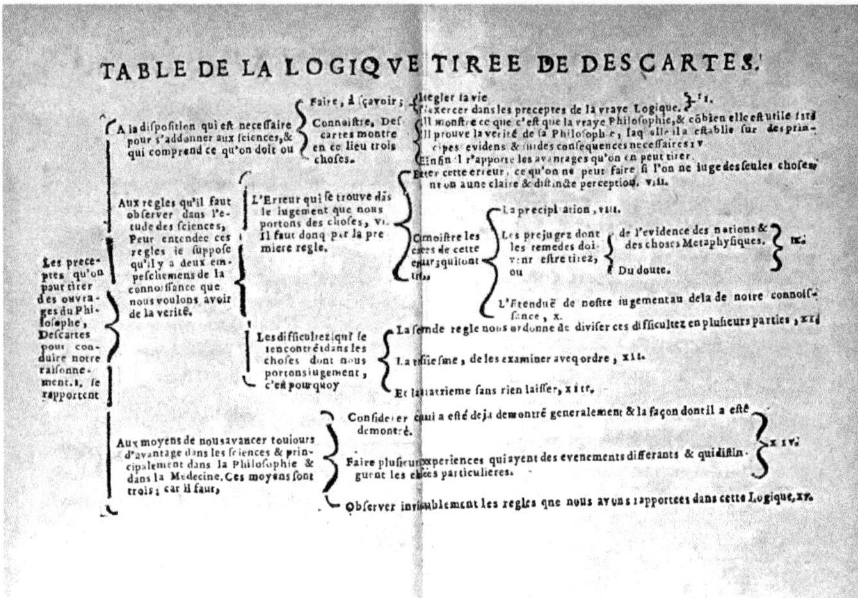

Fig. 3.4 Du Roure's Table of Descartes' Logic. (Bibliothèque Nationale de France)

the advantages they can bring us. These authors are Alsted, Zuingger, Timpler, Keckermann, Pacius, Zabarella, Abraham de Guise, Crassot, Saint Ange, and several others. … I note only that it is through ignorance or vanity that some people today attribute the invention of Tables to themselves. Tables are useful in that they serve judgment and memory, in that they contain in few words and in orderly fashion many things whose beauty they allow us to understand and appreciate."[37]

The fashion for Tables must have waned shortly thereafter, both with scholastics and Cartesians. One does not see any tables in Frassen's *Philosophiae Academica* in 1668, although one can glimpse an echo of Eustachius in Frassen's use of Eustachius' graphic display *Totius Corporeae Machinae ex 12 Coelis et 4 Elementis Compactae Descriptio* in late editions well into the 1740s. There are no tables in Antoine Le

[37] "Nous pouvons considerer en ce lieu les principaux Auteurs des Tables, et apres cela faire quelque reflexion sur ses avantages qu'elles nous peuvent apporter. Ces Auteurs sont Alstedius, Zuingger, Timplerus, Kecquerman, Pacius, Zabarella, Abraham de Guise, Crassot, Sainct Ange, et plusieurs autres. Je ne dis rien en ce lieu de ceux qui ont reduit en Tables toutes les autres sciences. Car on sçait que la Medecine par exemple a été publiée en cette façon par Vequer. Je remarque seulement, que c'est par ignorance, ou par orgueil, que quelques-un s'attribuent aujourd'huy l'invention des Tables. Leur utilité a en ce qu'elles servent au jugement et à la memoire, et qu'elles contiennent en peu de mots et par ordre beaucoup de choses dont elles nous font connoistre et aimer la beauté." Du Roure 1654, I, 18–19. The additional authors being referred to are: Johann Heinrich Alsted, *Compendium philosophicum* (1626), presumably Julius Pacius (or Pace), though I have not found any tables in his *Doctrina Peripateticae tomi tres* (1606), and Jacob Zabarella, *Tabulae Logicae* (1601). Vequer is most likely Jonannes Jacob Wecker, perhaps referring to his *Medicae Syntaxis* (1562) or to his *Medicinae Generalis* (1589), both of which contain Tables.

Grand and Pierre Sylvain Régis' multi-volume Cartesian textbooks from the 1690s. Tables, we have argued, were emblematic of the great changes occurring to the scholastic textbook at the beginning of the seventeenth century: letting go of the commentary tradition at minimum required some technique for keeping track of where one could find discussions of particular topics. The disappearance of tables toward the end of the seventeenth century is also the herald of another change, namely, the progress made by the next generations of Cartesian textbook authors in redefining the School curriculum in their own image.

Bibliography

Anonymous. 1650. *La philosophie particulière combattue par celle de l'escole. Ou l'on examine les Discours et les tables d'un philosophe de ce temps*. Paris: A. de Sommaville.

Ariew, Roger. 2015. La Logique de Port-Royal, les premiers cartésiens et la scolastique tardive. *Archives de philosophie* 78: 29–48.

Bouju, Théophraste. 1614. *Corps de toute la philosophie*. Paris: Vesue Marc Orry.

Conimbricenses. 1593. *In libros Ethicorum Aristotelis ad Nicomachum aliquot Cursus disputationes*. Lisbon: Ex officina Simonis Lopesij.

———. 1606. *Commentarii in universam dialecticam Aristotelis*. Coimbra: Ex officina Didaci Gomez Loureyro.

de Ceriziers, René. 1643. *Le philosophe français*. Paris: Chez Antoine de Sommaville et Augustin Courbé.

de Lesclache, Louis. 1651 [1650]. *La philosophie expliquée en tables, divisée en cinq parties*. Paris.

de Raconis, Charles François d'Abra. 1651. *Tertia pars Philosophiae, seu Physica*. Lyon: Sumptibus Irenaei Barlet.

de Rochemonteix, Camille. 1889. *Un Collège de Jésuites aux 17e et 18e siècles*. 4 vols. Le Mans: Leguicheux.

Descartes, René. 1996. *Œuvres de Descartes*, ed. Charles Adam and Paul Tannery, revised ed., 11 vols. Paris: Vrin.

Du Roure, Jacques. 1654. *La Philosophie divisée en toutes ses parties, établie sur des principes évidents et expliquée en tables et discours, ou particuliers, ou tirés des anciens et des nouveaux auteurs, et principalement des péripatéticiens et de Descartes*. Paris: Chez Thomas Jolly.

Dupleix, Scipion. 1620. *Corps de Philosophie contenant la Logique, la Physique, la Metaphysique, et l'Ethique*. Lyon: Simon Rigaud.

Eustachius a Sancto Paulo. 1629 [1609]. *Summa Philosophiae Quadripartita*. Cologne: Zetzner.

Frassen, Claude.1668. *Philosophia Academica, quam ex selectissimis Aristotelis et Doctoris Subtilis Scoti rationibus*. Paris: Apud Edmundum Couterot.

Garber, Daniel, and Michael Ayers, eds. 1997. *Cambridge History of Seventeenth Century Philosophy*, 2 vols. Cambridge: Cambridge University Press.

Gaultruche, Pierre. 1665. *Philosophiae ac mathematicae totius clara, brevis, et accurata institutio*, 5 vols. Caen.

Goudin, Antoine. 1726 [1668]. *Philosophia juxta inconcussa tutissimaque Divi Thomae dogmata*. Cologne: Cöllen & Huisch.

———. 1864. *Philosophie suivant les principes de Saint Thomas*, trans. T. Bourard. Paris: Poussielgue-Rusand.

Le Grand, Antoine. 1680. *Institutio philosophia, secundum principia Renati Descartes, nova methodo adornata et explicata, ad usum juventutis academicae*. London: J. Martyn.

———. 1972 [1694]. *Entire Body of Philosophy according to the Principles of the famous Renate Descartes*, 2 vols Repr. New York: Johnson.

Leibniz, G. W. 1875-1890. *Die philosophischen Schriften*, ed. C.I. Gerhardt, 7 vols. Berlin: Weidmann.

———. 1923– . *Sämtliche Schriften und Briefe*. Berlin: Akademie-Verlag.

Lohr, Charles H. 1988. *Latin Aristotle Commentaries, II: Renaissance Authors*. Florence: Leo S. Olschki.

Michaud, Joseph François, and Louis Gabriel Michaud. 1819. *Biographie Universelle, Ancienne et moderne*. Vol. 24. Paris: Michaud.

Petrescu, Lucian. 2018. Scholastic Logic and Cartesian Logic. *Perspectives on Science* 26: 533–547.

Toletus, Franciscus. 1572. *Commentaria una cum quaestionibus in universam Aristotelis logicam*. Venice: Apud Iuntas.

———. 1597. *Commentaria una cum quaestionibus in universam Aristotelis logicam: quibus accesserunt nonnulla additamenta Ludovici Carbonis*. Venice: Angelieri, Giorgio.

Chapter 4
Philosophical Cartography in Seventeenth-Century Paris

Susanna Berger

Abstract Susanna Berger's contribution considers summations of knowledge presented in philosophical broadsides. In particular, Berger examines broadsides designed by Martin Meurisse (1584–1644) and Jean Chéron (1596–1673) in collaboration with the engraver Léonard Gaultier (1560/61–1635), and the publisher Jean Messager (1572–1649). The prints of Meurisse, Chéron, Gaultier, and Messsager feature an innovative combination of word and image that encourages visual thinking among their observers: Not only do the broadsides use the space of the page to chart theoretical relationships, but they also offer visual commentaries that enrich philosophical concepts. These broadsides were employed as thesis prints for use in public examinations called disputations. In her chapter, Berger reflects on the ways in which these engravings relate to printed maps produced in this period. She argues that like maps, these philosophical engravings fabricate understandings of a field through the selective presentations of topics and a range of descriptive mechanisms that enhance the observer's comprehension, while advertising and advancing particular political interests. Her chapter includes a close analysis of a previously undiscussed section of a broadside summarizing natural philosophy and concludes with a consideration of the ways in which philosophical broadsides visualize the dangers of traveling beyond the known boundaries of learning.

Keywords Illustrated thesis prints · Maps · Aristotelian scholastic philosophy · Visual thinking · Philosophical broadsides

In his atlas *Theatrum orbis terrarum* (*Theater of the World*), the Flemish cartographer and humanist Abraham Ortelius (1527–1598) juxtaposes a world map with quotations of Seneca and Cicero (Fig. 4.1). The citation in the bottom right states: "I only wish that philosophy might come before our eyes in all her unity, just as the

S. Berger (✉)
University of Southern California, Los Angeles, CA, USA
e-mail: scberger@usc.edu

S. Berger, D. Garber (eds.), *Teaching Philosophy in Early Modern Europe*, Archimedes 61, https://doi.org/10.1007/978-3-030-84621-3_4

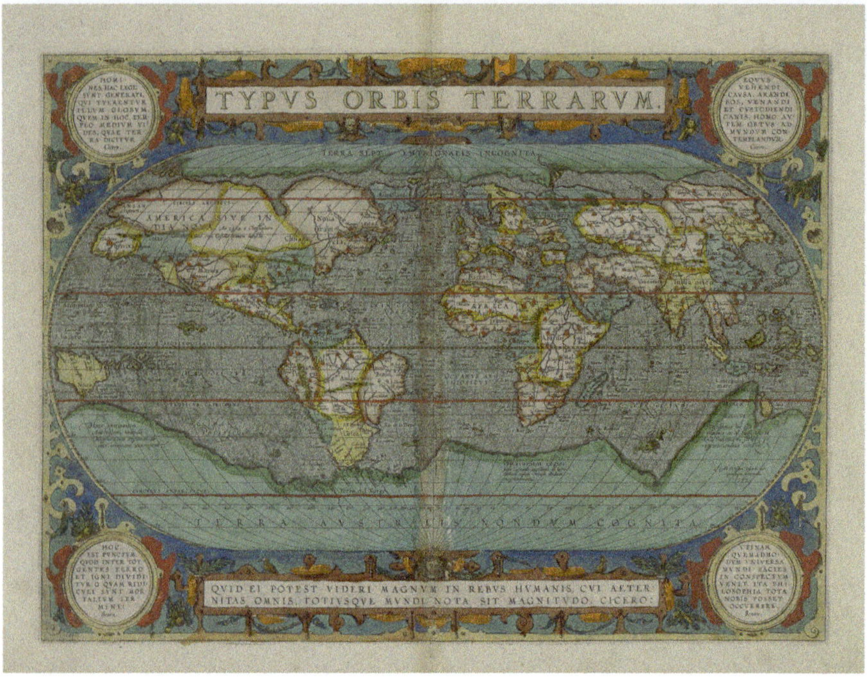

Fig. 4.1 Ortelius, *Theatrum orbis terrarum* (Antwerp, c. 1606), plate 1. Call #: STC 18855. (Used by permission of the Folger Shakespeare Library under a Creative Commons Attribution-ShareAlike 4.0 International License)

whole expanse of the firmament is spread out for us to gaze upon!"[1] The philosophical broadsides designed about a quarter of a century later by Martin Meurisse (1584–1644), a Franciscan professor of philosophy at the Grand Couvent des Cordeliers in Paris, as well as by the Carmelite philosophy professor Jean Chéron (1596–1673) might be said to have met this plea (Figs. 4.2 and 4.3). Between 1614 and 1622, in collaboration with the engraver Léonard Gaultier (1560/61–1635) and the publisher Jean Messager (1572–1649), Meurisse and Chéron produced a series of chorographic maps that visualize the disciplines of logic, natural philosophy, metaphysics, and moral philosophy across fictional realms.[2]

[1] Seneca 1979, vol. 4, ep. LXXXIX: "Utinam quemadmodum universa mundi facies in conspectum venit, ita philosophia tota nobis posset occurrere." Loeb classical library translation by Richard M. Gummere.

[2] The second century Greek astronomer and geographer Ptolemy distinguished "chorography" from "geography." Whereas the former denotes the practice of describing on a map particular regions, the later deals with descriptions of the earth in general. On the broadsides of Meurisse, Chéron, Gaultier, and Messager see Berger 2017. On Messager, see Selbach 2010, 35–51. For information on Gaultier, see Brugerolles and Guillet 2000, 1–24.

Fig. 4.2 Meurisse and Gaultier, *Clara totius physiologiae synopsis*, 1615. Engraving printed on paper, 25.5 × 18.5 in. (64.8 × 47 cm). Bibliothèque nationale de France, Cabinet des Estampes, Paris [AA4]. (Used by permission of the Bibliothèque nationale de France under a Creative Commons Attribution-ShareAlike 4.0 International License)

Fig. 4.3 Chéron and Gaultier, *Typus necessitatis logicae ad alias scientias capessendas*, 1622. Engraving on paper, 29.1 × 18.5 in. (74 × 47 cm). Princeton University Library Special Collections, Oversize Broadside 120

Meurisse and Gaultier created the *Artificiosa totius logices descriptio* (*Artful Description of Logic in Its Entirety*), hereafter *Descriptio*, in 1614.[3] The following year, they produced the *Clara totius physiologiae synopsis* (*Clear Synopsis of Physics in Its Entirety*), hereafter *Synopsis* (see Fig. 4.2).[4] In 1616 they made the *Laurus metaphysica* (*Laurel of Metaphysics*) and their fourth broadside, the *Tableau industrieux de toute la philosophie morale* (*Artificial Table of Moral Philosophy in Its Entirety*), hereafter *Tableau*, appeared in 1618.[5] Chéron and Gaultier created the *Typus necessitatis logicae ad alias scientias capessendas* (*Scheme of the Necessity of Logic for Grasping the Other Branches of Knowledge*), hereafter *Typus*, in 1622 (see Fig. 4.3).[6] These large-format broadsides, which feature a rich combination of word and image, were employed as illustrated thesis prints for students to use at oral examinations called disputations, held in the Franciscan and Carmelite convents that were affiliated with the University of Paris.[7]

We need only to think about alterations in, for instance, relative proportions of objects represented, coloring, or texture, to become cognizant of some of the ways in which maps create distortions. The images produced by cartographers do not replicate the world or heavens for us in a transparent or neutral fashion, but function to construct our understanding of these realms through selective representation of subject matter and a range of other descriptive means.[8] Similarly, these seventeenth-century philosophical maps are not simply illustrations of the theses that students were expected to discuss at disputations. They also offer interpretations of Aristotle's ideas that enrich and structure our understanding of his philosophy.

4.1 Mediated Descriptions

In addition to working on the philosophical broadsides of Meurisse and Chéron, Gaultier and Messager produced several chorographic maps of their own denoting actual sites. The *Vera et accurata descriptio situationis Magnae Cartusiae* (*True*

[3] For a close reading of the *Descriptio*, see Berger 2013a.

[4] On this broadside, see Berger 2013b. In this chapter, I will focus on details in the *Synopsis* that are not discussed in this *Art Bulletin* article.

[5] See Servus Gieben's account of the *Laurus metaphysica* (1990, 683–707).

[6] For an account of the *Typus,* see Berger, 2014. In producing these prints, Meurisse and Chéron also worked with one or several anonymous engravers of lettering. Meurisse and Chéron authored the texts inscribed on their works.

[7] Annotations reveal that these engravings were intended for use in disputations. The *Typus* even cites the names of the students who were expected to participate. Chéron 1622: "De his respond[ebunt] fratres P. Thomas Despres. I. Verbiale. N. Adam. V. La-ruele. H. Couplet. F. de Harne. R. Contansin. F. Fontaine. I. Boulin. etc. Anno. 1622." For sources on illustrated thesis prints, see Gieben 1993, 273–74n2; and Rice 1999, 165n1.

[8] Denis Cosgrove (1999, 2) stresses the artifice and epistemological import of mapping: "Acts of mapping are creative, sometimes anxious, moments in coming to knowledge of the world, and the map is both the spatial embodiment of knowledge and a stimulus to further cognitive engagements."

and Accurate Description of the Situation of the Convent of Grande-Chartreuse) was engraved by Gaultier and published by Messager in 1630 (Fig. 4.4). The adjectives "vera" and "accurata" invite viewers to regard this image as a truthful account of the monastery and its environs, when of course this achromatic, two-dimensional representation is necessarily a mediated description of a three-dimensional realm. In another chorographic map, Gaultier offers a bird's-eye view of Paris, one that he could have only created by imagining himself high above the ground (Fig. 4.5).[9] Producing maps such as this one required artists to rely on fanciful projection, in order to transmit credible likenesses of the world.

Titles often accentuate the mediation involved in early modern mapmaking. In both Ortelius's *Typus orbis terrarum* and the thesis print produced by Chéron, Messager, and Gaultier, the term "Typus," meaning "Scheme," designates these works as contrived or abstracted, in the manner of diagrammatic representations

Fig. 4.4 *Vera et accurata descriptio situationis Magnae Cartusiae,* engraved by Gaultier and published by Messager, 1630. Bibliothèque nationale de France, Cabinet des Estampes. (Used by permission of the Bibliothèque nationale de France under a Creative Commons Attribution-ShareAlike 4.0 International License)

[9] Jessica Maier stresses the artificiality of city maps for precisely this reason: "Early modern artists and viewers were earthbound, and—in an age before flight and aerial or satellite photography—there was no way for them to behold an entire city in a single glance." Maier 2015, 8. Maier has found that the term "ritratto" (portrait) is consistently used in titles for maps of cities produced in Renaissance Italy. Portraits of cities, as those of individuals, integrated an emphasis on likeness with artistic interventions. Maier 2015, 3–10.

Fig. 4.5 Gaultier's bird's-eye view of Paris, 1611. Bibliothèque nationale de France, Cabinet des Estampes. (Used by permission of the Bibliothèque nationale de France under a Creative Commons Attribution-ShareAlike 4.0 International License)

(see Figs. 4.1 and 4.3).[10] The map of the Convent of Grande-Chartreuse furthermore is characterized in the title cited above as a "descriptio" (description). The earliest philosophical broadside of Meurisse, Messager, and Gaultier, the *Artificiosa totius logices descriptio*, is also a "descriptio," one that is not natural but "artful" or "artificial." As Svetlana Alpers has stressed, the term "descriptio" was frequently invoked in the early modern period to characterize the activities of mapmakers, who offered conglomerative accounts of the world as opposed to presenting a single, homogeneous spatio-temporal slice of nature.[11] The *Descriptio* and the other maps of Meurisse, Chéron, Messager, and Gaultier skillfully use the space of the page to describe or map out the relationship of parts of Aristotelian scholastic philosophy to

[10] The metaphor of the theater of nature figures in the title of Ortelius's atlas, *Theatrum orbis terrarum*, as well as in the dedication of the *Synopsis*. Meurisse 1615: "...naturae theatrum ex(h)ibemus...." (We show this theater of nature). On this notion, see Berger 2013a, 269–93; and especially Blair 1997.

[11] Alpers 1983, 122 and 256n2.

each other and to the system as a whole. They are unified by conceptual relationships, as opposed to physical ones; they tend not to show a single moment in time, a perfectly cohesive space, or a single coherent action.[12]

In his treatise on pedagogy, Siegmund Jacob Apin (1693–1732), rector of the Aegidien school (in Braunschweig, Germany), also acknowledges the mediation involved in mapping. The treatise, titled *Dissertatio de variis discendi methodis memoriae causa inventis earumque usu et abusu* (*Dissertation on Various Methods of Learning, Invented for the Sake of Memory, and on their Use and Abuse*), was first published in 1725 and then again in a revised and augmented edition in 1731.[13] It lists and describes a range of early modern prints and illustrated books that were employed in the teaching of a variety of fields, such as philosophy, geography, grammar, history, astronomy, and law. Apin asserts that, thanks to the efforts of mathematicians, the discipline of geography has made considerable progress in his time, as compared with antiquity.[14] He goes on to discuss the aims of astronomers: "They wish to rival not only the heavens, but also the very structure of the sky, with a certain artful system."[15] Celestial cartographers, according to Apin, must show the "artful" or "artificial" arrangements of the heavenly realm.

After his account of early modern maps, Apin turns to the *Synopsis* and the *Typus*, two of the philosophical broadsides of Meurisse, Chéron, Messager, and Gaultier (see Figs. 4.2 and 4.3). He is particularly fond of the former, details of which will be analyzed in the next part of this chapter. The *Synopsis*, he writes, visualizes the world of natural philosophy with "such an abundance of invention that I cannot sufficiently admire the ingenuity of the author."[16] His reaction to the *Typus*, however, is far less positive. He asserts that its "images are horrible, absurd, and often have no similarity to the things they ought to indicate" (*imagines sunt horridae, absurdae, nullamque saepius cum rebus, quas debent indicare, habent similitudinem*).[17] Whereas Apin admired the ingenuity of Meurisse, Messager, and Gaultier, he is bothered by the artifice inherent to the map of Chéron, Messager, and

[12] In *The Art of Philosophy: Visual Thinking in Europe from the Late Renaissance to the Early Enlightenment*, I characterize the chorographic maps of Meurisse, Chéron, Messager, and Gaultier as "plural images" and contrast them with the modern model of the "singular image," or "tableau," in which the visual "field" is unified into a homogeneous whole through the implied presence of a fixed observer. Berger 2017.

[13] For an in-depth discussion of the *Dissertatio*, see Berger 2017, chapter 1.

[14] Apin 1731, 10: "In tantum vero ultimo aevo promoverunt hoc studium Mathematici, ut veteres ab orco revocatos, iuratos fore diceres, alium ab antiquo orbem esse nostrum: adeo priscos errores e tabulis suis expunxere, praeque illis longe nitidius pinxerunt ac curatitius." (In truth, Mathematicians have advanced this study in recent times so much, so that you would say that the ancients have been called back from the dead, under oath, [and] that another world from antiquity is ours: so they have erased the former errors from their maps, a long time before they painted them splendidly and carefully.)

[15] Apin 1731, 12: "Nec astra solum, sed & ipsam coeli compagem systemate quodam artificiali voluerent aemulari."

[16] Apin 1731, 18: "...tanta figmentorum ubertate, ut autoris ingenium non satis potuerim mirari."

[17] Apin 1731, 15.

Gaultier, by the gap between its images and what they denote. How are we to understand the artifice of these endeavors in philosophical mapmaking, an artifice that the broadsides' designers and critics recognized and stressed?

Historians of cartography have argued that across centuries and through a variety of means, maps promoted the attainment, dispensation, and codification of power for elite individuals or groups.[18] Did the broadsides of Meurisse, Chéron, Gaultier, and Messager also function as instruments of power, and if so, whose power did these maps of Aristotelian scholastic philosophy promote and over what?

To address these queries, we would do well to examine the two gates on the left and right edge of the bottom half of the *Typus*, where the powers of church and state are explicitly referenced (see Fig. 4.3). The gates' architraves are marked with inscriptions reminding viewers that Aristotelian scholastic logic is integral to the thinking of society's religious and civil officials.[19] Because scholastic philosophers had for centuries aimed at reconciling contradictory propositions in ancient texts of philosophy and theology, by the seventeenth century, the links between scholastic philosophy (in all of its varieties) and church doctrines were strong. In addition, fears of anti-Aristotelianism and Protestantism were particularly vivid in France following the assassination of Henri IV (1553–1610) and the religious wars.[20] Consequently, the promotion of Aristotelian scholastic philosophy in this period might have also struck viewers as a mode of strengthening the power of the church and state. The political functions of these philosophical maps also emerges through their dedications: the *Descriptio* is dedicated to the French statesman Jacques-Auguste de Thou (1553–1617), the *Synopsis* to King Louis XIII (1601–1643), the *Tableau* to Queen Anne of Austria (1601–1666), and the *Laurus metaphysica* to the First President in the Parlement of Paris Nicolas de Verdun (d. 1627). In their dedications, Meurisse and Chéron flatter and praise these eminent statesmen, strengthening their power.

4.2 Charting Natural Philosophy

How did philosophical cartographers juxtapose word and image to chart and comment on Aristotelian scholastic notions? I will explain how Meurisse and Gaultier formulate their considerations of the elements, generation, corruption, alteration, nutrition, growth, rarefaction, action, passion, and mixture, in carefully demarcated territories along the inside of the perimeter of a half-circle in the lower area of the

[18] Harley and Woodward 1987, 506. Harley 1988, 277–312.

[19] Chéron 1622: "The papacy is not wanting in logic" (*Pontificium non eget logica*); and "Civil law is not wanting in Logic" (*Jus civile non eget Logica*).

[20] There were of course plenty of Protestant scholastics. For a study of the natural philosophy of Philip Melanchthon (1497–1560) and his reinterpretation of Aristotelian natural philosophy, see Kusukawa 1995.

Synopsis (see Figs. 4.2, 4.6 and 4.7).[21] I will then continue with a discussion of the accounts of place, vacuum, motion, duration, time, and the continuum, in the region that surrounds the male natural body in the center of the half circle (Fig. 4.8).

Along the inside of the semicircular boundary, Meurisse follows Aristotle in asserting that the earth does not move, that it lies at the center of the cosmos, and that its shape is spherical.[22] Meurisse also notes that its "ambitus" (circumference)

Fig. 4.6 Detail of *Synopsis* pertaining to the elements, corruption, mixture, and the agent and patient. Bibliothèque nationale de France, Cabinet des Estampes. (Used by permission of the Bibliothèque nationale de France under a Creative Commons Attribution-ShareAlike 4.0 International License)

---

[21] There is no obvious overarching notion linking the concepts discussed in these spaces, although Aristotle does cite generation, destruction, increase, and alteration as four of six different kinds of change: Aristotle 1984, vol. 1, *Categories*, 15a14–15; and idem 1590, vol. 1, 20.

[22] Meurisse 1615: "Terra et aqua efficiunt unum globum sphericae figurae. est terra in medio mundi, immobilis, et centrum respectu coeli [*sic pro* caeli]." (Earth and water bring about one globe of a spherical form. There is earth in the middle of the world, unmovable, and a central point as concerns the sky.) Aristotle 1984, vol. 1, *On the Heavens*, 296b24–25 and 297a8. Idem 1590, vol. 1, 290.

Fig. 4.7 Detail of *Synopsis* pertaining to substantial generation, alteration, nutrition, growth, and rarefaction. Bibliothèque nationale de France, Cabinet des Estampes. (Used by permission of the Bibliothèque nationale de France under a Creative Commons Attribution-ShareAlike 4.0 International License)

measures 19,080 miles, its “diameter” 6070 miles, and its “semidiameter” (radius) 3035 miles. These numbers are nearly identical both to those noted as the most recent measurements by Christophorus Clavius (1538–1612) in his *Opera mathematica*, and to those cited by Eustachius a Sancto Paulo (1573–1640) in his popular philosophy textbook, which is the subject of Roger Ariew’s chapter in this volume.[23]

The rightmost area along the inner circular border interprets and describes the four elements (see Fig. 4.6). Meurisse, following Aristotle, notes that natural bodies

[23] The measurements of the diameter and radius given by Clavius and Eustachius are slightly more precise. For the earth’s diameter Clavius and Eustachius give 6070 10/11 miles and for its radius Clavius gives 3035 5/11 miles (1611, vol. 3, 115–16); and Eustachius 1609, vol. 2, 172. Meurisse 1615 also notes that the distance to the hollow space of the moon is 106,002 miles, but it is unclear where he found this measurement: “...ad concavum vero lunae 106002. milliaria...”

Fig. 4.8 Detail of *Synopsis* showing a male, natural body. Bibliothèque nationale de France, Cabinet des Estampes. (Used by permission of the Bibliothèque nationale de France under a Creative Commons Attribution-ShareAlike 4.0 International License)

are composed of elements, which are indivisible bodies.[24] He reiterates Aristotle's view that the sublunary elements can change into one another.[25] His textual explication of the elements is accompanied by four personifications of earth, water, air, and fire. The natural position of the earth is at the center of the cosmos. Above are the natural positions of water, air, and finally fire. Fire, who holds a burning candle and is surrounded by flames, is consequently shown flying above Air, who also flies,

[24] Meurisse 1615: "Datur elementum, estque illud in quod alia corpora dividuntur, quod inest potentia, aut actu, ipsum autem est indivisibile..." (The element is defined as that thing into which other bodies are divided, which is present in them potentially, or in actuality, but is itself indivisible...) Aristotle 1984, vol. 1, *On the Heavens*, 302a15–18. Idem 1590, vol. 1, 294. Abra de Raconis (1617, 513–14) cites Aristotle's definition of elements in his philosophy textbook.

[25] Meurisse 1615: "...quodlibet, elementum ex quolibet facilius...progignitur." (any element at all comes about from any other rather easily.) Aristotle 1984, vol. 1, *On Generation and Corruption*, 331a14–331b37. Idem 1590, vol. 1, 319.

with his legs submerged in clouds.[26] Below Air is Water, who leans on an urn from which flows a stream of water containing two swimming fish. Reeds, the traditional headdress of river gods and nymphs, are positioned on Water's head and she holds an oar with her right hand. Adriaen Collaert (1560–1618) personifies water with the same attributes as Meurisse and Gaultier: Collaert's Water leans on an urn and wears a headdress of reeds (Fig. 4.9). To the right and a little below the *Synopsis*'s Water sits Earth. She holds a cornucopia and rests on the ground, leaning against a cow.[27] The same year that he cut the *Synopsis*, Gaultier also engraved representations of the four elements on a frontispiece to a French translation of *Les hieroglyphiques* by Pierio Valeriano (1477–1558) (Fig. 4.10). On the frontispiece, Water also leans on an urn spilling water, and the posture of Earth resembles the stance of Earth in the *Synopsis*: in both representations, Earth leans on her left shoulder, while holding flowers up with her right arm.

Other pedagogical broadsides also visualize the elements, though sometimes in less detail. The Roman philosophy professor Philander Colutius (d. 1627) describes the elements in his *Physica seu naturae theatrum in typum totius philosophiae naturalis* (*Physics or the Theater of Nature, according to the Scheme of Natural Philosophy in Its Entirety*), hereafter *Physica*.[28] The keystone of an arch that is inscribed with an account of the elements exhibits them as well: earth is at the center, surrounded by water, then air, and finally fire (Fig. 4.11). The elements are also visualized in the *album amicorum* (friendship album) of Petrus Hondius (1589–1610).[29]

In the scholastic Aristotelian tradition, the most fundamental type of change is substantial change, whereby a new substance comes into existence, or an old substance ceases to exist. The text and image on substantial generation occupy the leftmost space under the circular border (see Fig. 4.7). The opening of Meurisse's account of substantial generation, based on a passage from Aristotle's *On Generation and Corruption*, emphasizes that nothing that can be perceived by the senses persists throughout substantial change: "Generation is defined as substantial; and is the changing of everything totally, with no perceptible subject persisting."[30] Whereas prime matter persists throughout substantial changes, it is imperceptible, and form,

[26] Cesare Ripa (1560–1622) states in his *Iconologia* that "Fuoco" (Fire) holds "un bel vaso pieno di foco" (a beautiful vase full of fire) in both hands. He characterizes "Aria" (Air) as "sedendo sopra le nuvole" (sitting above the clouds). Ripa 1603, 120–21.

[27] Ripa also mentions a cornucopia in his description of "Terra" (Earth) (1603, 122).

[28] On the *Physica*, see Berger 2016, 33–44. Matthäus Buschweiler (fl. 1611–1620) is cited on the sole surviving impression of this broadside as either its engraver or, more probably, its publisher. This impression is held by the HAB in Wolfenbüttel and is dated 1 September 1611. It is likely a second edition of the print, as it is filled with mistakes.

[29] For the images, see the album in the Cabinet des Manuscrits of the BRB, II 2254, fols 161^v (fire), 162^v (air), 163^v (water), and 164^v (earth).

[30] Meurisse 1615: "Datur generatio substantialis; estque mutatio totius in totum, nullo subiecto sensibili remanente." Aristotle 1984, vol. 1 *On Generation and Corruption*, 319b14–17. Idem 1590, vol. 1, 309. Eustachius cites this same definition from Aristotle in his textbook (1609, vol. 2, 195).

Fig. 4.9 Adriaen Collaert's 1575–1618 engraving *The Four Elements, Aqua*, London, BM, Prints and Drawings, 1870,0625.669. © The Trustees of the British Museum

Fig. 4.10 The four elements in a frontispiece engraved by Gaultier for Pierio Valeriano's *Les hieroglyphiques* (Lyon: Paul Frellon, 1615). Paris, Bibliothèque nationale de France, Cabinet des Estampes, Ed. 12 Rés. (Used by permission of the Bibliothèque nationale de France under a Creative Commons Attribution-ShareAlike 4.0 International License)

which is perceptible, does not persist throughout substantial changes. Meurisse and Gaultier accompany this discussion of generation with an image of a baby in a crib, two chickens with a nest, and a tree. The baby exemplifies the substantial generation of humans and the tree of plants. The chicken to the right of the tree sits on eggs and stands for the generation of animals.

Fig. 4.11 Detail showing elements from *Physica*. Wolfenbüttel, Herzog August Bibliothek, IE2. (Used by permission of the Herzog August Bibliothek under §51 UrhG)

Corruption is charted on the other side of the circular border, to the left of the elements (see Fig. 4.6). Meurisse explains corruption as the removal of substantial form from matter: "Corruption is change from the substantial being of form in matter to the non-existence of the same thing."[31] He stresses the link between corruption and generation: "The corruption of one thing is the generation of another.... In it resolution right down to prime matter comes about."[32] He visualizes corruption as a corpse lying on a bed of hay behind a skeleton holding a scythe. Throughout the sixteenth and seventeenth centuries, the human skeleton was frequently depicted to symbolize death; it usually held a scythe to denote life's being cut off. One of many

[31] Meurisse 1615: "Corruptio est mutatio ab esse substantiali formae in materia, ad non esse eiusdem."

[32] Meurisse 1615: "Corruptio unius est generatio alterius.... in ea fit resolutio usque ad materiam primam." Eustachius's (1609, vol. 2, 198) textbook contains almost the exact same phrase as the title of "Quaestio VII" in "Seconda partis Physicae": "An Corruptio unius sit generatio alterius." This instance of near equivalence between both genres suggests a close conceptual relation among the philosophical maps of Meurisse, Chéron, Messager, and Gaultier and textbooks from the period.

examples of this trope appears in the bottom right corner of a world-map made in 1596–97 by Jodocus de Hondt (1563–1612).[33]

Above and slightly to the right of the territory pertaining to generation is a plot of land visualizing alteration (see Fig. 4.7). Meurisse follows Aristotle in characterizing alteration as change in a thing from one property or quality to another opposing property.[34] Aristotle draws a distinction between substantial change and alteration, noting that whereas in substantial generation and corruption the persisting *substratum* is not perceptible, in alteration it is, although its properties change.[35] Aristotle cites water changing into air as an example of a substantial change.[36] Meurisse and Gaultier comment on alteration with a representation of a mother and child observing steam rising from a kettle of water over a flame. In the *Commentarii Collegii Conimbricensis...in libros de generatione et corruptione* the distinction between generation and alteration is discussed at length and Aristotle's characterization of water changing into air as substantial change is contested with reference to humors, such as the vapors shown emerging from the kettle in the representation of Meurisse and Gaultier.[37] Presumably, the visual commentary offered by the *Synopsis* alludes to these sorts of challenges to the Aristotelian doctrine.

To the right of the discussion of alteration is a territory visualizing nutrition, growth, and rarefaction (see Fig. 4.7). Nutrition, or the process of providing and accepting food, permits animated substances to live; Meurisse notes that food is subject to substantial conversion before it is able to nourish plants and animals.[38] He and Gaultier represent nutrition with a child feeding at its mother's breast. Growth is an increase in the size or quantity of the living being.[39] Meurisse explains that

---

[33] Reproduced in Scheurleer 1975, 250–51. On visual commentaries on mortality and the transience of worldly goods in *alba amicorum*, see Berger 2017, 167–68.

[34] Meurisse 1615: "Alteratio est motus physicus qui inter contrarias qualitates passibiles versatur..." (Alteration is a physical motion, which occurs among contrary perceptible qualities liable to change or decay...) Aristotle 1984, vol. 1, *On Generation and Corruption*, 319b32–320a1. Idem 1590, vol. 1, 309. See also Eustachius 1609, vol. 2, 201; and Abra de Raconis 1617, 583.

[35] Aristotle 1984, vol. 1, *On Generation and Corruption*, 319b3–320a3. Idem 1590, vol. 1, 309.

[36] Aristotle 1984, vol. 1, *On Generation and Corruption*, 319b16. Idem 1590, vol. 1, 309.

[37] Conimbricenses 1606, lib. 1, cap. 4, qu. 18, art. 4, p. 137: "...dicendum, cum ex aqua aer gignitur, corrumpi aquae humorem non per se, sed per accidens ad corruptionem subiecti." (it must be said, when air is generated from water, the humor of the water is not corrupted *per se*, but *per accidens* according to the corruption of the subject.) The Conimbricenses were Jesuits from the University of Coimbra. I am grateful to Paul Richard Blum for drawing my attention to this passage.

[38] Meurisse 1615: "Nutritio est productio partialis substantiae viventis ex conversione substantiali alimenti." (Nutrition is the bringing-forth of a partial living substance out from the substantial conversion of food stuff.) Aristotle emphasizes that the nutritive power is necessary for sustaining life: Aristotle 1984, *On the Soul*, 416b17–20; and idem 1590, vol. 1, 391. For accounts of nutrition from early modern Parisian textbooks, see Eustachius 1609, vol. 2, 294–98 and Abra de Raconis 1617, 723–27.

[39] Meurisse 1615: "augmentatio est acquisitio majoris quantitatis ex substantiali conversione alimenti." (Growth is the acquisition of a greater quantity from the substantial conversion of food stuff.) Aristotle 1984, *On Generation and Corruption*, 319b32–33. Idem 1590, vol. 1, 309.

whereas nutrition persists as long as the body lives, growth does not.[40] He also states that growth occurs throughout the entire body, which remains equivalent in number while increasing in size.[41] To explain growth, he and Gaultier show two infants on a turning wheel. The breastfeeding woman holds a string attached to the wheel's handle. By placing this nutritive source in command of the wheel's rotations, Meurisse and Gaultier visualize growth's dependence upon nutrition: without nourishment, growth is impossible.

This wheel of nutrition recalls the popular iconography of the wheel of fortune. Derived from classical and medieval philosophy, the iconography denotes the arbitrary character of Fate. In his *Liber de intellectu*, Charles de Bovelles (1475–1566) shows *Fortuna*, seated across from *Sapientia* and holding a wheel (Fig. 4.12). A king sits triumphantly at the top of the wheel, while three less fortunate men cling to its sides and bottom. By connecting growth to a sort of Fortune's wheel, Meurisse and Gaultier suggest that as natural bodies grow over time, their circumstances will be subject to the whims of Fortune.[42]

Whereas growth is a quantitative change brought about from the adding of substance (through nourishment) to a living being, rarefaction is a change that is not explained by an increase in the amount of substance, but rather by an increase in extension.[43] Meurisse characterizes rarefaction as an increase in quantity (through extension) and condensation as a loss.[44] He and Gaultier visualize rarefaction with a young boy blowing bubbles (see Fig. 4.7). The *Adages*, a collection of ancient proverbs written by Desiderius Erasmus (1466–1536) with commentary in 1508, includes the proverb "Homo bulla" (Man is but a bubble), to indicate the vulnerability of life:

> The lesson of this proverb is that there is nothing so fragile, so fleeting and so empty as the life of man. A bubble is that round swollen empty thing which we watch in water as it grows and vanishes in a moment of time.[45]

---

[40] Meurisse 1615: "...nutritio toto tempore vitae durat, accretio non." (nutrition persists for the whole time of life, the accretion does not.) Aristotle 1984, *On Generation and Corruption*, 322a23–25. Idem 1590, vol. 1, 311. Meurisse notes that nutrition is instantaneous, whereas growth is not: "differunt quia nutritio est instantanea, et accretio successiva..."

[41] Meurisse 1615: "Manet semper idem numero quod augetur, una quaeque pars aucti augeri debet." (That which is increased always remains the same in number, and each and every part of the increased thing must be increased.)

[42] On Fortune in *alba amicorum*, see Berger 2017, 167.

[43] Duns Scotus 1639, vol. 8, *Ordinatio*, 4. 12. 4, n. 22, p. 767–68. For an account of Scotus's understanding of rarefaction and condensation, see Cross 1998 159–66. See also Eustachius 1609, vol. 2, 328–29.

[44] Meurisse 1615: "Rarefactio fit per acquisitionem novae quantitatis, et condensatio per deperditionem..." (Rarefaction comes about through the acquisition of new quantity, and condensation through the loss.)

[45] Erasmus 1621, II.iii.48, p. 482: "Proverbium hoc admonet, humana vita nihil esse fragilius, nihil fugacius, nihil inanius. Est enim bulla tumor ille inanis qui visitur in aquis momento temporis enascens simul & evanescens." Translated by Roger Aubrey Baskerville Mynors in Erasmus 1991, 156.

Fig. 4.12 *Fortuna*, in Charles de Bovelles's *Liber de intellectu....* (Paris: Henri Estienne, 1510), 116v. Cambridge University Library, Rare Books Room, Acton.b.sel.32

The "Homo bulla" proverb was fashionable among early modern artists, and the *Synopsis*'s representation of rarefaction resembles prints of the proverb, such as the *Quis Evadet?* made by Hendrik Goltzius (1558–1617) in 1594.[46] Not only does Meurisse's figuration of rarefaction visualize this philosophical notion, but it also reminds viewers of the transience of human existence.

In the fourth region from the right under the curved boundary is a territory pertaining to Aristotle's analysis of the agent and patient in regards to motion and change (see Fig. 4.6).[47] Aristotle argues that in any transitive motion or change the

[46] On the *Homo bulla*, see Knipping 1974, vol. 1, 86–88. For representations of this proverb in *alba amicorum*, see Berger 2017, 168–69.

[47] The textbooks of Eustachius (1609, vol. 2, 187–94) and Abra de Raconis (1617, 566–83) both include sections entitled "De Actione et Passione."

agent's action and the patient's passion are a single activity, as opposed to two distinct processes; Aristotle explains that the instructor's teaching is the same activity or motion as the student's learning.[48] Meurisse notes that the action belongs to the agent and the passion belongs to the patient, and echoes Aristotle by equating the motions of agent and patient.[49] Meurisse's discussion of action and passion is accompanied by an image of a schoolteacher whipping the bare buttocks of his pupil. The *Synopsis* is citing Aristotle's example of teaching and learning, since presumably this teacher punishes his student in order to teach him a lesson. Representations of teachers whipping students were common throughout the Middle Ages and Renaissance.[50] For instance, a sculpture of Grammar holding birch rods above her students was integrated around 1145–1150 into the south portal of Chartres cathedral's west façade.[51] And among the ink sketches executed in 1515 by Hans Holbein the Younger (1497–1543) in the margins of an edition of Erasmus's *Moriae Encomium* (*The Praise of Folly*) is a representation of a master whipping a student.[52] Holbein critiques the master by showing a fool's cap with bells, the traditional garment of a court jester, dangling down his back.

In the third territory from the right under the curved boundary, mixture is defined as the unification of mixed and altered things, and is differentiated from generation (see Fig. 4.6).[53] Meurisse articulates the Scotist position that in mixtures of the elements, it is unnecessary to assume the existence of the elements' substantial forms.[54] The accompanying representation shows a man pouring a liquid into a bowl and stirring it with a stick, presumably mixing the liquid with another substance that is already inside the bowl.

---

[48] Aristotle 1984, vol. 1, *Physics*, 202b19–22. Idem 1590, vol. 1, 213.

[49] Meurisse 1615: "Actio est in agente et passio in patiente ut in subiecto.... Idem potest agere in se ipsum et in passum..." (The action is in the agent and the passion is in the patient as in the subject.... The same thing can act in its very self and in the thing that suffers.)

[50] Grammar's attributes included the *flagrum* (whip) and the *machaera* (single-edged sword). Wirth 1983, 279.

[51] For a discussion of this and other medieval personifications of Grammar, see Cleaver 2009, accessed on 8 July 2010.

[52] This book is housed in the Kupferstichkabinett of the Basel Kunstmuseum. For a facsimile of the *Moriae Encomium* with the original illustrations see: Schmid 1931, fol. M4v.

[53] Meurisse, 1615: "Datur mixtio, estque mixtorum alteratorum unio: distincta a generatione." (Mixture is defined as the unification of things mixed and altered: it is distinct from generation.) Meurisse's definition is based on Aristotle's: Aristotle 1984, vol. 1, *On Generation and Corruption*, 328b23–24; and idem 1590, vol. 1, 317. See also Eustachius 1609, vol. 2, 208–09; and Abra de Raconis 1617, 608–11. Aristotle also distinguishes between mixing and substantial change: Aristotle 1984, vol. 1, *On Generation and Corruption*, 327b7–9; and idem 1590, vol. 1, 316.

[54] Members of the Franciscan order promoted the ideas of the Franciscan philosopher and theologian John Duns Scotus (1265/66–1308). Meurisse 1615: "Elementa non manent in mixto secundum suas formas substantiales..." (Elements do not remain in a mixture according to their substantial forms.) King 2003, 53. King refers to the following passage in Duns Scotus 1950–, vol. 19, *Lectura*, 2. 15. un., nn. 38–43, p. 149–52. Eustachius (1609, vol. 2, 209) entitles a *quaestio* in his textbook "An elementa maneant in mixto."

Behind the representation of the male, natural body at the center of the half-circle is a portico with a triangular pediment (see Fig. 4.8). The architrave carries text stating, "Every entity exists in place in either a repletive, definitive, or circumscriptive fashion."[55] Meurisse and Gaultier have positioned the man within the portico to give him a place, because, as Meurisse explicitly notes, every body exists in place.[56] On the triangular pediment is an Aristotelian definition of place: "Place is defined as some shared, some specific, and is said to be the first, motionless extreme boundary of the body containing it."[57] In an arch in the *Physica*, Colutius likewise cites Aristotle's account of place.[58] Additionally, the arch's keystone represents an arcade in order to capture the notion of place through an architectural structure (Fig. 4.13).

A definition of vacuum is inscribed inside the portico of the *Synopsis*, but it is not accompanied by an image, since a vacuum cannot be represented. Whereas Aristotle believed that it is impossible for a vacuum to exist, late thirteenth-century theologians worried that this thesis undermined God's omnipotence.[59] Meurisse, like Scotus, defends the possibility of a vacuum created by God and states, in concurrence with most seventeenth-century scholastics, that motion can take place in the vacuum.[60] Colutius summarizes Aristotle's conception of a vacuum in the arch to

---

[55] Meurisse 1615: "Omne ens est in loco vel repletive, vel deffinitive, vel circumscriptive." Schoolmen characterized presence in a place as "repletive," when discussing God's mode of presence in a place, as "definitive," in reference to the mode of presence of non-bodily spiritual beings, and as "circumscriptive," for the occupation of space in the manner of material bodies. Aristotle also asserts that every body is in place. Aristotle 1984, vol. 1, *Physics*, 213b33. Idem 1590, vol. 1, 222. For a discussion of Scotus's consideration of whether a body can exist without a place, see Lewis 2003, 71.

[56] On the portico's left column is an inscription stating, "Unum corpus in pluribus locis...esse non repugnat." (One body...is not averse to existing in many locations.) In a similar vein, Abra de Raconis (1617, 326–36) entitles part of a question in his textbook "an idem numero corpus possit esse in pluribus locis." (whether a body the same in number can exist in more places.) Abra de Raconis approaches this question with reference to the Eucharist, or the consecrated bread and wine, in which the body and blood of Christ is believed to be physically present. The Eucharist could offer an example of the ways in which one body might exist in numerous locations. The portico's right column is marked with text on the relation of accidents to the body. Meurisse 1615: "Quaedam accidentia competerent corpori. utrobique existenti sed non omnia." (Some accidents are suitẹd to a body existing in one place or another, but not all of them [i.e. accidents].)

[57] Meurisse 1615: "Datur locus, alius communis, alius proprius, et dicitur extrema superficies ambientis corporis, immobilis prima." Aristotle 1984, vol. 1, *Physics*, 212a20–21. Idem 1590, vol. 1, 221. Aristotle's conception of place is also cited by Abra de Raconis (1617, 325) and Eustachius (1609, vol. 2, 86).

[58] Colutius 1611: "Locus est terminus primus im[m]obilis continentis Arist[oteles] 4 phys[icorum] tex[tu] 41." (Place is the first boundary of a motionless container [see] Aristotle [book] 4 of the *Physics*, *textus* 41.)

[59] Aristotle 1984, *Physics*, 214a16–b11. Idem 1590, vol. 1, 223. Cross 1998, 202.

[60] Meurisse 1615: "Dari potest vacuum non quidem naturali virtute nec Angelica sed dumtaxat divina in quo motus fieri posset etiam successivus." (The vacuum can be defined not indeed by either a natural, or an Angelic virtue, but only by a divine one, in which [vacuum] motion is able to come about even in a continual form.) For an explanation of Scotus's arguments for the possibility of a vacuum, see Cross 1998, 202–08; and Lewis 2003, 71–74. On seventeenth-century accounts

Fig. 4.13 Detail from *Physica*. Wolfenbüttel, Herzog August Bibliothek, IE2. (Used by permission of the Herzog August Bibliothek under §51 UrhG)

the right of that inscribed with an account of place, but does not positively assert the existence of a vacuum.[61] The *Physica* visualizes a vacuum as a black circle, shown in the keystone of the arch (see Fig. 4.13).

The winged spheres beneath the male natural body in the *Synopsis* are annotated with texts pertaining to motion (see Fig. 4.8). The wings indicate that the spheres are moving. The wing on the sphere to the right is marked with the standard Aristotelian definition of motion: "Motion is the act of an entity in potentiality, insofar as it is in potentiality."[62] Colutius cites this same definition of motion in one of the central arches of the *Physica* (see Fig. 4.13).[63] In the keystone at the top of this arch, motion is visualized through a crown gear and verge escapement, as in a clock.[64] Both Meurisse and Colutius note that Aristotle's conception of motion designates not only motion in place, but also other modes of motion or change.[65] The wing of the left sphere of the *Synopsis* is inscribed with text pertaining to the possible modes of change: "There is motion towards these three predicates, i.e. where, quantity, and quality."[66] Aristotle uses the term "locomotion" to denote motion in

of the vacuum and motion in it, see Ariew and Gabbey 1998, 436. Eustachius (1609, vol. 2, 92) entitles a *quaestio* in his textbook "An motus in vacuo fieri posset" (Whether motion is able to come about in a vacuum.)

[61] Colutius 1611: "Vacuum quod esse locum corpore privat<i>um affirmant ut .4 phys[icorum] tex[tu] 6. et 65 Arist[oteles] scribit[.]" (Vacuum is that which they assert to be place deprived of body, as Aristotle writes in [book] 4 of the *Physics* at *textus* 6 and 65.) Although Colutius does not positively argue for the existence of a vacuum, he seems to disagree with Aristotle, as he concludes, "...sed nihil probat..." (but he proves nothing.).

[62] Meurisse 1615: "Motus est actus entis in potentia ut in potentia." Aristotle 1984, vol. 1, *Physics*, 201a11–12. Idem 1590, vol. 1, 211. This definition is also cited by Eustachius (1609, vol. 2, 99) and Abra de Raconis (1617, 404).

[63] Colutius 1611: "Motus...est actus entis in potentia prout in potentia..." (Motion...is the act of an entity in potentiality, insofar as it is in potentiality.)

[64] On this mechanism, see Usher 1954, 312–13.

[65] Aristotle 1984, vol. 1, *Physics*, 226a24–25. Idem 1590, vol. 1, 233.

[66] Meurisse 1615: "Ad tria praedicamenta est motus ad ubi, quantitatem, et qualitatem." Colutius 1611: "motus species essendiales [*sic pro* essenti-] sunt tres[;] ad quantitatem, ad quantitem [*sic*

respect of place; "increase" or "decrease" to denote motion in respect of quantity; and "alteration" to denote motion in respect of quality.[67] Meurisse inscribes further distinctions pertaining to motion on both spheres.[68]

The *Synopsis*'s male natural body holds an enormous hourglass inscribed with text on duration, which Meurisse explains as "the continuous existence of a thing" (see Fig. 4.8).[69] He cites three kinds of duration commonly identified by scholastic philosophers: *aeternitas* (duration experienced by God), *aevum* (the mode of duration of angels and celestial Intelligences), and *tempus* (duration experienced by man).[70] Meurisse quotes Aristotle's explanation of time as "a number of motion in respect of before and after."[71] With this representation of the moving man holding the hourglass, Meurisse and Gaultier emphasize visually that the Aristotelian understanding of time requires the existence of a body in motion. In his annotations, Meurisse offers further details on the nature of duration.[72] Colutius describes time in the second arch from the right in the *Physica*, but he does not discuss duration (see Fig. 4.13). Colutius also cites Aristotle's account of time, interpreting "numerus" (number) as "measurement."[73] In the keystone of the arch of time, he presents Old Father Time with his traditional attributes, the long beard, wings, scythe, and hourglass.[74]

Behind the male natural body in the *Synopsis* is an oblique line annotated with an explanation of continuum, which begins, "A continuum consists of always divisible parts that actually exist within it. [Consequently, the continuum does] not consist of indivisible points, though they exist in the continuum."[75] The opening of this

pro qual-], et ad locum." (There are three essential species of motion: that with regard to quantity, to quality, and to place.) See also Abra de Raconis 1617, 367; and Eustachius 1609, vol. 2, 105.

[67] Aristotle 1984, vol. 1, *Physics*, 226a25–31. Idem 1590, vol. 1, 233.

[68] On the left sphere Meurisse discusses motion in relation to the moveable thing and the boundaries of motion. The right sphere is marked with three potentially controversial theses, perhaps included to provoke debate during the disputations, although their precise purpose and significance is unclear.

[69] Meurisse 1615: "Datur duratio estque existentia rei continuata."

[70] Abra de Raconis (1617, 350–57) cites these three modes of duration.

[71] Meurisse 1615: "...tempus, quod est numerus motus secundum prius et posterius." Aristotle 1984, vol. 1, *Physics*, 219b1. Idem 1590, vol. 1, 227. According to some scholars, Aristotle understood "number" to signify a magnitude assigned to motion's duration, such as a minute, or an hour. Others hold instead that Aristotle employs the word "number" to differentiate the enumeration of one motion from another. See Lewis 2003, 98 and Sorabji 1983, chapter 7. Abra de Raconis (1617, 361) quotes Aristotle's definition of time.

[72] Meurisse (1615) follows Scotus in asserting that time "is not one instant": "...non est unum instans." Lewis 2003, 85–86.

[73] Colutius 1611: "Est mensura seu numerus motus Secundum prius et posterius..." (It is the measure or number of motion in respect of before and after.)

[74] Ripa describes "Tempo" (Time) as a winged old man (1603, 483).

[75] Meurisse 1615: "Continuum constat partibus actu in ipso existentibus semper divisibilibus non constat ex punctis indivisibilibus, quae tamen sunt in continuo." For a discussion of the conceptions of continuity among seventeenth-century schoolmen, see Ariew and Gabbey 1998, 444–47.

definition is based on Aristotle's explanation of a continuum: "Now a continuum is that which is divisible into parts always capable of subdivision."[76] Meurisse writes that indivisible points do exist in the continuum, but they do not *constitute* it. Aristotle likewise states the impossibility of indivisibles constituting a continuum.[77] It is presumably because Aristotle identifies a line as a continuous quantity that Meurisse and Gaultier inscribe their account of a continuum onto a line.[78] Meurisse and Gaultier chose to represent a continuum near their explanations of time, place, and motion, because Aristotle characterizes all of these as continua.[79] It is conceptual affiliations that determine the geography of this and many of the other details of the *Synopsis*.

4.3 Monsters of Ignorance

The previous section explored some of the ways in which the cartography of Meurisse and Gaultier advanced Aristotelian scholastic precepts. I would now like to consider whether these philosophical maps offer insights into the ramifications of venturing outside known or accepted territories. To address this question, I will begin by discussing how early modern geographical cartographers visualized encounters with unknown or obscure regions of the globe.

Among the most captivating components of medieval and Renaissance maps are the wondrous and frightening sea monsters that swim through turbulent waters, often to the horror of vulnerable seamen.[80] These creatures, which tended to be enormous in scale, signaled for European viewers the perils of travel to less well-known regions, while simultaneously advertising the marvels of newly "discovered" and faraway reaches of the globe. We need only look at the oversized sea monsters in Ortelius's world maps to understand how these decorative elements served to entertain, inspire fear, and incite wonder (see Fig. 4.1). In the Indian Ocean of his *Typus orbis terrarum*, for example, we encounter a sea monster, whose head is topped by twin waterspouts and whose body is many times larger than all of the islands nearby (Fig. 4.14). The creature's strange anatomy and size accentuate the dangers and wonders of this ocean.

---

[76] Aristotle 1984, vol. 1, *On the Heavens*, 268a6–7. Idem 1590, vol. 1, 266.

[77] Aristotle 1984, vol. 1, *Physics*, 231a20–231b6. Idem 1590, vol. 1, 237. Scotus also denies that indivisibles make up a continuum. He holds that if a quantity were to consist of indivisibles, it would also have to contain gaps. Scotus does argue, however, that the limits of a continuum are indivisibles. For an explanation of Scotus's arguments against atomism, see Cross 1998, chapter 7.

[78] Aristotle 1984, vol. 1, *Categories*, 4b36–5a1, and *On the Heavens*, 268a8. Idem 1590, vol. 1, 12 and 266.

[79] Aristotle 1984, vol. 1, *Categories*, 5a9–10, *Physics*, 232b23, and *On Generation and Corruption*, 337a22–23. Idem 1590, vol. 1, 12, 238, and 324.

[80] Van Duzer 2014; Nigg 2013.

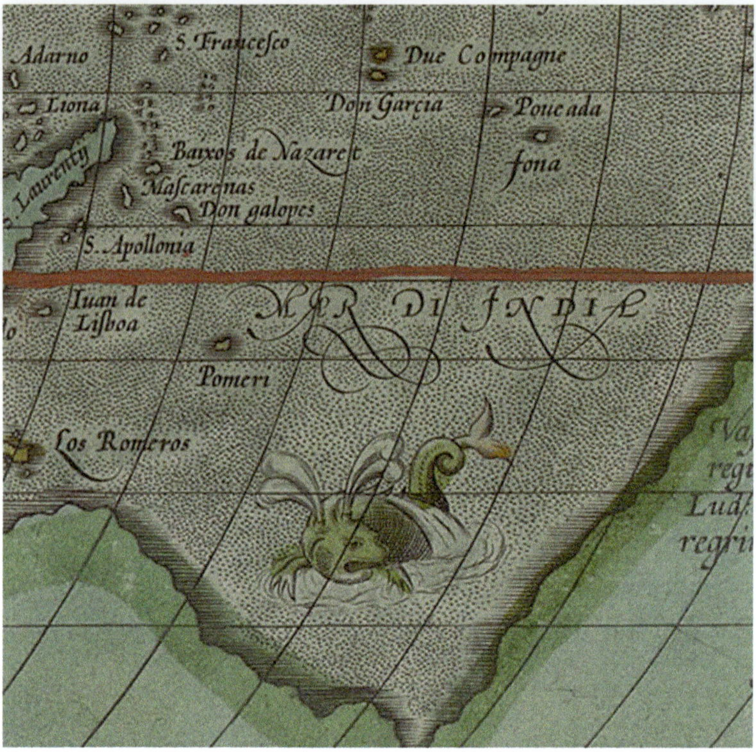

Fig. 4.14 Sea monster in Ortelius's map of the Indian Ocean from his *Typus orbis terrarum*. Call #: STC 18855. (Used by permission of the Folger Shakespeare Library under a Creative Commons Attribution-ShareAlike 4.0 International License)

The gigantic dimensions of the monsters populating early modern cartography are particularly conspicuous when these creatures are juxtaposed with foundering ships carrying frightened sailors.[81] In Nikolaos Sophianos's *Totius Graeciae Descriptio* of 1545, spear-wielding sailors in a galley appear miniature in comparison to the colossal, open-mouthed sea monster that is on the verge of gulping them up.[82] Ortelius's map of the Holy Land from his atlas shows in the bottom left corner a figure that was in fact swallowed by a sea monster, Jonah (Fig. 4.15). We find the Hebrew prophet as he is being thrown out of the ship and towards the mouth of an enormous whale.

The philosophical broadsides of Meurisse, Chéron, Messager, and Gaultier also display threatening sea monsters and endangered seamen. Instead of marking

[81] Perhaps the most threatening and famous sea monster of the early modern era is Hobbes's Leviathan, who was famously visualized in an etching created by Hobbes in collaboration with Abraham Bosse (1602/4–1676). This monster is also gigantic and like the maps of early modern cartographers, functioned as an instrument of power and control. See Berger 2017, chapter 5.

[82] Reproduced in van Duzer 2013, 39.

Fig. 4.15 Ortelius's map of the Holy Land from his *Theatrum orbis terrarum*, plate 111. Call #: STC 18855. (Used by permission of the Folger Shakespeare Library under a Creative Commons Attribution-ShareAlike 4.0 International License)

regions of the globe that were obscure and little-known to Europeans, however, these monsters denote a want of knowledge more generally. The association between sea monsters and ignorance is most explicit in the *Tableau*, where a blindfolded creature with raised arms is characterized as a "monster of ignorance" (*monstre d'ignorance*) (Fig. 4.16). Above this scaly philistine, a ship that struggles to stay afloat is related in an inscription to "the fear of shipwreck" (*la crainte du naufrage*). On the right edge of the *Tableau*, a "Siren of concupiscence" (*Syrene de concupiscence*) combs her hair and admires her reflection in a mirror (Fig. 4.17). She too is a source of danger; an annotation warns readers, "Concupiscence along with her distractions, as in this depicted Siren, reduces up to a point the liberty of the Will."[83] Chéron and Gaultier are likely to have been inspired by this detail as well by decorative creatures illustrated in maps from this period, such as the two sirens with these same attributes that swim near the Straight of Magellan in a map of the Americas, made in 1562 by Diego Gutiérrez.[84]

[83] Meurisse 1618: "Ceste Syrene de concupiscence, avec ses perturbations, diminue aucunement la liberté de la vonlonté."

[84] Reproduced in van Duzer 2013, 39. The association between sirens and unorthodox, sensuous pleasures can already be found in book twelve of Homer's *Odyssey*, when Odysseus crafts an

Fig. 4.16 Monster of ignorance in the *Tableau*. Paris, Bibliothèque nationale de France, Cabinet des Estampes, Reserve QB-201 (21). (Used by permission of the Bibliothèque nationale de France under a Creative Commons Attribution-ShareAlike 4.0 International License)

In the *Typus* we discover more sea monsters that symbolize the dangers of ignorance. On a path on the right edge of the broadside, a man flees a blindfolded siren and cries "Blessed is the man that findeth wisdom" (*Beatus homo qui invenit sapientiam*) (Fig. 4.18).[85] Near the top of the print, to the left of a temple of wisdom, a man who has slipped on his approach to wisdom and ventured beyond the confines

elaborate ruse to escape these creatures unscathed. Homer 1998, vol. 2, pp. 408–409.

[85] Book of Proverbs 3:13.

Fig. 4.17 Siren of concupiscence in the *Tableau*. Paris, Bibliothèque nationale de France, Cabinet des Estampes, Reserve QB-201 (21). (Used by permission of the Bibliothèque nationale de France under a Creative Commons Attribution-ShareAlike 4.0 International License)

of Aristotelian scholastic logic tries to flee a sea monster (Fig. 4.19). He exclaims, "The sea saith: It is not with me" (*Mare ait non est mecum*).[86] The menacing head of a second creature rises above the waters a bit further behind.

Below, a seascape incorporates imagery of seafaring and military battle (Fig. 4.20). Three large boats denote the three kinds of syllogisms: those that result in knowledge, opinions, and errors. The syllogisms that result in knowledge are summarized on the central galleon, a large multi-decked sailing ship that could withstand a major ocean crossing. Those that result in error are summarized on the left-most boat and those that result in opinions are described in the boat on the right.

[86] Job 28:14.

Fig. 4.18 Traveler in *Typus*. Princeton University Library Special Collections, Oversize Broadside 120

Figures drown in the water near the boat associated with errors. On the left, a man uses both arms to lift himself back up the rocky cliff towards the path to wisdom; a figure looking in his direction from the top of the hill remarks, "The greatest misery is not knowing to where you are heading."[87] The mast of the boat of the sophists has broken; the men in this vessel will soon join their peers in the water. Like sailors who venture to the outermost reaches of the known world, those students who do not stay within the scope of Aristotelian scholastic philosophy run the risk of perishing in turbulent waters.

Far from neutral accounts of fields of learning, the cartography of Meurisse, Chéron, Messager, and Gaultier promoted the theories of Aristotle and his scholastic interpreters through a rich and vivid integrations of word and image. The philosophical maps produced by these pedagogues and their artistic collaborators enabled students to understand and contemplate complex fields of learning in their entirety. As they might physically grasp a map of the world in their hands and gain an

[87] Chéron 1622: "Summa miseria est nescire quo tendas." This quotation is taken from Isidore of Seville's *soliloquiorum seu synonymorum de angustia & miseria hominis, libri duo ex vetustissimo codice recogniti* (1552, G4r).

Fig. 4.19 A man who has slipped on his approach to wisdom in *Typus*. Paris 1622, Princeton University Library Special Collections, Oversize Broadside 120

Fig. 4.20 Three large boats that denote the three kinds of syllogisms in the *Typus*. Paris 1622, Princeton University Library Special Collections, Oversize Broadside 120

understanding of the world or universe, students could also grasp the philosophical maps of Meurisse, Chéron, Messager, and Gaultier to comprehend entire realms of thought. By advertising the Aristotelian scholastic perspective, these innovative theoretical maps also promoted the interests of the Catholic church and the French state. Yet the political import of these philosophical maps should of course not be exaggerated, as the audiences for these prints in the Franciscan and Carmelite convents of Paris were likely to have been for the most part already convinced of the legitimacy of these institutions.

Bibliography

Abra de Raconis, Charles François d'. 1617. *Tertia pars philosophiae seu physica*. Paris: Denis de la Noue.

Alpers, Svetlana. 1983. *The Art of Describing: Dutch Art in the Seventeenth Century*. Chicago: University of Chicago Press.

Apin, Siegmund Jacob. 1731. *Dissertatio de variis discendi methodis memoriae causa inventis earumque usu et abusu, recognita et aucta*. Braunschweig: Ludolphus Schroederus.

Ariew, Roger, and Alan Gabbey. 1998. The Scholastic Background. In *The Cambridge History of Seventeenth-Century Philosophy*, ed. Daniel Garber and Michael Ayers, 425–453. Cambridge/New York: Cambridge University Press.

Aristotle. 1590. *Operum ... omnium longe principis, nova editio, Graece et Latine*, ed. Isaac Casaubon. Lyons: Guillelmus Laemarius.

———. 1984. *The Complete Works of Aristotle*, ed. Jonathan Barnes. Princeton: Princeton University Press.

Berger, Susanna. 2013a. Martin Meurisse's Garden of Logic. *Journal of the Warburg and Courtauld Institutes* 76: 203–249.

———. 2013b. Martin Meurisse's Theater of Natural Philosophy. *The Art Bulletin* 95: 269–293.

———. 2014. The Invention of Wisdom in Jean Chéron's Illustrated Thesis Print. *Intellectual History Review* 24: 343–366.

———. 2016. Philander Colutius and the Visualization of Natural Philosophy. In *Jean Michel Massing's Festschrift*, ed. Philip Lindley and Mark Stocker, 33–44. Turnhout: Brepols.

———. 2017. *The Art of Philosophy: Visual Thinking in Europe from the Late Renaissance to the Early Enlightenment*. Princeton: Princeton University Press.

Blair, Ann. 1997. *The Theater of Nature: Jean Bodin and Renaissance Science*. Princeton: Princeton University Press.

Brugerolles, Emmanuelle, and David Guillet. 2000. Léonard Gaultier, graveur Parisien sous les règnes de Henri III, Henri IV, et Louis XIII. *Gazette des Beaux-Arts* 135: 2–24.

Chéron, Jean. 1622. *Typus necessitatis logicae ad alias scientias capessendas*. Paris: Jean Message.

Clavius, Christophorus. 1611. *Opera mathematica*. Vol. 3. Mainz: Reinhard Eltz.

Cleaver, Laura. 2009. Grammar and Her Children: Learning to Read in the Art of the Twelfth Century. *Marginalia* 9, http://www.marginalia.co.uk/journal/09education/cleaver.php. Accessed 8 July 2010.

Colutius, Philander. 1611. *Physica seu naturae theatrum in typum totius philosophiae naturalis*. Speyer: Matthäus Buschweiler.

Conimbricenses. 1606. *Commentarii Collegii Conimbricensis Societatis Iesu, in libros de generatione et corruptione Aristotelis Stagiritae*. Mainz: Joannes Albinus.

Cosgrove, Denis. 1999. Introduction: Mapping Meaning. In *Mappings*, ed. Cosgrove, 1–23. London: Reaktion Books.

Cross, Richard. 1998. *The Physics of Duns Scotus: The Scientific Context of a Theological Vision*. Oxford: Clarendon Press.

Duns Scotus, John. 1639. *Opera omnia*, ed. Luke Wadding. Lyon: Durand.

———. 1950–. *Opera omnia*, ed. Carolus Balic. Vatican City: Vatican Polyglot Press.

Erasmus, Desiderius. 1621. *Colloquia Familiaria*. Amsterdam: Johannes Janssonius.

———. 1991. *Collected Works of Erasmus. Adages II.i.1 to II.vi.100*, trans. Roger Aubrey Baskerville Mynors. Toronto: University of Toronto Press.

Eustachius a Sancto Paulo. 1609. *Summa philosophiae quadripartita, de rebus dialecticis, moralibus, physicis, & metaphysicis*. Paris: Carolus Chastellain.

Garber, Daniel. 1988. Descartes, the Aristotelians and the Revolution That Did Not Happen in 1637. *The Monist* 71: 471–486.

Gieben, Servus. 1990. Il "Lauro della metafisica" di Martino Meurisse: Foglio di tesi, inciso da Leonardo Gaultier nel 1616. *Collectanea Franciscana* 60: 683–707.

———. 1993. Iconografia antoniana in due fogli di tesi del Museo Francescano di Roma. *Il Santo* 2: 273–298.

Harley, J.B. 1988. Maps, Knowledge, and Power. In *The Iconography of Landscape: Essays on the Symbolic Representation, Design and Use of Past Environments*, ed. Denis Cosgrove and Stephen Daniels, 277–312. Cambridge, UK: Cambridge University Press.

Harley, J.B., and David Woodward. 1987. Concluding remarks. In *The History of Cartography, Vol. 1: Cartography in prehistoric, ancient, and medieval Europe and the Mediterranean*, ed. Harley and Woodward, 502–509. Chicago: University of Chicago Press.

Homer. 1998. *Odyssey*, ed. Jeffrey Henderson, trans. A.T. Murray, rev. George E. Dimock. Cambridge, MA: Harvard University Press.

Isidore of Seville. 1552. *...soliloquiorum seu synonymorum de angustia & miseria hominis, libri duo ex vetustissimo codice recogniti*. Salzburg: Bauman.

King, Peter. 2003. Scotus on Metaphysics. In *The Cambridge Companion to Duns Scotus*, ed. Thomas Williams, 15–68. Cambridge/New York: Cambridge University Press.

Knipping, John B. 1974. *Iconography of the Counter Reformation in the Netherlands*. Nieuwkoop/Leiden: B. de Graaf.

Kusukawa, Sachiko. 1995. *The Transformation of Natural Philosophy: The Case of Philip Melanchthon*. Cambridge, UK: Cambridge University Press.

Lewis, Neil. 2003. Space and Time. In *The Cambridge Companion to Duns Scotus*, ed. Thomas Williams, 69–99. Cambridge/New York: Cambridge University Press.

Maier, Jessica. 2015. *Rome Measured and Imagined: Early Modern Maps of the Eternal City*. Chicago: University of Chicago Press.

Meurisse, Martin. 1614. *Artificiosa totius logices descriptio*. Paris: Jean Messager.

———. 1615. *Clara totius physiologiae synopsis*. Paris: Jean Messager.

———. 1616. *Laurus metaphysica*. Paris: Jean Messager.

———. 1618. *Tableau industrieux de toute la philosophie morale*. Paris: Jean Messager.

Meyer, Véronique. 2002. *L'illustration des thèses à Paris dans la seconde moitié du XVIIe siècle: peintres, graveurs, éditeurs*. Paris: Commission des Travaux Historiques de la Ville de Paris.

Nigg, Joseph. 2013. *Sea Monsters: A Voyage Around the World's Most Beguiling Map*. Chicago: University of Chicago Press.

Rice, Louise. 1999. Jesuit Thesis Prints and the Festive Academic Defense at the Collegio Romano. In *The Jesuits: Cultures, Sciences, and the Arts, 1540–1773*, ed. John W. O'Malley, Steven J. Harris, Gauvin Alexander Bailey, and Thomas Frank Kennedy, 148–169. Toronto: University of Toronto Press.

Ripa, Cesare. 1603. *Iconologia*. Rome: Lepido Facii.

Rosenthal, Earl. 1971. *Plus Ultra, Non Plus Ultra*, and the Columnar Device of Emperor Charles V. *Journal of the Warburg and Courtauld Institutes* 34: 204–228.

Scheurleer, Theodor H. Lunsingh. 1975. Un amphithéâtre d'anatomie moralisée. In *Leiden University in the Seventeenth Century: An Exchange of Learning*, ed. Guillaume H.M. Scheurleer and Posthumus Meyjes, 211–277. Leiden: Brill.

Schmid, Heinrich Alfred. 1931. *Erasmi Rotterdami Encomium Moriae*. Basel: Oppermann.

Selbach, Vanessa. 2010. L'activité de l'éditeur d'estampes Parisien Jean Messager (1572–1649): L'affirmation de la gravure française du premier quart du XVIIe siècle, au carrefour des influences flamandes et italiennes. *In Monte Artium* 3: 35–51.

Seneca, Lucius Annaeus. 1979. *Ad Lucilium epistulae morales*, trans. Richard M. Gummere. Loeb Classical Library. Cambridge, MA: Harvard University Press.

Sorabji, Richard. 1983. *Time, Creation, and the Continuum: Theories in Antiquity and the Early Middle Ages*. Ithaca/London: Cornell University Press.

Usher, Abbott Payson. 1954. *A History of Mechanical Inventions*. Cambridge, MA: Harvard University Press.

van Duzer, Chet. 2014. *Sea Monsters and Medieval and Renaissance Maps*. London: The British Library.

Wirth, Karl-August. 1983. Von mittelalterlichen Bildern und Lehrfiguren im Dienste der Schule und des Unterrichts. In *Studien zum städtischen Bildungswesen des späten Mittelalters und der frühen Neuzeit: Bericht über Kolloquien der Kommission zur Erforschung der Kultur des Spätmittelalters 1978 bis 1981*, ed. Bernd Moeller, Hans Patze, and Karl Stackmann, 256–370. Göttingen: Vandenhoeck & Ruprecht.

Chapter 5
The Mathematics Theses Defended at *collège de Clermont* (1637–1682): How to Guard a Fortress in Times of War

Domenico Collacciani and Sophie Roux

Abstract To capture some of the early modern teaching practices while studying the interaction between ancient and new doctrines, this chapter focuses on the thesis booklets defended at the Jesuit *collège de Clermont* from 1637 to 1682. First, it examines what this exceptional corpus (almost one thesis in mathematics per year is available) reveals about the role that the Parisian Jesuits assigned to the defense of mathematical theses in the context of their complex relationship with the University of Paris. Then, it focuses on the optics theses that were defended under Pierre Bourdin: they borrowed from the Jesuit theoretical optics of the 1610-1620s, while making room for extraordinary devices such as the *Specillum Fidei*. Finally, it shows that the optics of Ignace-Gaston Pardies, who held the chair of mathematics some twenty years later, incorporated the main lesson of the *novatores*: it is not a theory of vision, but a theory of light, which is based on mechanics defined as the science of motion in general. While Bourdin preserved the structure of Aristotelian knowledge, Pardies overturned this structure by considering mechanics as the key to all physics.

Keywords Early modern science · Jesuits · Theses · Optics · Light · Eye · Pierre Bourdin · Ignace-Gaston Pardies · Descartes

---

Mathesis, République des Savoirs, ENS, Collège de France, CNRS, PSL Research University, 75005, Paris, France. This article took a very long time to complete. Above all, we would like to thank Dan Garber and Susanna Berger for their patience with us; Susanna deserves special thanks for the care with which she revised our article and helped us improve our English. We would also like to thank our colleagues in Berlin, Bologna, Brussels, Groningen, Paris, Princeton and Rome who, by their questions and comments, gave us new impetus when we were stuck in the middle of the ford.

D. Collacciani (✉)
CRISES Université Paul-Valéry Montpellier 3; Mathesis, République des savoirs (USR 3608, ENS, CNRS, collège de France), École normale supérieure, PSL University, Paris, France

S. Roux
Mathesis, République des savoirs (USR 3608, ENS, CNRS, collège de France), École normale supérieure, PSL University, Paris, France
e-mail: sophie.roux@ens.psl.eu

S. Berger, D. Garber (eds.), *Teaching Philosophy in Early Modern Europe*, Archimedes 61, https://doi.org/10.1007/978-3-030-84621-3_5

5.1 Introduction

Early modern theses that were defended in colleges and universities have recently attracted historical attention. They were first studied by historians of art and of the book: the illustrations that some theses included allowed scholars to reconstruct their material and social production as well as the ceremonies associated with their public defense.[1] More recently, among other documents, their doctrinal content has been of interest for intellectual historians studying the transition from the so-called "philosophy of the Schools" to the modern philosophy, especially, but not only, in Jesuit colleges.[2] An impressive book has thus been devoted to theses in logic, natural philosophy, metaphysics, ethics, and politics that were defended between 1555 and 1648 at the University of Dilingen, the first Jesuit University north of the Alps.[3] Yet theses in mathematics defended in Jesuit colleges remain understudied. Often only a few printed mathematics theses remain—we do not know if the other mathematics theses were never printed or they were printed but then lost.[4]

This chapter studies mathematics theses defended during a period of fifty years at the collège de Clermont in Paris, the most important French Jesuit college. It begins at the end of the thirties, around the time when Descartes's *Discourse on method* and *Essays* (1637) appeared; this moment also corresponds to when these theses began to be defended or at least to be published. The chapter ends in 1682, when the *collège de Clermont*, having received the patronage of Louis XIV, became Louis-Le-Grand.[5] Our first objective, which belongs to social history, is to capture some of the teaching practices of the early modern period. Our second objective, which falls within intellectual history, is to explain how the ancient and the new doctrines interacted, more specifically how French Jesuits reacted to Descartes and to other *novatores*. To reach these two aims without writing an excessively long paper, we will focus on optics and leave aside other disciplines tackled in these theses, including military architecture, mechanics, and cosmology.[6]

---

[1] Meyer 1993; eadem 2002; eadem 2017; and Rice 1999; eadem 2004; eadem 2007a, b.

[2] Brockliss 1981; idem 2002; idem 2006; Gatto 1994; Ariew 1999; Hellyer 2005; Klein 2017; Raphael 2017.

[3] Leinsle 2006.

[4] On mathematics theses defended in Antwerp and Louvain, see Vanpaemel 2008; De Bruycker 2009; Dhombres and Radelet-de Grave 2009. According to the *Ratio studiorum*, the printing of theses was not compulsory. Lukács 1986, 375.

[5] The list of these theses is given in Bibliography, section 2, which is organized by the presiding professor and for each presiding professor, listed chronologically under the name of the student defending. References to the theses in the notes are given in **boldface** to distinguish them from other bibliographical entries. Most of them were for the first time identified in Le Dividich 1996, xxxiii–xxxvi; see also Dupont-Ferrier 1921–1925, vol. III, Appendix I, 273–85.

[6] Military architecture, especially Bourdin's, is studied in d'Orgeix 2005, 91–105; Romano 2006; De Lucca 2012, 96–104, 118, 189, 217, 229, and 332. Topics in cosmology and natural philosophy in theses defended in the sixties are studied in Roux 2017. The discipline of mechanics in these theses has not yet been studied.

We begin with a general presentation of the corpus, which is exceptional insofar as there is almost one thesis per year, so that one can identify structural features common to all of the theses, but also variations from professor to professor and changes over time (II). Then, we focus on the optics theses that were defended during the long period in which Pierre Bourdin was the professor of mathematics at *collège de Clermont* (III). The brief time when Ignace-Gaston Pardies held the same chair some twenty years later on gives us the occasion to discuss an optics that fully incorporated the lessons of the *novatores*, even if it claims to respect the doctrine of the Ancients (IV).

5.2 An Overview of the Mathematics Theses Defended at the *collège de Clermont*

Before coming to the optical doctrines defended in these theses, we will explain why our focus is on mathematics theses rather than on philosophy theses, give some indications about what is meant by "theses," and, finally, enumerate the professors who were teaching mathematics at the *collège de Clermont*. Our corpus itself will give us many clues, but we will not hesitate to complete them by information known from other contexts.

1. French Jesuit philosophy theses are typically 2–3 feet high posters printed on paper, canvas, or silk; they are sometimes richly illustrated and often dedicated to great figures.[7] By contrast, the Jesuit theses in mathematics examined in this chapter are *quarto* format books ranging from 2 to 20 pages in length, rarely preceded by a dedication, more rarely still by a letter developing this dedication.[8] The physical differences between philosophy and mathematics theses produced by Jesuits in France are striking when the same student successively defended theses in philosophy and mathematics, as was the case of Chrétien-François de Lamoignon on 14 and 15 June, 1663, or of Jean-Baptiste Colbert on 29 and 30 August, 1668. The mathematics thesis booklet of Jean-Baptiste Colbert (1651–1690), the eldest son of the Minister of the same name, is a small treatise of mixed mathematics that does not include fancy figures (Fig. 5.1). His philosophy thesis print is a magnificent poster engraved by François de Poilly (Fig. 5.2).

In terms of their theoretical content, Jesuit philosophy thesis prints generally present "Conclusions (*conclusiones*)" that reiterate the Aristotelian orthodoxy

[7] On the dedications and on the illustrations of theses, and on their material production, see Meyer 1993; eadem 2002; eadem 2017.

[8] Exceptions include the thesis that Charles Potier dedicates to Claude Lestandart, holder of the *abbaye du Val Secret* near Château-Thierry where Potier came from (1640), the thesis that Jacques Truel de Cohon dedicates to Timoleon Le Roy, *premier commis* to the War Minister Michel Le Tellier (1648), and the thesis that Jules Louis Bolé de Champlay dedicates to François Michel Le Tellier, who succeeded his father as War Minister (1668).

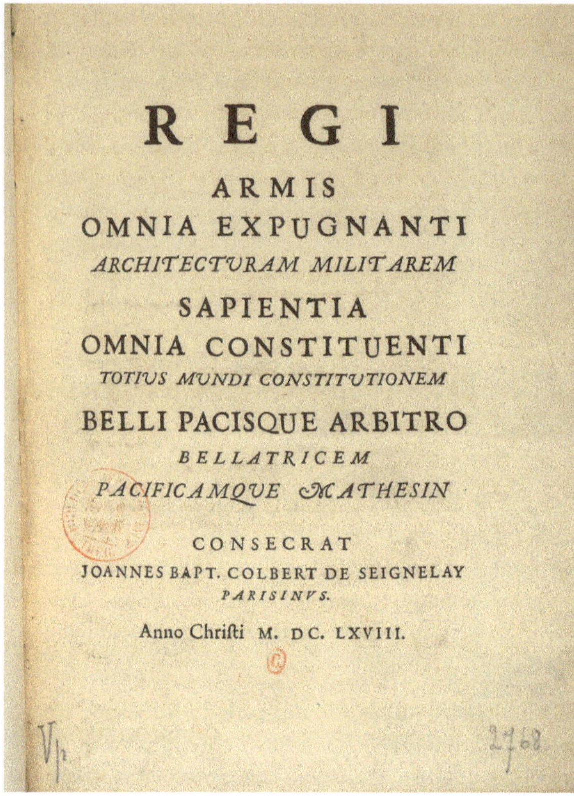

REGI

ARMIS

OMNIA EXPUGNANTI

ARCHITECTVRAM MILITAREM

SAPIENTIA

OMNIA CONSTITUENTI

TOTIVS MVNDI CONSTITVTIONEM

BELLI PACISQUE ARBITRO

BELLATRICEM

PACIFICAMQVE MATHESIN

CONSECRAT

JOANNES BAPT. COLBERT DE SEIGNELAY

PARISINVS.

Anno Chriſti M. DC. LXVIII.

Fig. 5.1 The front page and the first page of Colbert's mathematics thesis booklet in-4 (30 August 1668). Bibliothèque Sainte-Geneviève (Paris)

established by Francisco Suárez, Francisco de Toledo, and the Coimbrans without providing any explanation, objection, discussion, or comment. During the period examined here, they assert for example that natural philosophy is the speculative science of the natural body inasmuch as it is natural; that the three principles of natural bodies are matter, privation, and form; that motion is the actualization of the potential, insofar as it is potential; that quantity is an accident really distinct from matter, etc.[9] They do not present much textual content that would be of interest to intellectual historians, except insofar as they confirm the received view according to which the professors of the Schools belonged to a distant and immobile landscape with respect to which the *novatores* moved forward.

By contrast, the mathematics thesis booklets introduced here discuss various subjects, occasionally of pure mathematics (elementary arithmetic and geometry), but usually of mixed mathematics (optics, astronomy, mechanics, fortifications, and military architecture), or, as an additional handwritten title indicates in 1676, of

[9]Ariew 2014, 15–17.

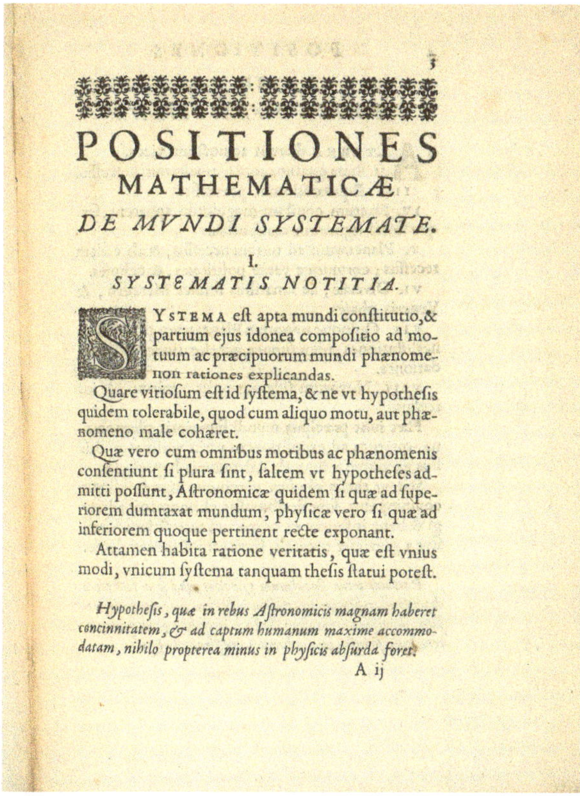

3

POSITIONES
MATHEMATICÆ
DE MVNDI SYSTEMATE.

I.
SYSTEMATIS NOTITIA.

SYSTEMA eſt apta mundi conſtitutio,& partium ejus idonea compoſitio ad motuum ac præcipuorum mundi phænomenon rationes explicandas.

Quare vitioſum eſt id ſyſtema, & ne vt hypotheſis quidem tolerabile, quod cum aliquo motu, aut phænomeno male cohæret.

Quæ vero cum omnibus motibus ac phænomenis conſentiunt ſi plura ſint, ſaltem vt hypotheſes admitti poſſunt, Aſtronomicæ quidem ſi quæ ad ſuperiorem dumtaxat mundum, phyſicæ vero ſi quæ ad inferiorem quoque pertinent recte exponant.

Attamen habita ratione veritatis, quæ eſt vnius modi, vnicum ſyſtema tanquam theſis ſtatui poteſt.

Hypotheſis, quæ in rebus Aſtronomicis magnam haberet concinnitatem, & ad captum humanum maxime accommodatam, nihilo propterea minus in phyſicis abſurda foret.

A ij

Fig. 5.1 (continued)

"physico-mathematics."[10] The questions that are tackled involve the latest scientific developments, or at least, events that were discussed in the learned circles. For example, the book *Epistolae duae de motu impresso a motore translato* (*Two letters on the motion impressed by a moving* mover), in which Gassendi associated Copernican cosmology with a demonstration of Galileo's law of falling bodies having recourse to indivisibles, launched in 1642 what has been called "the second Galileo affair." [11] Dominique de Vic defended a thesis entitled *Dissertationes contra Galilæum* (*Dissertations against Galileo*) in 1643. Torricelli's experiment having been reproduced and amplified in France for the first time in October 1646 by Pierre Petit and Pascal father and son, Blaise Bouthier defended a thesis giving a significant place to Torricelli's experiment in July 1647, which means that he had been

[10] On the Sorbonne copy of **Anonyme 1676**, the title is completed by the following manuscript annotation: "Theses mathematicae seu potius physicomathematicae" (mathematics, or rather physic-mathematics theses).

[11] Galluzzi 1993.

Fig. 5.2 Colbert's philosophy thesis print (29 August 1668, 103 x 76 cm. The King, descending from a chariot driven by Mercury and surrounded by allegorical figures representing the arts and sciences, is brought by Minerva to the list of theses to be defended. BnF (Paris), département Estampes et photographie

taught on these issues during the preceding winter.[12] In 1651, Pierre de la Villette takes up verbatim the first nine paragraphs of Cavalieri's *Exercitationes Geometricae Sex* (*Six geometrical dissertations*), which four years beforehand outlined the notion of indivisibles and set out the principles of the two methods that today's historians

[12]When Pascal published his *Expériences nouvelles touchant le vide* in October 1647, the only writings to mention the Rouen experiment before were Pierius 1648 and Petit's letter to Chanut of November 1646, which was published one year later in Petit and Magni 1647.

of science call the distributive method and the collective method.[13] Last, but not least, in 1665 some theses were devoted to the comets that enlightened the European skies that very year.[14] Of course, this does not mean that the Jesuit professors accepted scientific novelties. De Vic lists the absurd consequences that follow from Galileo's law of falling bodies and from his explanation of the tides.[15] Thierry argues that the space at the top of Torricelli's tube, far from being a vacuum, is filled with an "aerial spirit, distinct, by its most powerful part, from the common air and from other airs."[16] De la Villette refuses to label Cavalieri's doctrine as science because of the paradoxes that it generates.[17] The theses on comets were followed by a series of theses condemning not only Descartes' opinion on comets, but his natural philosophy more generally.[18] Such discussions deserve our attention. Much more than philosophy theses, mathematics theses are an indicator of the demarcation that the Jesuits made between the novelties that could be accepted, because they did not challenge the Aristotelian framework, and the novelties that could not, because they endangered this framework.

To account for the difference between philosophy and mathematics theses, it is helpful to turn to the *Ratio studiorum*. While the doctrine to be taught and the textbooks to be used were determined in philosophy and theology, there were no specifications concerning mathematics. Similarly, the rules for the various disputations in philosophy and in theology are detailed, whether they concern the times and places of the events, their number and their rhythm, the pomp with which they were to be organized, or the functions of the different protagonists (the prefect, the rector, the teachers, the examiners, the defendants, the objectors, etc.), but nothing is said about the public exercises in mathematics, except that every month, or at least every second month, the professor "should have one of the students solve a famous mathematical problem in the presence of many students of philosophy and theology.[19] Being less controlled than philosophy disputations, mathematics disputations allowed for more variety and for more original developments.

---

[13] Compare **Villette 1651**, 13–4, § 37–41 and Cavalieri 1647, 3–4, § 1–5, 6, § 9. Cavalieri published his method in Cavalieri 1635; Cavalieri 1647 is an answer to the objections formulated in Guldin 1635–1641.

[14] **Tarteron 1665**; **Prou 1665a**; and **Ragayne de la Picotière 1665a**.

[15] **De Vic 1643**, 14, § 2. **Gedoin 1644**, 4.

[16] For the quotation, see **Thierry 1648**, 11, § 29, and 12, § 33: "Vocetur aut a potiori parte Spiritus aërius, a communi tamen aëre distinctus." See also **Truel de Cohon 1648**, 11, § 33–4. Both defend the position of Étienne Noël (1581–1659), who was then rector of the *collège de Clermont*, and who is now remembered for having been mocked by Pascal.

[17] **Villette 1651**,14–5, § 43–6.

[18] **Prou 1665b, idem 1665c**; **Ragayne de la Picotière 1665b**; **Decombes 1665**; and **de la Bletonniere 1665**.

[19] Lukács 1986, 374–7, 402. The monthly disputation in mathematics was introduced in 1599. According to Cosentino (1970, 212) if no rules were formulated to regulate mathematics disputations, contrary to what Clavius had wished, it was because the enactment of rules would have obliged all colleges to create a chair of mathematics, whereas only the most important colleges could afford one. On Clavius' fight for the teaching of mathematics through the different versions of the *Ratio*, see Romano 1999, 111–32; on the first chairs of mathematics in France, see *ibid.*, 393–414.

Another explanation for the difference between philosophy and mathematics theses lay in the bitter and long-lasting rivalry that developed between the Jesuits and the Sorbonne. [20] Since the Parisian Jesuits were denied the authorization to award diplomas, they tried to push their advantages into the educational market by demonstrating their supremacy in mathematics, a field in which the Sorbonne professors were incompetent and over which they had no control.[21] Moreover, the Jesuits used disputations to spread what they considered to be scientific orthodoxy throughout France. Today, the philosophy thesis prints defended at the *collège de Clermont* are found only in the *bibliothèque de la Sorbonne*, in two collections of theology and philosophy theses defended mainly at the University, and occasionally at external colleges.[22] As evidenced by the collections held today by libraries outside of Paris and by correspondences of the time, mathematics thesis booklets of the *collège de Clermont* were by contrast circulated throughout France, and sometimes even beyond. The two most striking examples of such a widespread dissemination are the thesis booklet defended in 1663 by Chrétien-François de Lamoignon, the eldest son of the President of the *Parlement* of Paris, which is now to be found in many libraries all over Europe and even in America, and the anti-Cartesian thesis booklet defended in July 1665 that was sent by Parisian scientists to their correspondents abroad.[23]

2. For simplicity, we speak generically of "theses" in this chapter. It is however only at the end of the period under consideration that this term is systematically used in the titles of our documents, even if beforehand they were sometimes referred to as such.[24] In the years 1643–1655, they are almost always entitled *agones panegyrici* or *agones mathematici*, "panegyric (or mathematical)

[20] On the rivalry between the Jesuits of Paris and the Sorbonne, see Jourdain 1862–1866, 60–7, 83–7, 94–9, 107–9, and 150–8; Dupont-Ferrier 1921–25, vol. I, 26–28. Compère and Julia 1984–2002, vol. III, 359–407 argue that the monarchy stimulated this rivalry both for motivating the two institutions to give their best and for being granted the role of an arbitrator between them.

[21] For a similar situation in Antwerp and Louvain, see De Bruycker 2009, 139.

[22] Bibliothèque interuniversitaire Sorbonne (Paris), OBL 32–1 (105, 130 bis); VCM 6=6680 (1–2). For an exception, see the thesis print of Coret 1661, now to be found in many European libraries. On 12 December 1661, in the midst of the Jansenist crisis, this thesis print defended the infallibility of the Pope, including on questions of fact.

[23] Petit to Huygens, 7 August 1665, in Huygens 1888–1950, vol. V, 433; Oldenburg to Boyle, 14 July 1665, in Oldenburg 1965–1973, vol. II, 431–2. For some mathematics thesis booklets, there remain however only one copy preserved in the library of the Conservatoire des arts et métiers, where, after the Revolution, treatises on applied mathematics and on fortifications coming from the libraries of the religious orders were deposited. Romano (2006, 362–5) discusses on the circulation of manuscripts, especially those of Bourdin's.

[24] Bourdin himself uses the term *theses* only in the title of the 1640 thesis defended by Pierre Gaillard, but Paul Le Mercier introduced his manuscript transcriptions of the 1638 thesis and of the 1639 thesis respectively as "Theses mathematicae seu Musaeum mathematicum" and as "Aliae theses mathematicae … seu conclusiones mathematicae" (BNF, Ms. Lat. 17862, resp. f. 906 and 966) and Descartes speaks of the "Jesuit optical theses" (to Mersenne, 9 February 1639, in Descartes 1996, vol. II, 499). Oldenburg to Boyle, 14 July 1665, in Oldenburg 1965–1973, vol. II, 432–3 uses the terms "thesis" and "disputation" interchangeably.

jousting."[25] After the interruption following Bourdin's death, the privileged term is *positiones* in the years 1663–1669. Finally, from 1669 onwards, our documents are dubbed *theses mathematicae*.[26] Beyond the use of the word *theses*, the nature of these documents should be clarified.

First, early modern theses were not supposed to introduce new results, methods or interpretations, but were the occasion for students to establish their competence in a given academic field by defending a set of established positions. Second, a distinction existed between *theses pro gradu*, that led to the awarding of diplomas, and *theses sub preside*, that were more or less formalized exercises through which the students who had acquired some skills had an opportunity to demonstrate them in front of an audience.[27] Since the Jesuits never obtained the right to award diplomas, which remained a privilege of the Sorbonne, the theses defended at the *collège de Clermont* were at best *theses sub preside*.[28] Nevertheless, in the context of their rivalry with the Sorbonne, the Parisian Jesuits used to mimic the ceremonies surrounding the defense of *theses pro gradu* to capture the symbolic benefits normally reserved for the intellectual magisterium of the Sorbonne.

Jesuits organized magnificent public celebrations in June or July to mark the end of each school-year. In our documents, the use of the word *agones* in the titles during the period 1643–55 is an allusion to this ceremonial context. Considering what we know of previous periods and of other colleges, the significant phenomenon at the *collège de Clermont* is that, from the late 1630s onwards, these ceremonies included festivities totally devoted to mathematics.[29] The front page of each thesis booklet mentions the place and date of the defense, but also the name and origin of the defendant. It is almost always followed by a program written in the future announcing the various events scheduled for these mathematical festivities, which took place at the end of the week, including on Sunday afternoons.[30] The celebra-

[25] The Greek term from which *panegyricus* derives refers to an eulogy, to the assembly in which this eulogy is pronounced, and, finally, to the festive event itself.

[26] Similar terminological variations are noted in Leinsle 2006, 41.

[27] On this difference and the exercises in Jesuit establishments, see Leinsle 2006, 26–9.

[28] Jourdain 1862–1866, 150–8, Compère and Julia 1984–2002, vol. III, 364.

[29] For a description of the annual celebrations at the *collège de Clermont* and at college of *La Flèche*, see resp. Dupont-Ferrier 1921–25, vol. I, 245–6; and de Rochemonteix 1889, vol. IV, 149–56. For a description of the public, see Meyer 2002, 38–42; and Dhombres and Radelet-de Grave 2009, 27; it should be noted, however, that the testimonies of ceremonies available to us concern the defenses of important people, like the eldest sons of a Minister and of the President of the *Parlement de Paris*, and that we do not know how an ordinary defense was conducted. For the place of mathematics in these celebrations during the late sixteenth and early seventeenth century, see Romano 1999, 491–511.

[30] In our corpus, there are several exceptions to this rule. First, there were Sunday exercises; as we argued in Collacciani and Roux (2017b, 92–93), they probably corresponded to monthly disputations that were not public, but involved only the small group of mathematics students gathered in the *aula mathematica*. Second, there were theses that were defended in unusual times of the year because of extraordinary events such as the passage of a comet (**Tarteron 1665** and **Prou 1665a**, both defended on 29 January 1665; Vernon to Oldenburg, 10 April 1672, in Oldenburg 1965–1973,

tions were opened by a solemn speech (*solemnis prolusio*) delivered by the professor of mathematics; students then performed mathematical exercises for two or three consecutive days. Such programs likely functioned as advertising leaflets that were distributed beforehand, to make the public know of the event, but also, possibly, to allow the *virtuosi* to prepare comments, questions, and objections.[31]

The authors of the theses were not the defendants, but their teachers, whose names are not mentioned on our documents. The professors could submit the same set of problems to a number of students and the same thesis could be defended on the same day by several students, who sometimes were rich enough to print their own volume in their own name (this is, for example, the case of Louis Ragayne de la Picottière and Louis Prou in 1665). Sometimes students shared the costs by printing a single thesis booklet that mentions their respective names (for example Pierre de Cornouaille and Jacques Manchon in 1638 or Hyacinthe de Ruffec and Jean Richomme in 1655).[32] In the last years of the period under scrutiny, in 1668, 1669, and 1671, thesis booklets were also produced that did not mention the names of all of the defendants; as for the 1672 thesis, which was defended by Louis Le Mazier on 24 June, by "two of the Society of Jesus" on 25 June, and by "one of the Society of Jesus" on 26 June. Because theses were written by teachers, the same paragraph, the same set of paragraphs, or even the same set of theses altogether, can be found in theses defended one or several years apart from one another: for example, a thesis entitled *De corporum caelestium motibus* (*On the motions of celestial bodies*) was defended in July 1667 by Jean-Baptiste and Pierre-Antoine de Castagnère de Chasteauneuf and in March 1668 by Jean-Baptiste Colbert de Segnelay.

3. Most of the professors who held the chair of mathematics at the *collège de Clermont* are known through the Jesuit archives:

   1634–1653: Pierre Bourdin (1595–1653)
   1653–1658: Unknown
   1658–1661: Georges Fautrel (Unknown–1687)
   1661–1664: Nicolas d'Harouys (1622–1698)
   1664–1669: Michel Beaussier (Unknown–Unknown)
   1670–1673: Ignace-Gaston Pardies (1636–1673)
   1673–1676: Claude-François Millet de Chales (1621–1678)

vol. IX, 14). Last, for some unknown reasons, some defenses took place during the week, such as Guillaume Benard de Rezay's defense (**Benard de Rezay 1666**) on Thursday 29 July 1666, Jean-Baptiste and Pierre-Antoine de Castagnere de Chasteauneuf's defense (**Castagnere de Chasteauneuf 1667**) on Wednesday 13 July 1667, Jean-Baptiste Colbert de Seignelay's defense (**Colbert de Seignelay 1668**) on Thursday 30 August 1668, or Alexandre Milon's (**Milon 1671**) defense on Monday 6 July 1671.

[31] Le Dividich 2003, 56–61 for leaflets announcing courses in mathematics; Rochemonteix 1889, vol. IV, 153; Meyer 2002, 38–9; and eadem 2017, 36–8 for leaflets announcing the defense of theses.

[32] The strategy of sharing the costs of printing is noted by Leinsle 2006, 37; Dhombres and Radelet-de Grave 2009, 26; Meyer 2017, 20. How the time was shared when there were several defendants is not known.

1676–1685: Jean de Fontaney (1643–1710).[33]

The temporal distribution of thesis booklets allows us to distinguish periods, which correspond to the activity of some teachers and to the inactivity of others, or even to the fact that the chair of mathematics may have been vacant in the mid-fifties. Sixteen theses in mathematics produced between 1638 and 1651 survive, that is, almost one for every year during the fifteen years (1638–1653) in which Pierre Bourdin taught at the *collège de Clermont*. In the ensuing decade or so, only one thesis was defended in 1655, a year in which we do not know who was the mathematics teacher, assuming there was one. During the eight years (1661–1669) when Nicolas d'Harouys and Michel Beaussier were professors, there are twenty-one theses; this frenzy of thetic activity is explained by the extraordinary passing of two comets in a row in 1664–65, but mostly by the antagonism that developed between the Jesuits and the Cartesians. Under the impulse of Ignace-Gaston Pardies, the normal rhythm of one thesis per year was reintroduced.

As a quick glance at these thesis booklets shows the disciplines favored in the thirties and forties, when Bourdin was the mathematics teacher, were optics and military architecture. In the sixties, because of the comets, the establishment of an observatory, and the interests of those who were then mathematics teachers, most of the theses were devoted to astronomy and cosmology, the main task, but not the only one, being to compare the different systems of the world.[34] Finally, in the few theses that were defended when Pardies was a professor, the place of optics is again important. In what follows, we will focus on optics to contrast Bourdin's teaching and Pardies' teaching.

5.3 Pierre Bourdin's Collections of Optical Problems: Spectacular Devices Enlisted to Defend a Traditional Theory of Sight

Pierre Bourdin (1595–1653) began his career by teaching grammar and humanities at the college of La Flèche (1618–1623). He was then professor of rhetoric at Rennes (1627), Rouen (1628), and Bourges (1629–1632). Two years after his return to La Flèche as a professor of mathematics (1633), he was transferred to the same

[33] We follow the indications given in Dainville 1954, 110–11.

[34] On the 1664–65 comets, see Roux 2017. On the observatory of the *collège de Clermont*, see Bigourdan 1918; Dupont-Ferrier 1921–25, vol. I, 189–90; and Dainville 1954, 44. Nicolas d'Harouys built machines for teaching the different astronomical systems that are alluded to in **Ragayne de la Picottière 1665b**, 4, § 7. Garnier (1678, 118) describes these machines and explains that, because of their number and magnitude, they could not be stored in the rooms of the Library, but had to be kept in a large room of their own, where they could be seen by eminent travelers; see for example Huygens' *Journal*, in Huygens (1888–1950, vol. XXII, 545), André de Graindorge to Huet, 9 May 1665 and 5 August 1665, in Graindorge 1942, resp. 267 and 303; and Brice 1685, vol. II, 51.

position at the *collège de Clermont*. Since he held this title for almost twenty years, from 1634 to his death in 1653, and organized a defense almost every year, we have an impressive number of his thesis booklets at our disposal.[35] To capture their gist, we will first approach them from a formal perspective; we will then show that they had recourse to engravings and optical devices; finally, we will ask ourselves what was the place of the ancients and the moderns in Bourdin's optical theory.

1. From a formal perspective, Bourdin's thesis booklets belong to a "genre" that can be easily characterized. They are often presented as encyclopedias ordered according to disciplines (*Philosophica, Optica, Acoustica, Historica, Cosmographica, Theologica, Astronomica, Arithmetica*, or *Ex philosophia*, *Ex optica*, etc.).[36] Bourdin, who had been a professor of rhetoric and who probably knew the art of memory, did not confine himself to the order of disciplines: the 1638 thesis enumerates the different collections of a cabinet of curiosities, the 1639 thesis, the rooms of a palace, the 1646 thesis, the accomplishments of mathematics. Each of these headings includes, rather than a series of continued propositions on a given theme, a collection of problems that are independent from one another. For example, in the 1640 thesis, a problem under the heading *Militaris* asks to show that straight lines are more suitable than curved lines for building defense walls; a problem under the heading *Sacra* asks to compute the speed of the skies, knowing that during the time to say "Pater," the stars travel 800 leagues; or a problem under the heading *Philosophica* asks to demonstrate geometrically that quantity is infinitely divisible in actually finite parts, and, this, in two ways, the Aristotelian and the Euclidean.[37] The "collection of questions" aspect was also present, albeit less markedly, in the theses defended under Jean Leurochon in Pont-à-Mousson in the twenties and it was part of the genre of mathematical recreations that appears in Leurochon's *Récréations mathématiques* or in Marin Mersenne*'s Questions inouies*.[38] Bourdin's emphasis on problem solving, that appears when paragraphs begin with expressions such as *explicare* (or *aperire, indicare) qua possit* (or *quo fit, cur, quo pacto, unde, qua arte fiat*), is however typical for him.

Bourdin's thesis booklets present yet another specificity that is not found in the thesis booklets of any other Jesuit professor of the *collège de Clermont*: some of

[35] On Bourdin, see Sommervogel 1890–1932, vol. II, 29–30; on his career, see Dainville 1954, 110; and Romano 1999, 563–4. On his relations with Descartes, see Ariew 1995; idem 1997; idem 1999, 5, 24–9, 156, 173, 193–6, and 203–4; and Collacciani and Roux 2017b. On his treatise on fortifications, see d'Orgeix 2005, 91–105; Romano 2006; De Lucca 2012, 96–104, 118, 189, 217, 229, and 332. On his theses and more generally his teaching at the *collège de Clermont*, see Le Dividich 1996, 39–44, 47–8, 180–8, 234–43, and 272–6; and Collacciani and Roux 2017b.

[36] Leinsle (2006, 20) notes that some theses are a series of exercises independent of each other, while others constitute small treatises on a given subject.

[37] **Potier 1640**, 3, § 4, and 7, § 1–2.

[38] Leurochon 1622; and idem 1629. On the diversity of these theses, see Romano 1999, 501–3. See also (ibid., 618–21) on the reproduction of mathematics thesis booklets defended in Paris in 1622. Concerning mathematical recreations, see Chabaud 1994.

them include pictures. In the thesis booklet of Jacques Pallu and Jacques Touchelée, whom Bourdin made defend a mathematics thesis in 1635 when he was a professor at La Flèche, there are no illustrations; in 1638, Pierre de Cornouaille and Jacques Manchon's included two engravings in their thesis booklet (one representing the Copernican hypothesis, the other a reinforced fortification); and from 1639 onward, Bourdin developed a collaboration with engravers to illustrate his thesis booklets. Figure 5.3 reproduces copper-plate engravings made by Alexandre Boudan that appear in a 1640 thesis booklet. While the plate on the right illustrates various topics, the plate on the left is entirely devoted to optical phenomena, optics being indeed the discipline that most interested Bourdin. These illustrations raise unanswered questions. We do not know why some thesis booklets feature printed images, whereas other impressions of the same sets of theses that were supposedly defended at the same dates contain fewer images, drawings rather than engravings, or no images. We also do not know what shares the engravers and Bourdin had in the production of these images, nor what type of contract bound them. Some things are known however. The signature inscribed on some of the engravings identifies his main—and possibly his only—collaborator as Alexandre Boudan (1600–71), an illuminator, copper-plate engraver, and dealer of etchings established in rue Saint-Jacques at the *Corne de cerf* around 1633, then at the *Image Saint-Maur* from 1643 onward, who also worked with other Jesuits.[39] Second, some of these engravings are also included in the manuscript notes taken from Bourdin's teaching, which indicates that they were likely also distributed in class.[40] Moreover, they resurface in various editions of his *Cours de mathématique*, which is presented as a "figured mathematics" (*figurata mathesis*) and organized according to a "map" (*ichnographia*). Bourdin's textbook was a visual encyclopedia of mathematics where, in accordance with a more general pedagogical tendency of the Jesuits, pictures were used as mnemonic support for students.[41] It is therefore no exaggeration to conclude that images were essential in Bourdin's pedagogical practice.

Another particularity of Bourdin's thesis booklets is that some paragraphs begin with "*exhibere (or proferre) experimenta*," (to exhibit or to show experiments). One wonders if the experiments in question were carried out during the defense or if they were only presented with engravings. It is possible, that, as some of the theses ask, the defendants had to trace ellipses or to explain the functioning of a book wheel, which was able to open several books at the right page.[42] But it is unlikely that they could, as other theses demand, display fountains and machines capable of lifting

[39] Jones (1947, 119–20) and Ariew 1995, 212; idem 2014, 21, date Boudan's engravings in Bourdin's thesis booklets back to 1631, but Collacciani and Roux (2017b, 110–11) prove that they were made in 1637, at the time when Bourdin was beginning to teach mathematics in Paris.

[40] **Petit 1639**, in BNF, Ms. Lat. 17861, p. 879; **Culant 1639**, in BNF, Ms. Lat. 17861, p. 881; Bourdin 1641; idem 1645; idem 1661. Le Dividich (1996, 41; and 2003, 61–3) mention the use of these engravings in Bourdin's classes.

[41] Collacciani and Roux 2017b, 106–11.

[42] **Henri 1639**, 6 § 3, 7 § 1, and 6 § 1. The machine to display books is certainly a book wheel similar to the one presented in Ramelli 1588.

Fig. 5.3 Potier, *16*40, in-4, Bibliothèque Sainte-Geneviève, Paris, cliché N. Boutrous. While the right plate presents a variety of topics (astronomical observations, magnetic phenomena, geographical maps, and fountains), the left plate is entirely devoted to optical phenomena: anatomy and physiology of the eye (A, B, C), reflection (D), refraction (F), laws of perspective (G, H), anamorphosis (E, L, M), and, at the center, the *specillum* (O) which is put in parallel with Galileo's telescope (P)

Fig. 5.3 (continued)

enormous burdens.[43] Unambiguous evidence that real experiments were performed comes from the program of 1639, which distinguishes between the time devoted to discourses and to experiments.[44] Thus, Bourdin's thesis booklets displayed a

[43] **Henri 1639**, 7 § 1–4; and **Cornouaille and Manchon 1638**, 7 § 4–5.

[44] **Henri 1639**, 2. The programs of 1640, 1648 and 1651 also mention real experiments.

science that was literally spectacular: there were spectators around and they were attending spectacular performances. This is unsurprising in the context of public ceremonies that were intended to convince the public of the magnificence of Jesuit science.

2. Relevant here is an optical device called the Lens of Faith (*Specillum Fidei*) that is presented as new and unexpected in several theses. The most precise indications about it appear in the 1639 thesis, which describes the marvelous effects of this device without explaining how it works (Fig. 5.4).

These effects were the following: when seen with the naked eye, a given plate would depicts the Host surrounded by angels; however, when seen through the Lens of Faith, the same picture revealed, not the Host, but the Lord of the angels, that is Jesus Christ. The Lens of Faith showed a representation of transubstantiation.[45]

Bourdin was already interested in anamorphoses when he arrived in Paris, but it is probably through Jean-François Niceron that he first heard of this dioptrical device.[46] Although it was invented as early as 1628 by Charles du Lieu, a Jesuit from Lyon (1609–1678), it was not described in a publication before Niceron's *Perspective curieuse* (1638).[47] It appeared in Jesuit works of the mid-seventeenth century, from Jean du Breuil (1649) to Nicolas Forest Duchesne (1650) to Gaspar Schott (1659); a century later, it was still known: it was used by Charles-Amédée van Loo to paint a portrait of Louis XV in 1742.[48] The Museo Galileo in Florence and the Teyler Museum in Harlem having preserved different pieces of the whole apparatus—a painting in Florence, an optical tube in Harlem—, it is possible to get an idea of its functioning and to construct a replica.[49] The *specillum* was a tube that included faceted mirrors: each of these facets refracting a fragment of the picture that was seen with the naked eye. What was seen through the *specillum* was a new picture composed from several fragments of the initial picture (Figs. 5.5 and 5.6).

The skill of the artist consisted in painting two pictures that were not only coherent, but also associated through a symbolic relation, so that the picture seen through the lens revealed the meaning of the picture seen with the naked eye. According to the examples mentioned in the literature of the day, the *specillum* allowed one to see Urban VIII as a substitute of several popes, Ferdinando II de' Medici (or Louis XIII)

---

[45] **Henri 1639**, 6. It is also mentioned in **Foucquet 1646**, 6 and, 11 § 4.

[46] Collacciani and Roux 2017b, 99–100. The most complete article on this device is Malcolm 1998; concerning Niceron, see Truci 1976; and Bessot 2005; concerning Bourdin, see Collacciani and Roux 2017b, 95–100.

[47] Niceron 1638, 101 and du Breuil 1649, "Instruction sur le traite VII," sigs. Zz1v–2r. On Charles du Lieu, see Malcolm 1998, 219–21.

[48] Du Breuil 1649, 161–2; Forest-Duchesne 1650, 226–7; Schott 1657, 470–1 and pl. 23, fig. 7, 453.

[49] For a photograph of the Florentine apparatus, see Bonelli 1959, 38; the optical tube was lost in the 1966 Arno flood, see Righini-Bonelli 1976, 197–96; for its present state, see https://catalogue.museogalileo.it/object/OpticalToy.html. Only the optical tube remains in the Teyler Museum, see https://www.teylersmuseum.nl/en/collection/instruments/fk-0316-1-5-anamorphoscopes-5. Hunt and Sharp 2011 propose a reconstruction of the apparatus.

Fig. 5.4 Henri, 1639, in-4, BnF (Paris), département Littératures et art. This is the only copy that includes this frontispiece in which the rectangles and circles are printed while the images inside them are drawn by hand. It is entirely devoted to optical transformations; the three pictures in the top cartouches represent the *specillum*

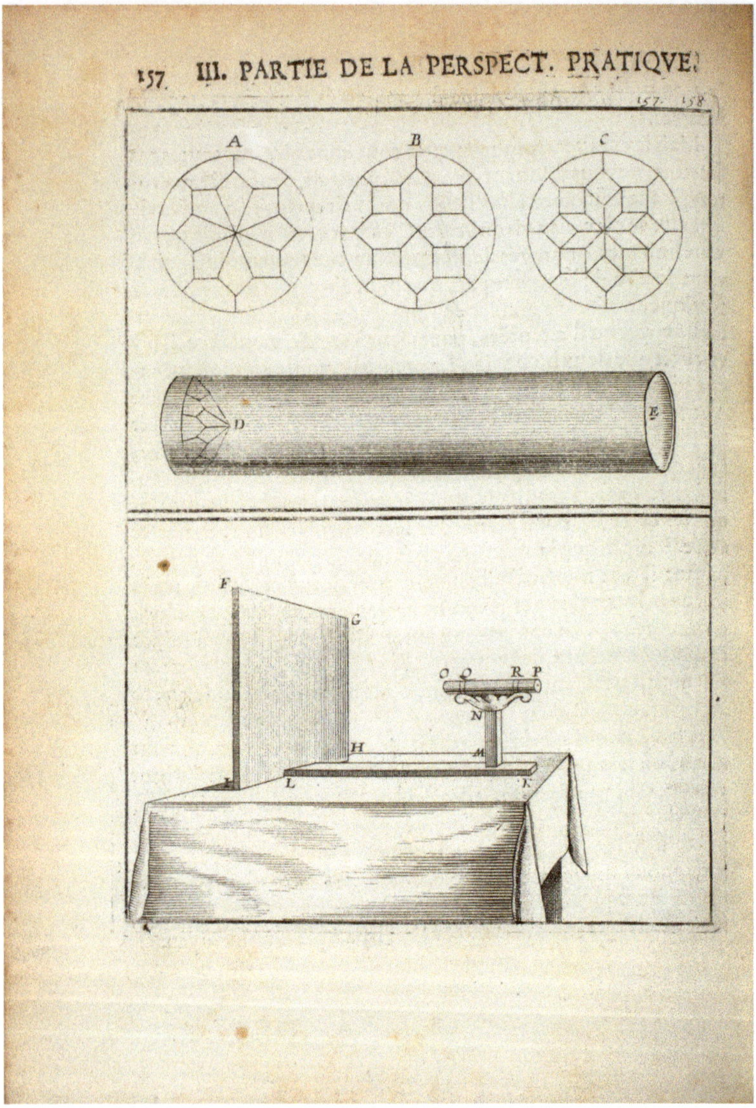

Fig. 5.5 The optical device according to du Breuil, 1649, 157, in-4 (left, Bibliothèque Sainte-Geneviève, Paris, cliché N. Boutrous) and to Niceron, 1638, in fol., plate 23 (right, Bibliothèque de l'Institut National d'Histoire de l'Art, Collections Jacques Doucet, 4 RES 876)

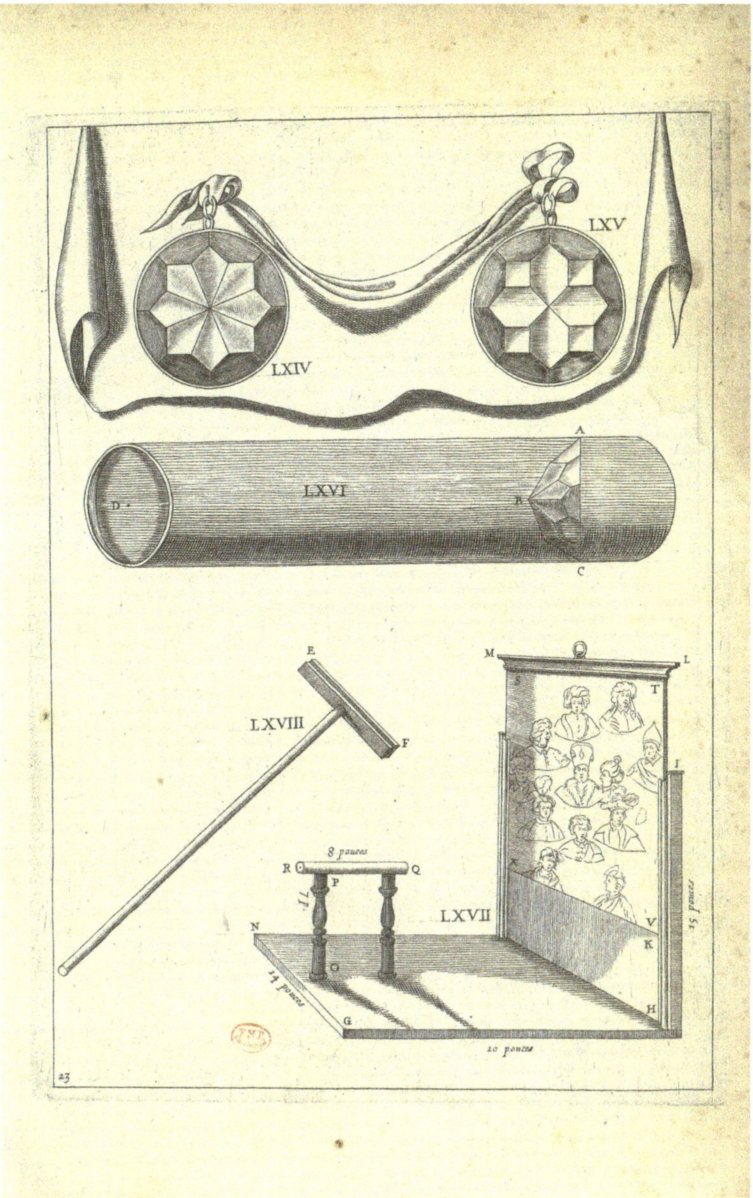

Fig. 5.5 (continued)

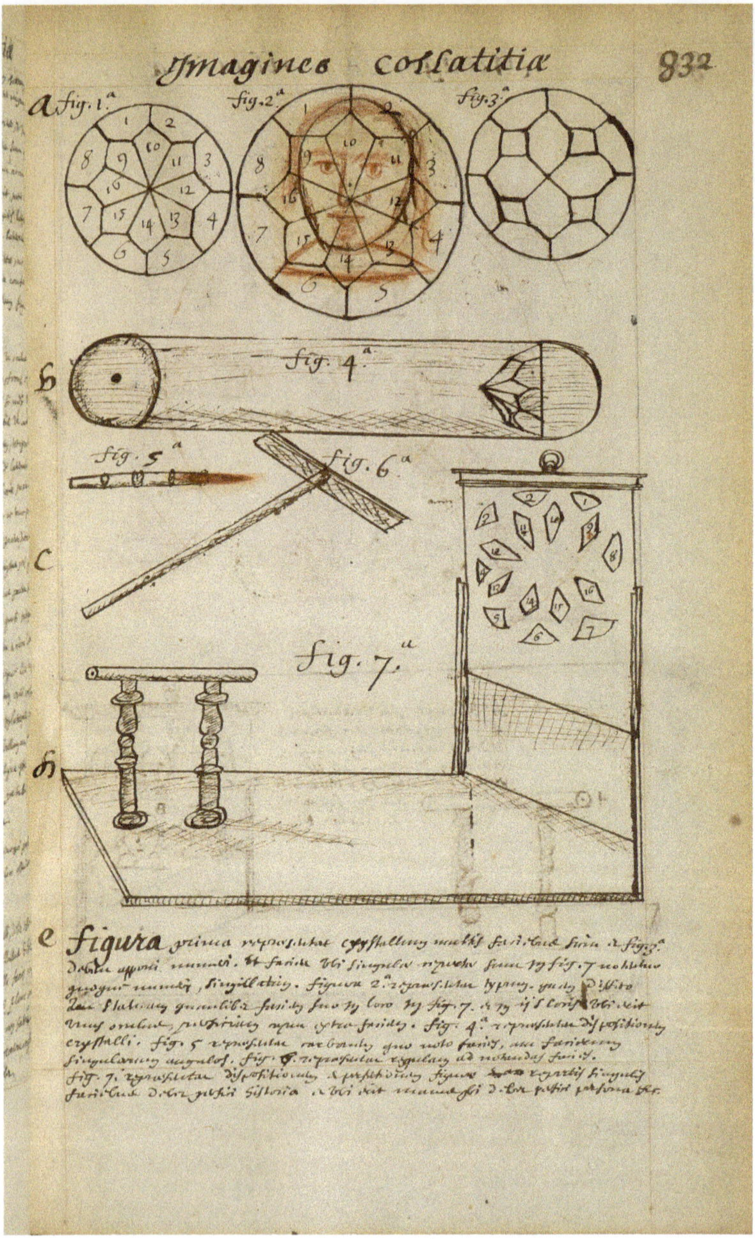

Fig. 5.6 A drawing explaining the *specillum* in BnF (Paris), Département des manuscrits, Ms Lat. 17862, f. 932, in-4

Fig. 5.7 Portraits of Ignatius and other saints hidden in a representation of angels and cherubs visible through the *specillum* according to du Breuil 1649, 166, in-4, Bibliothèque Sainte-Geneviève, Paris, cliché N. Boutrous

rather than several Mohammedan sultans, Louis XIV in the place of his parents, a New Testament truth instead of several Old Testament prophecies.[50] As for the *Specillum Fidei* mentioned in Bourdin's theses, as we have seen, it made one see Jesus Christ in the place of the Host (Fig. 5.7). To look at these pictures through the appropriate lens revealed the truths hidden behind the painted appearances; it was not only a scientific device, but a moral device, even an apologetic one when the truth in question was that of the Eucharist.

3. If Bourdin displayed a spectacular device at the cutting edge of his time, his optical theory was inspired by Jesuits of the previous generation like François de Aguilón (1567–1617) and especially Christoph Scheiner (1575–1650), who accepted Kepler's discovery that vision is made on the retina, rather than on the crystalline lens, but refused to give up the existence of species transporting the resemblance of things to the brain. From the analogy between the eye and the *camera obscura*, the pupil being the hole through which the light is transmitted, the crystalline lens being the lens by which a picture is formed, and the retina being the screen on which it is painted, Kepler had deduced in the *Paralipomena ad Vitellionem* that opticians had to study the geometric transmission of light rays without resorting to *species*, while natural philosophers had to devote them-

[50] Niceron 1638, 106, 115–8; and Du Breuil 1649, 162 and 165.

selves to the investigation of what was happening beyond the retina, in the dark caverns of the brain.[51] In his *Oculus hoc est fundamentum opticum* (*The Eye, that is the optical principle*), Scheiner, who is Bourdin's main source of inspiration, took up Kepler's analysis of the different functions of the parts of the eye and argued that the retina is the formal instrument of vision.[52] He went even further thanks to delicate anatomical observations and established that the optic nerve is not in the axis of the eye, but also that some pictures can be seen in the eyes of dead men and animals.[53] But, unlike Kepler, to ensure that we know things as they actually are, he kept the notion of *species*, which he applied equally to images emanating from objects, to visual rays, to retinal images, and to brain images.[54] Although he admitted that all rays joining the visible object to the eye are to be considered visual rays, he asserted that only one of them is sensed by the eye and deserves as such to be called the "principal, primary, and immediate or formal ray."[55]

Bourdin's thesis booklets take up the division of parts and functions of the eye proposed by Scheiner in the wake of Kepler.[56] The crystalline lens is established by nature to gather the visible species on the retina: vision is distinct when each point on the retina corresponds to a point of the object; it is confused when the crystalline lens is damaged and there is no one-to-one correspondence between the points of the object and the points of the retinal image. As for the retina, it is the formal organ of vision: it is the place where the visible *species* of the objects (*species objectorum*) that have passed through the cornea and have been collected by the crystalline lens are joined to the visual spirits (*spiritus visorii*) that come from the brain through the optic nerve.[57] This stresses a point that was already noted by Ibn al-Haytham but had been reinforced by Kepler: while visual rays coming out of the eye (*radii emissi*) are imaginary, incoming rays carrying the images of the objects (*radii immissi*) are real.[58] But for all that, Bourdin's optics, exactly like Scheiner's, did not abandon the notion of *species*. Like Scheiner, Bourdin states that *species*, light (*lux*),

---

[51] Kepler 1604, 37–56, 168. On Kepler's optics, see Lindberg 1976, 178–208; Sabra (1981); Simon 1988 and idem 2019; Smith 2015, 322–75.

[52] Scheiner 1619, 124–5, 137–8, and 197–9. See also Scheiner 1630, $106a_{30}$–$116b_{9}$. On Scheiner's compromise, see Pantin 2008, 263–7; Gal and Chen-Morris 2010 203–6; and Smith 2015, 374–5.

[53] Scheiner 1619, 17–18; and Scheiner 1630, $110b_{39}$–$112a_{3}$.

[54] Scheiner 1619, 128–38; Pantin 2008, 263–7.

[55] Scheiner 1619, 73–4.

[56] Aguil6n and Scheiner are among the very few authors to be named, see **Bouthier 1647**, 12 § 1–2.

[57] **Cornouaille and Manchon 1638**, 15 § 2, 4–5; **Henri 1639**, 14 § 1; **Culant** 1639, [1] § 2; *Potier 1640*, 9 § 1–2 and 4; **Gaillard 1640**, [1] § 1–2; **de Vic 1643**, 11 § 4–5; **Gedoin 1644**, 13 § 29; **Bouthier 1647**, 12 § 2; **Thierry 1648**, 8 § 19–20; and **de Bourneuf**, 1650, 10–11 § 27, and § 30. See also BNF, Ms. Lat. 17861, f. 900 and 905 and Ms. Lat. 17862, f. 906–9.

[58] **Culant 1639**, [3] § 1; **Henri 1639**, 14 § 2; **Potier 1640**, 7 § 5; **Gaillard 1640**, [1] § 6; **Bouthier 1647**, 12 § 2. On the arguments against the emission of visual rays, see Lindberg 1976, 66–7, 178-85; Smith 2015, 185–8, 256–75; and Simon 2019, 43–44, and 127–28.

illumination (*lumen*), and color are the same albeit in different respects.[59] The most developed argument on *species* is given in the 1642 thesis:

> I. To explain vision, *species* are needed in the medium. These are not (except perhaps metonymically), as is commonly believed, images of objects, but a very simple entity of the same kind for absolutely all visible things, whose function is to join what is looked at to the agent when these powers are distant. Likewise, the Peripateticians believe that, by light, as by a vehicle, celestial things produce various effects on sublunar things.... II. The species of visible things impressed by powers and expressed by them are really similar to objects and are their natural images, and outside of them there are no other images of things, or formal *species*, not even in mirrors, but neither when, through an opening, external objects appear on a screen in a closed room.[60]

Scheiner is not Bourdin's only source: it is likely that he learned from Vospicus Fortunatus Plempius the strange fact reported by Platerus (Felix Platter), Rungius (Johannes Runge), Smetius (Henrich Smet), Vagerus (Matthaeus Wagger), Kepler, and Plempius, that "while they were deprived of the use of their eyes, they were able to see by their nose and to distinguish correctly colored objects through their nostrils."[61] However, Scheiner's importance is revealed by the engraving "Oculus" that was first used in the years 1638–1640 during Bourdin's classes as well as in his thesis booklets and then regularly republished in his books until the 1661 edition of the *Cours de mathématique* (Fig. 5.8). In the lecture notes taken by Paul Le Mercier, part of this engraving was copied, which indicates that copying figures by hand constituted a significant part of Bourdin's teaching (Fig. 5.9). Even leaving aside Bourdin's appetite for pedagogy through images, using engravings and drawing pictures was relevant in anatomy: visualizing the spatial structure of the eye and the relationship among its parts was more efficient than describing it through words. The images labeled 4 and 15 in the "Oculus" engraving might have been inspired by Felix Platter's anatomical images, but, since these images had already been reproduced in Kepler's *Ad Vitellionem paralipomena*, it is impossible to know if Bourdin was inspired by Platter or by Kepler (see Figs. 5.8 and 5.10).[62] Moreover, other images in the "Oculus" print are borrowed from Scheiner: this is the case for the

---

[59] **De Bourneuf 1650**, 10 § 15–16.

[60] **Despont 1642**, 10 § 1–2: "Ad explicandam visionem necessariæ sunt species in medio. Hæ non sunt, ut vulgo creditor, (nisi forte metonymice) imagines obiectorum, sed entitas quædam simplicissima, et uniusmodi respectu omnium omnino visilium, cuius sit munus Aspectabilia potentiis distantibus coniugere ad agendum, ut de lumine censent Peripatetici, quo veluti vehiculo, cœlestia in sublunaribus varios producunt effectus... II Species rerum visilium impressæ potentiis et ab iisdem expressæ sunt vere similes obiectis, et illorum imagines naturales, ac præter eas nullæ sunt aliæ rerum imagines, aut species formales, ne in speculis quidem, sed nec in charta dum clauso in conclavi per foramen externa obiecta apparent." For other mentions of species, see **Thierry 1648**, 8 § 17–18; and **de Bourneuf 1650**, 10 § 25–6. **de Vic 1643**, 10 § 1 is discussed *infra*, p. XXX.

[61] **Potier 1640**, 5 § 5: "... cùm essent privati oculorum usu, naso viderent, et obiectos colores naribus aptè discernerent, ut referunt probatae fidei Platerus, Rungius, Smetius, Vagerus, Keplerus, et Fortunatus Vopiscus Plempius." This is the last problem discussed in Plempius 1632, 310–12, who refers exactly to the same group of authors.

[62] We would like to thank Tawrin Baker for the clues on anatomical representations of the eye that he shared with us. For more, see Baker 2016.

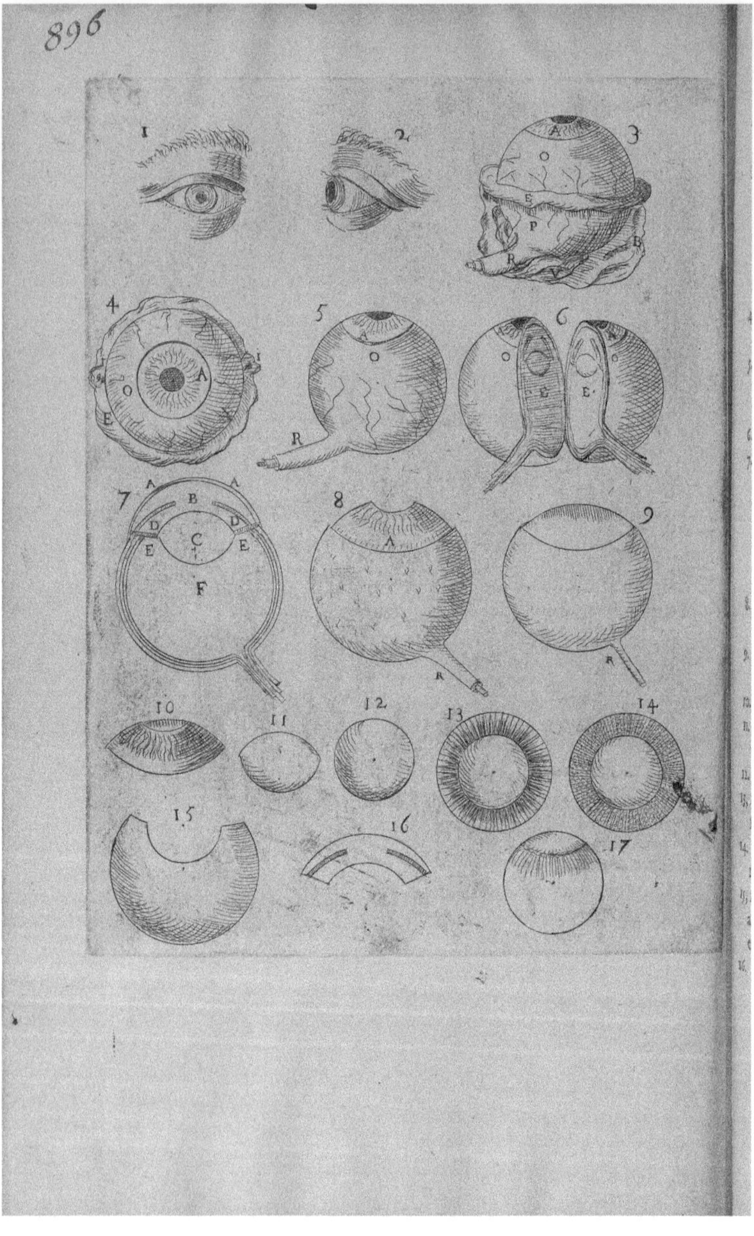

Fig. 5.8 The engraving "Oculus," in-4. BnF, Département des manuscrits, Ms. Lat. 17861, f. 896; Cornouaille and Manchon, 1638, 15; Henri, 1639, Bibliothèque Municipale de Lyon, [n.p.]; Gaillard, 1640, [5]

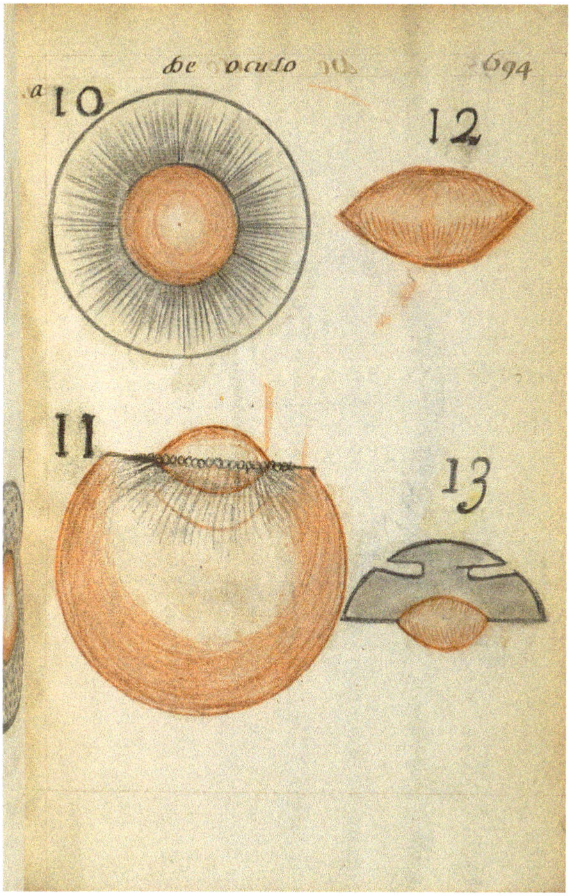

Fig. 5.9 Some of the corresponding drawings in BnF, Département des manuscrits, Ms. Lat. 17862, f. 693–5. The drawing that shows the inclination of the optical nerve (right) is copied from Scheiner (Figure 11, left)

image labeled 3 as well as images 5 to 9, in which the optic nerve is inclined with respect to the central axis of the eyeball, or for image 16, in which the aqueous humor appears isolated from its neighboring parts, which is a rare representation since it could not be obtained by dissection (Fig. 5.11).[63]

As engravings begin to appear in the lecture notes of Bourdin's students in 1637–1638 and in his thesis booklets in 1638, one wonders whether the pictures of Descartes' *Dioptrique* inspired him. Some of Bourdin's and Descartes' pictures have a family resemblance, because they use the same graphic conventions: they combine anatomical and optical elements on a single picture, so that the path of rays

[63] Scheiner 1619, 17–18. The inclination of the optic nerve with respect to the central axis of the eyeball is also taken into account in Descartes' *Dioptrique*, see Descartes 1996, vol. VI, 106.

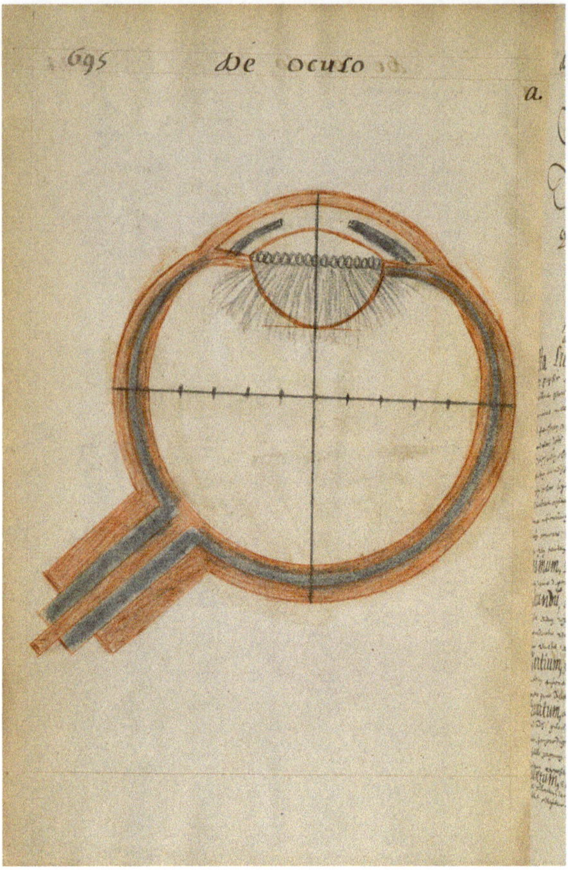

Fig. 5.9 (continued)

is traced not only in the medium between the object and the eye, but in the eye itself; they represent not a singular ray, but pencils of rays that form double cones (from each point of the visible body, many rays issue forth, which are recombined after their passage through the tunics of the eye to form a single point on the retina); they use oriented objects such as arrows, daggers or swords to make the inversion of the image on the retina sensible. But it cannot be concluded that Bourdin and the artist(s) with whom he collaborated borrowed these conventions directly from Descartes, since they already existed in Kepler and Scheiner.[64] Moreover, it is unlikely that Bourdin and his collaborator(s) developed overnight an image-based pedagogy from scratch in reaction to Descartes' pictures. Nevertheless, it is not

[64] On these conventions, see Shapiro 2008, 282–6. Concerning pictures in Descartes' natural philosophy, see Lüthy 2006; Zittel 2009; and idem 2011.

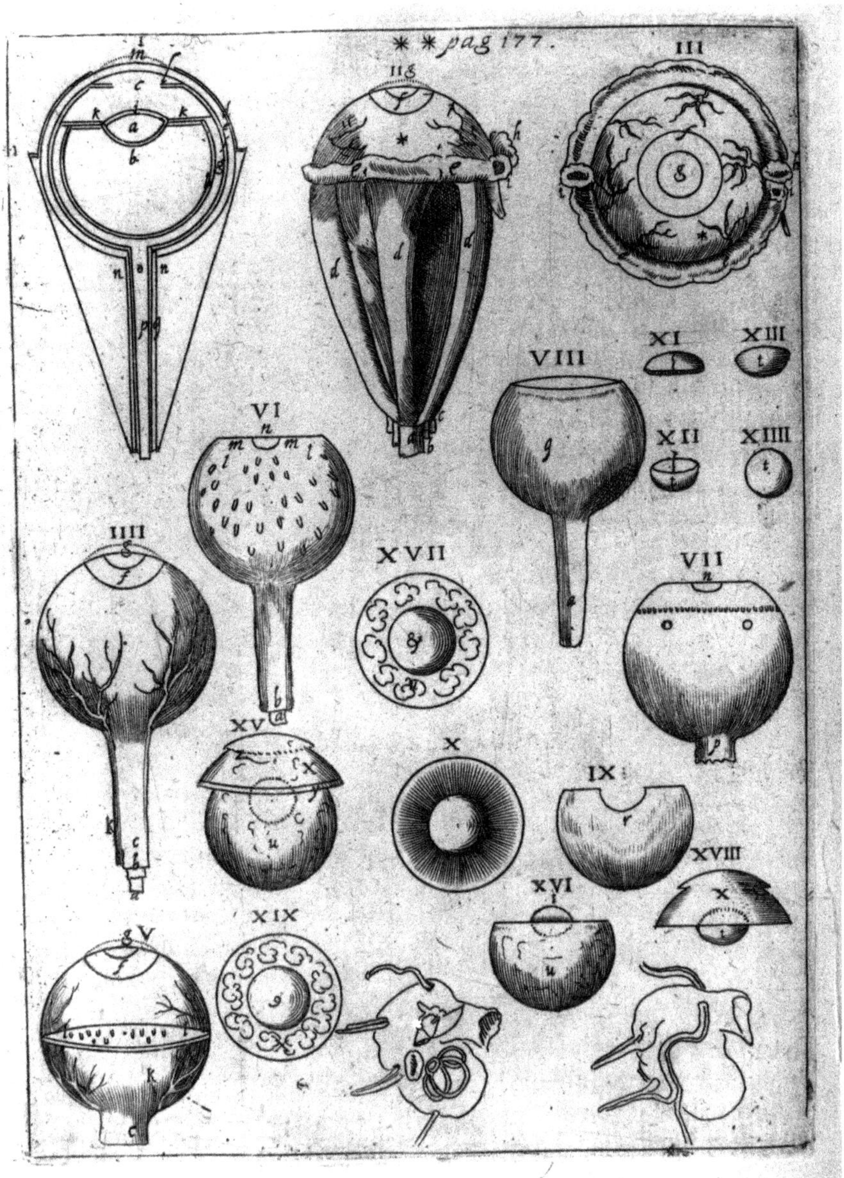

Fig. 5.10 Platter, 1583, tab. 49, as reprinted in Kepler, 1604, 177 (public domain from commons.wikimedia.org)

impossible that the seduction exerted by Cartesian pictures prompted Bourdin and Boudan to invest in their own pictures (Figs. 5.12 and 5.13).

Bourdin's reaction to the doctrine contained in the *Dioptrique* was unknown before Charles Potier's thesis (1640). In this work Bourdin criticizes Descartes' law

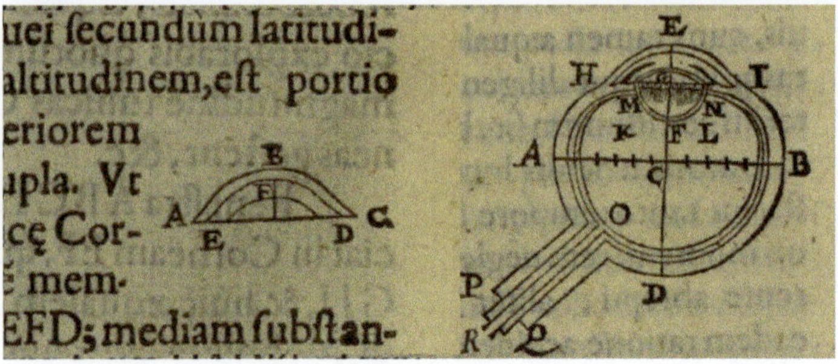

Fig. 5.11 Scheiner, 1619, 14 (the aqueous humor EFD between the crystalline and the vitreous humor) and 17 (the eye with its inclined optical nerve)

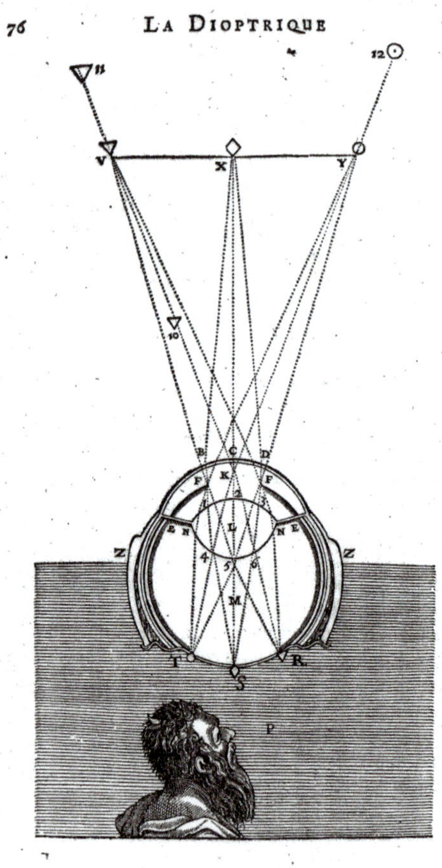

Fig. 5.12 Graphical conventions of Descartes' *Dioptrique*, in Descartes, 1964–1974, vol. VI, 106

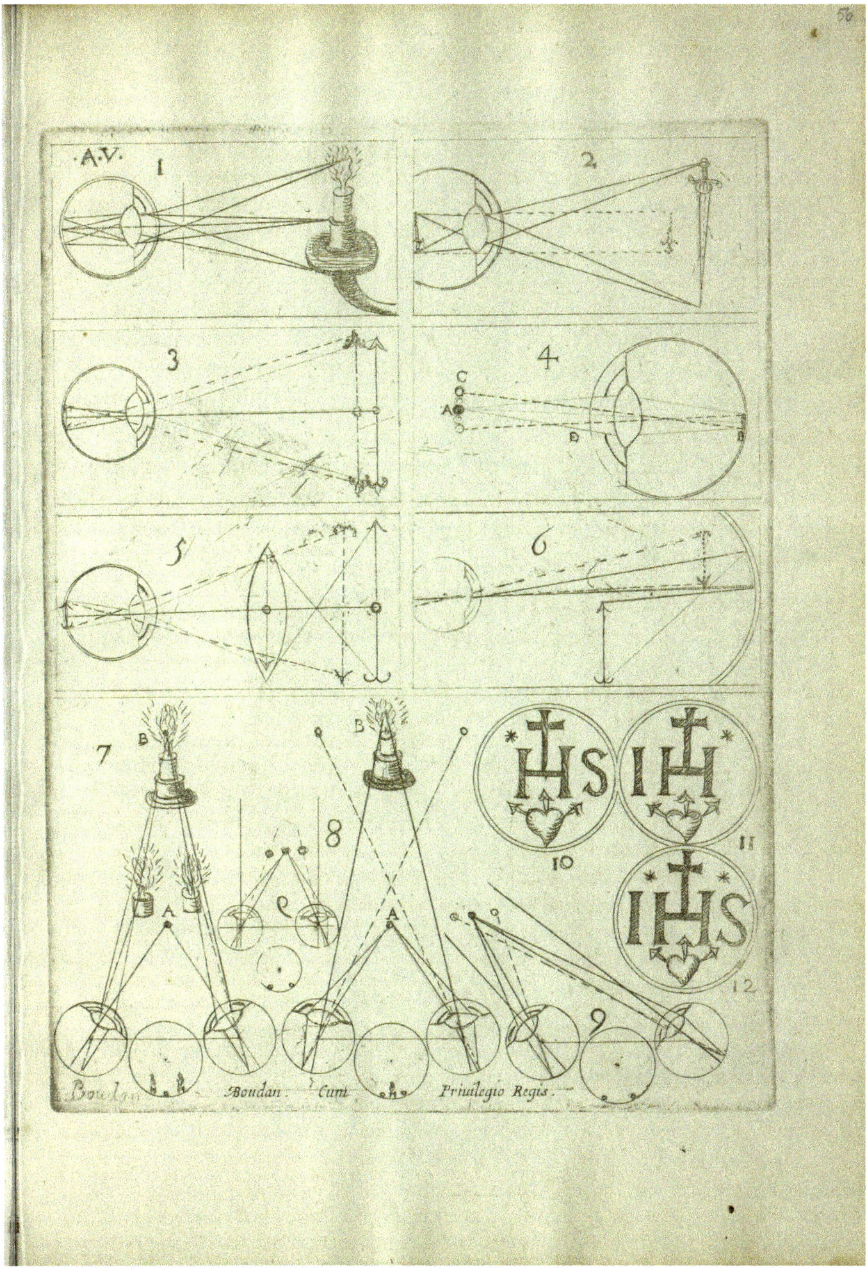

Fig. 5.13 Graphical conventions of the engraving "Visio absoluta," in-4, BnF, Département des manuscrits, Ms. Lat., 17862, f. 925; Henri 1639 [n.p.], Bibliothèque Municipale de Lyon

of refraction, but also his use of subtle matter to explain light and colors.[65] His first argument against subtle matter is that it is only an imaginary entity.[66] Moreover, far from making the explanations more simple, it makes them more complex:

> To explain the action of light and colors on the eyes by the motion of some subtle matter diffused through the pores of the air and of transparent bodies, which luminous bodies set in motion and through which they affect the eyes in different ways, which is quite different than by intentional *species*, amounts to healing a scar with a new injury, and to bothering for nothing with the same difficulties, and with some new ones on top of that. To show this and to demonstrate the inanity of subtle matter.[67]

Bourdin presents explicitly as two rival explanations of light and color the Cartesian explanation, which uses subtle matter, and the Aristotelian explanation, which uses intentional *species*. We hope that it is not an over interpretation to attribute to Bourdin the idea that the mechanization of light made possible by Descartes' subtle matter made vanish into thin air scholastic entities such as intentional *species*. Bourdin was attacking subtle matter to defend scholastic entities. Without examining the transformation of Bourdin's and Descartes' optical dispute into a metaphysical controversy and their reconciliation thanks to Jacques Dinet, it can be noted that in 1643, while their reconciliation was in progress, Dominique de Vic defended a thesis that included subtle matter among the entities that could explain vision:

> For vision, it is necessary that something is produced by the visible object in the transparent medium: whether it is a *species*, as Peripateticians want, a motion of a subtle medium that acts like a stick, according to the opinion that Simplicius attributes to Aristotle, or something that is transmitted through pores or voids as Democritus says.[68]

This does not mean that Bourdin believed in the existence of subtle matter and in his capacity to explain light and colors: it is only one hypothesis among others. In the

---

[65] Collacciani and Roux 2017a, 59–65. According to Bourdin, the law of reflection and the law of refraction are the same, insofar as the less inclined a light ray is, the less reflected or refracted it is, see **Pallu and Touchelée 1635**, 10 § 7; **Potier 1640**, 9 § 6; and BNF, Ms. Lat. 17861, f. 908. His conflict with Descartes was linked to the fact that they were not referring to the same angles by the terms "angle of incidence" and "angle of refraction."

[66] **Potier 1640**, 11 § 3.

[67] **Potier 1640**, 15 § 3: "Lucis et colorum actionem in oculos explicare per motum materiæ cuiusdam subtilis fusæ per aëris, et transparentium corporum poros, quàm lucida corpora moveant, eáque tangant oculos modis variis, atque omnino aliter, quam per species intentionales, est cicatricem curare novo vulnere, et gratis implicare sese iisdem, et insuper novis difficultatibus. Has afferre, et inanitatem subtilis illius materiæ demonstrare."

[68] **De Vic 1643**, 10 § 1: "Ad visionem necesse est ut à visili in medio diaphano producatur aliquid, sive illud sit species, ut placet Peripateticis; sive motio medii subtilis ad instar baculi, ut Aristoteli affingit Simplicius; sive aliquid aliud transmissum per poros, aut vacuum, ut innuit Democritus." Matton (2018) argued against Collacciani and Roux (2017a, 71–2) that this passage does not contain any allusion to Descartes, but is only an allusion to Simplicius' exposition of Aristotle's doctrine, that Bourdin might have known through Gassendi (1642, 173–4). This seems improbable to us, because Bourdin never discusses Gassendi, while he had been involved in a long controversy with Descartes. In 1644, Bourdin's and Descartes' reconciliation was so complete that Bourdin was entrusted with the task of distributing copies of *Principia philosophiae* to the French Jesuits (Descartes 1996, vol. IV, 143 and 160).

last two theses Bourdin had Jean de Bourneuf (1650) and Pierre de la Villette (1651) defend, he challenged the atomist and Cartesian hypotheses concerning the nature of light, returning to what he called the "common and royal way," namely the Aristotelian way of explaining the light:

> Different [philosophers] explain the nature of light in different ways: some by a perpetual flow and emission of corpuscles from the luminous body, others by the motion produced by the luminous or colored body on corpuscles placed in the transparent medium. Both explanations are ingenious, but not at all appropriate, if you look at the details. Hence, we have still the common and royal way. This is why we say that light is a form in a suitable subject, but without penetration of the subject. This subject is the transparent and another would be suitable for another thing. The nature (*indoles*) of light, which is sufficiently perceived by the one who sees, is not perceived at all by the one who was born blind.[69]

To summarize, Bourdin's thesis booklets were encyclopedic collections of problems taken from different parts of mixed mathematics, the resolution of which sometimes required the manipulation of engravings or even the performance of real experiments. Bourdin's optical teaching was founded on Scheiner's *Oculus* (1619), but that did not prevent him from making his students produce extraordinary effects such as the appearance of a picture of Christ thanks to the *Specillum Fidei*. More generally, two levels can be distinguished in Bourdin's theses and in his teaching: on the one hand, a stable structure of knowledge corresponding to the Jesuit science of the years 1610–1620, on the other, new experimental elements coming from the scientific actuality that added a touch of novelty and excitement to this structure of knowledge without contradicting it. As for Descartes, Bourdin perceived him as a *novator* among many others, who had to be corrected when he went astray—that is what Bourdin did concerning subtle matter –, but who did not pose a real threat to the traditional ways of teaching nor induce a transformation of the Aristotelian structure of knowledge.

5.4 Ignace-Gaston Pardies: A New Theory of Light Claiming to Be True to the Ancients

Ignace-Gaston Pardies (1636–1673) was educated in the Jesuit colleges of Pau and Toulouse. He began his career as a preceptor of humanities at the *collège de la Madeleine* in Bordeaux (1656–1660), where he studied theology (1660–1663). He then spent one year as a professor of humanities in Bordeaux (1664) and one year

[69] **De Bourneuf 1650**, 8, § 17–8: "Luminis naturam varii vario modo exponunt: alii per corpusculorum perpetuum fluxum et eiaculationem à lucido: alii per corpusculorum perspicuo in medio positorum motum factum à lucido, vel colorato. Ingeniosè quidem utrique; at minus commodè, singula si spectes. Nobis igitur via restat, communis et regia. Quare lumen dicimus esse formam in subiecto apto, sine tamen subiecti penetratione. Subiectum illud perspicuum est, et aliud quidem alio magis. Lumine vero indoles ea quæ percipitur satis ab oculato, à cæco nato minimé." See also **Villette 1651**, 12 § 34–5. The mention of a blind person is probably an allusion to *Dioptrique*, in Descartes 1996, vol. VI, 85–6.

of probation in Pau (1665); at this point he had already some reputation as a mathematician: he was in correspondence with Athanasius Kircher (1602–1680) and he was a friend of Emmanuel Maignan (1601–1676), he offered his expertise on a project to make the river that passes through Pau navigable and published two books on comets.[70] After spending a few years as professor of philosophy and mathematics in La Rochelle (1666–1668) and in Bordeaux (1668–1670), he was appointed professor of mathematics at the *collège de Clermont* at the age of thirty-four. Only three years after his arrival in Paris, he died prematurely from a fever, contracted while tending the poor.[71]

Due to his short life and career, Pardies only authored a few thesis booklets. At the *collège de la Madeleine* in Bordeaux, Étienne Roux and André Duhamel defended in 1669 theses devoted to mechanics. At the *collège de Clermont*, Alexandre Milon and another student defended in 1671 theses on optics, while Louis Le Mazier and other students defended in 1672 theses on cosmology.[72] Jean-Baptiste Joseph-Ignace de Mesgrigny and Charles-Bénigne Hervé defended theses devoted to different parts of mathematics in 1670; the sections on optics and on military architecture in their thesis booklets were borrowed from thesis booklets printed in 1666, 1668, and 1669, that is, when Pardies was not in Paris. The sections on optics and military architecture were likely written by his predecessor and Pardies probably arrived in Paris shortly before the start of the school-year in October 1671.[73] The only thesis booklet devoted to optics at the *collège de Clermont* under Pardies is the booklet dated 1671. We will first show that it differs from those defended under Bourdin because of its presentation, but also because of the intellectual context in which it was defended; we will then make explicit the principles of Pardies' optics that surface in it; finally we will insist that it is not without reason that Pardies was accused of Cartesianism.

1. Theses of the kind that Bourdin developed disappeared after him. Mathematics thesis booklets defended at the *collège de Clermont* in the second half of the century resemble more standard *disputationes*, *dissertationes,* or *theses* that feature a set of propositions to be debated: they do not ask the defendant to solve elementary problems belonging to different parts of mixed mathematics using engravings or performing real experiments. Whereas Bourdin's thesis booklets present collections of questions and propositions that jump from one topic to another unrelated one, the mathematical thesis booklets produced later in the seventeenth century by Pardies and others present more coherent progressions

[70] Pardies 1665a; idem 1665b.

[71] On Pardies's life and works, see Anonymous 1726; Sommervogel 1890–1932, vol. VI, 201–8; and Ziggelaar 1971.

[72] The thesis booklets on military architecture that the catalogue of the Bibliothèque municipale de Lyon dates from 1671 without giving any reason for such a date are left aside.

[73] The section on optics of **de Mesrigny 1670**, 7–14 is borrowed from **Anonymous 1666**, 3–13; **Lamoignon de Basville** 1666, 3–13; or **Benard de Rezay 1666**, 3–13; its section on military architecture is borrowed from **Bolé de Champlay** 1668, 10–11 or from **Anonymous 1669**, 14–15 § 3–6. We were not able to see **Hervé 1670** but our guess is that it is identical to **Mesrigny 1670**.

from one topic to another that in certain cases appear as mini-treatises specialized in one domain. The only trace of the type of thesis booklet developed by Bourdin that remains in the second half of the seventeenth century is that paradoxes are often formulated, sometimes at the end of each paragraph or of each page, as is the case in the thesis that Chrétien François de Lamoignon defended in 1663, sometimes gathered at the end of the thesis, as exemplified by the list of "Paradoxes that follow from the principles of Descartes' philosophy" at the end of the thesis booklet defended by Louis Ragayne de la Picottiere and others in 1665.[74] Here, paradoxes are not logically unacceptable or self-contradictory statements; rather, they are statements that run counter to received opinions and that, as such, are supposed to arouse astonishment, admiration, or repulsion in an audience—recall that these theses were defended during ceremonies that brought together a public that expected something spectacular, at least rhetorically.[75]

These formal differences are however slight compared to the difference in intellectual context between the first and the second half of the seventeenth century. As we have seen, Bourdin considered Descartes as a *novator* among many others; if Descartes did not succeed in making the Jesuits adopt his *Principia philosophiae* in their classrooms, he succeeded at least in being considered by them as a kind of fellow traveler with whom they could exchange polite letters and honest words. In the sixties however, while the Cartesians were busy publishing the works of Descartes that he had not published, writing Cartesian works on subjects he had not dealt with, and giving lessons in Cartesian philosophy to anyone who wanted to listen to them, the Jesuits and more generally the partisans of the philosophy of the Schools judged that, as far as natural philosophy was concerned, the main challenge that they had to face was to prevent the Cartesian party from continuing to spread a natural philosophy that was perceived as a threat to the old philosophy. This polarization of the field between the new philosophy advocated by the Cartesians and the old philosophy defended by the Jesuits affects all of France, but manifests itself particularly vividly at the *collège de Clermont*, as is evident from the thesis booklets devoted to cosmology.[76]

Throughout our period, the comparison of hypotheses concerning the structure of the world was a *topos* for those who taught cosmology. In his lectures, Bourdin said for example that Ptolemy's hypothesis was irreconcilable with new observations, Frascator's hypothesis, too complex, Copernicus' hypothesis, ingenious but implausible, and moreover contrary to Scripture, so much so that he came to favor

---

[74] Concerning paradoxes in **Lamoignon (1663)** and **Ragayne de la Picottiere 1665b**, 15–16, see Roux 2017 114–5. The 1665 theses were defended by many students, see **Ragayne de la Picottiere 1665a**; **Prou 1665b**; **Ragayne de la Picottiere** 1665b; **Prou 1665c**; **Decombes 1665**; and **de la Bletonniere 1665**: in what follows, we only give reference to Ragayne de la Picottiere.

[75] For a study of Renaissance paradoxes, see Colie 1966. That thesis booklets presented paradoxes is noted in Leinsle 2006, 40, 45–6; and Dhombres and Radelet-de Grave 2009, 173.

[76] On this polarization of the French natural philosophy, see Roux 2013a.

a combination of Tycho Brahe's and Ptolemy's hypotheses.[77] There was then no mention of Descartes, which is only natural since he had not yet published his *Principia philosophiae*. It is still the case in 1663: Chrétien François de Lamoignon's thesis print is devoted to a criticism of Copernicus' hypothesis, presented as a fortress that should be taken by assault, while Joseph Le Meilleur's thesis print compares the hypothesis of the concentric circles ("ridiculous"), the Ptolemaic hypothesis ("intolerable"), the Copernican hypothesis ("unlikely"), the Tychonian ("almost chimeric in practice"), and the semi-Copernican ("combining all the disadvantages of the Copernican and the Tychonian hypotheses") before introducing "the new mathematical hypothesis of Clermont," which is supposed to take the best of all the previous ones without any of their defaults.[78] In 1664, an anonymous defendant criticized, but only in passing and without naming its author, the Cartesian explanation of magnetism for being ingenious, but too poetic and too mechanical: "insofar as it is poetic, it is rejected by philosophy, which is the enemy of all fiction; since it is too mechanical, it is not accepted by nature, to which it is up to command the laws of human art, not to obey them."[79] This corresponds exactly to two fundamental methodological criticism that will be addressed to Descartes in the 1665 thesis booklet: he mistakes his fictitious mathematical hypotheses for true theses explaining the reality of things and he confuses the analysis of artificial and natural things. In both cases, the problem is that the traditional division of labor between mathematics and natural philosophy is not respected.[80] But in 1664 this is only a criticism made in passing; the passage of two comets in 1664–1665 occasioned a confrontation between the Cartesians and the Jesuits that quickly led to an unprecedented attack of Cartesian natural philosophy.

This confrontation, along with Descartes' works placement on the Index of Prohibited Books *donec corrigantur* on November 20, 1663, precipitated a change of intellectual context: not only was Descartes now explicitly mentioned, but his system of the world was condemned more harshly than any other. In 1666, Lamoignon de Basville noted:

> If it [the Cartesian hypothesis] differs in anything from the Copernican one, it adds to it a lot of absurd things of its own making and it shakes up an otherwise ingenious hypothesis. Certainly the Copernican hypothesis sins against the laws of Wisdom, but the Cartesian hypothesis sins much more seriously [against them], since it attributes everything to the power of God, so that it seems to leave nothing to his wisdom; the former depicts a world much greater by its mass than it was before, the latter makes it immense; the former forbids

[77] BNF, Ms. Lat. 17862, f. 501–13.

[78] **Lamoignon 1663**; **Le Meilleur 1663**, 12–13 § 3–8. On the political context of Lamoignon's thesis, see Lerner 2001, 534–7. Nicolas d'Harouys was the author of the "new mathematical hypothesis of Clermont" which is also presented in **Lamoignon de Basville 1666**, 19 but not in **Benard de Rezay 1666**.

[79] **Anonymous 1664**, 6 § 4: "Nimium tamen et poetica videtur esse et mechanica; quatenus est poetica, reiicitur a philosophia omnis inimica figmenti; ut mechanica nimius est, non admittitur a natura, cuius est artis humanæ legibus imperare, non subiici."

[80] **Ragayne de la Picottière 1665a**, 3 § 3, 4 § 4, 11 § 39; and **Ragayne de la Picottière 1665b**, 5–6 § 14–15, § 19; commented in Roux 2017, 114.

> the graves to go down in a straight line, the latter orders them to go up, or at least to endeavor to do so. The Copernican glories in the simplicity of its motions, the Cartesian in their multiplicity, since it imparts an infinity of motions to infinitely infinite parts.... The Copernican, though not true, is nevertheless possible; the Cartesian is neither true nor possible.[81]

Similarly, the theses defended in 1667 and 1669 criticized Descartes' hypothesis for adding its own vices to those of the Copernican hypothesis.[82] In addition, for reasons that we will not detail here, Louis XIV decided in 1671 to ban the public teaching of Descartes's philosophy, which resulted in what was sometimes called the persecution of Cartesianism.[83] In short, while Descartes and the Jesuits were playing a subtle game of intellectual seduction and mutual manipulation when Bourdin was professor at the *collège de Clermont*, Pardies made his students present his optics publicly at a time when there was an all-out-war between his order and the Cartesians.

2. Pardies was not entirely unattuned to the Jesuit's wonderful optics that we were able to glimpse in the Lens of Faith. In 1665, Pardies wrote to Kircher that he admired his *Ars magna lucis et umbrae* (1645) so much that he had hung his portrait in his room and that, inspired by his anacamptic clocks, he had built a marvelous clock (*horologium thaumaticum*), which produced a rainbow while indicating the time in different places.[84] In 1666, he described for the *Journal des sçavans* two beautiful rainbows that he observed while traveling in the south of France.[85] Still, his main contribution was theoretical and, as we will now explain, his theory was a theory of light rather than a theory of sight. From the very first paragraph, Pardies' optics appears as a post-Keplerian optics: the light propagates in straight rays, which do not obstruct one another; when these rays reflect on opaque bodies, they detach the colors and transport them to the eye, which functions as a *camera obscura*, so that pictures of the bodies are reproduced

---

[81] **Lamoignon de Basville 1666**, 17: "Si quid a Copernicæa differt, maxime quam absurda de suo multa conferat, et hypothesym cæteroquin ingeniosam labefacet. Peccat quidem contra Sapientiæ leges Copernicæa, sed multo gravius Cartesiana, quæ sic Dei potentiæ tribuit omnia, nihil ut sapientiæ videatur relinquere: mundum illa jam mole quam par era longe maiorem expresserat, hæc immensum exhibet: prohibet prima ne recta deorsum ferantur gravia; eadem sursum ire certe jubet altera. Gloriatur Copernicæa motuum simplicitate, multiplicitatem consectatur Cartesiana dum infinities infinitis partibus motus pariter infinitos imprimit.... Denique Copernicæa quamvis non vera, possibilis tamen est: Cartesiana nec vera, nec possibilis." See also **Anonymous, 1666**, 15 § 6.

[82] **Castagnere de Chasteauneuf 1667** (or **Bolé de Champlay 1668**), 4 § 3. For other criticisms, see idem, 5 § 5, 8 § 11, 9 § 11, and 11, § 16; **Anonymous 1669**, 8 § 5. While the other theses present one system after another, **Castagnere de Chasteauneuf 1667** and **Bolé de Champlay 1668** present one celestial phenomenon after another and, for each of them, compare the different systems.

[83] On the official condemnations of Cartesianism, see Armogathe (1969, 1977); Ariew (2013) and Roux 2019.

[84] Pardies to Kircher, 20 August 1665 and Pardies 1673a; and Ziggelaar 1971, 30–6.

[85] Pardies to Payen, 27 October 1666, in *Journal des sçavans*, 7 February 1667, 45–8.

upside down on the retina, which is itself compared to a cloth.[86] Like Scheiner and Bourdin before him, Pardies adopts this Keplerian optics without sacrificing *species*:

> Certainly, there are *species* emitted by objects, but these cannot be distinguished from the rays themselves that reflect, intersect, and attain their term… Vision occurs through reception, not transmission. The *species*, once they have attained their term on the retina, determine the faculty of seeing by means of the small filaments of the optic nerve. Objects appear all the larger that their images on the retina are larger; vision is all the stronger that more rays penetrate the eye; it is all the more distinct that the rays coming from one point gather in a narrower area of the retina.[87]

While Jean-Baptiste Joseph-Ignace de Mesgrigny and Charles-Bénigne Hervé still devoted most of the sections on optics to vision a year earlier, Milon does not expand any further on vision, but focuses instead on the nature of light. He begins by refuting both the Epicureans' opinion that light is a body and Descartes' opinion that light is explained by a pressure (*contensio*) of the celestial matter which, from the sun and luminous bodies, would endeavor to move (*conari ad motum*) and, compressing the eye, determine the faculty of seeing to produce sensation.[88] To attack the opinion of the Epicureans, whic include Gassendi, Pardies notes that Aristotle never approved of it. To counter the opinion of Descartes, he points out two of its unfortunate consequences: "According to this hypothesis, we would see the sun and luminous bodies even when we turn our backs to them or when we are locked in a room, and a ray of sunlight transmitted through a pore in a sheet of paper placed obliquely would refract towards the perpendicular."[89]

The first objection is that if, as the *Principia philosophiae* wants, light was a pressure propagating through an omnipresent subtle matter, this pressure would be transmitted through the corpuscles of subtle matter in the interstices of the walls or in the space behind us, so that we would see through the walls and behind our backs. The Cartesian hypothesis cannot account for the difference between opaque and transparent bodies, nor for the difference between what is in our field of vision and what is not.[90] The second objection is that, if, as the *Dioptrique* says, the propaga-

[86] **Milon 1671**, 3 § 1–2.

[87] **Milon 1671**, 4 § 3: "[E]sse aliquas Species emissas ex objectis, quæ tamen non distinguantur ab ipsiis radiis reflexis, decussatis, terminatis…. Visionem fieri per receptionem, nullatenus per emissionem. Species terminatas in Retina determinare videndi facultatem ope tenuium nervi optici capillamentorum. Objecta tanto majora apparere quanto major est eorum in Retina imago: visionem eo fortiorem quo plures radii ad oculum penetrant: eo distinctiorem quo radii ex uno procedentes puncto magis ad se se accedentes colliguntur ad Retinam in angustiori loco."

[88] Pardies neatly summarizes the explanation of light proposed in Descartes 1996, vol. VIII-1, 108 sqq. By comparison **Lamoignon de Basville 1666** and **Anonymous 1666**, 3 § 2 are vague.

[89] **Milon 1671**, 4 § 4: "In ea hypothesi, solem lucidáque corpora etiam aversi aut intra cubiculum occlusi videremus; et radius solaris per foramen obliquè jacentis chartæ transmissus ad perpendicularem infringeretur."

[90] Other similar objections appear in Paradoxes 2 and 3, in **Milon 1671**, 7: light would be reflected as well by the earth or by a rough piece of wood, as by a mirror; the light of the Sun or of a burning torch would be as intense at large distances as at small ones.

tion of light was comparable to the transmission of a ball, the refraction of a ray of light that passes from a less dense medium to a denser one, as is the case when it passes from air to a cloth, would cause its deviation towards the perpendicular (*ad perpendicularem*). However, it is the opposite which is actually true: the refraction of a ray of light passing from a less dense medium to a denser medium does not cause its deviation towards the perpendicular, but its deviation away from the perpendicular (*a perpendiculari*).[91]

The hypothesis that light is a body being rejected, whether in its Cartesian version or in its Epicurean version, Milon puts forward with his own thesis, which is that light, being not a body, is a quality:

> We can perhaps form a conjecture about its production and propagation by comparison with sound: it is indeed quite rightly that the Sound has been called the Monkey of Light. It is evident that sound is produced, formed, and propagated by certain vibrations of the undulating air or of some other equivalent body; so that, the waves advancing from different places, when they make contact, go further and cross each other without hindering each other, and then, penetrating the organ of hearing, they excite sensations of sound which vary with the variety of the commotions or tremors of the air. Thus, it may be that light is excited, that it is propagated and that it affects the organ of vision by the numerous waves and frequent vibrations, not of the air, but of a more subtle substance.[92]

The four arguments given in favor of this conjecture are that: 1. Manifest experiments have shown the existence of a tenuous substance called ether.[93] 2. The reason why the Sun, fire and other luminous bodies excite the vibrations of this substance can be found.[94] 3. This conjecture can explain the phenomena of light, colors, and vision.[95] 4. Objections can be answered by referring to the example of sound.

---

[91] This objection is also at the root of Paradox 12, in **Milon 1671**, 8. This is at the time a common criticism of Descartes 1996, vol. VI, 93–105; it is for example to be found in Grimaldi 1665, prop. 19, 173–6. There might be an echo of this criticism in **Anonymous 1666** (or **Benard de Rezay 1666**), 6 § 10, which is identical to **Lamoignon de Basville 1666**, 6 § 11. Later on, it will be found in Pardies 1670, 86–7 § 37 and Ango 1682, book I, prop. 72, 85–8.

[92] **Milon 1671**, 5 § 5–6: "[C]onjecturam aliquam facere fortasse possumus ex comparatione ad sonum: jure enim optimo Luminis Simius appellatus Sonus est. Manifestum autem est sonum produci et formari et propagari per certas quasdam aëris undulantis aut alterius æquivalentis corporis vibrationes: adeo ut variæ ejusmodi undulationes ex varii locis procedentes contagione quodam ulterius procedentes, se se impunè trajiciant, ac deinde in aurem illapsæ organum auditus verberent, excitentque varios sonorum sensus pro varietate commotionum seu tremorum aëris. Ita fortasse posset ut lumen excitaretur et propagaretur organumque visus afficeret, per certas quasdam et crebras, non quidem aëris, sed subtilioris alicujus substantiæ undulationes et frequentes succussiones."

[93] This is an allusion to the interpretation of the barometric experiments that were given in Jesuit colleges. In the case of the *collège de Clermont*, see **Thierry 1648**, 10–15 § 26–33; **Truel de Cohon** 1648, 10–15 § 33–44; and **De Bourneuf** 1650, 13 §§ 37–38. These are commented on by Collacciani and Roux 2017b, 103.

[94] This is what **Milon 1671**, 6 § 7 is about.

[95] This is what **Milon 1671**, 6–7 § 8–9 is about. The theory of colors advanced here by Pardies is similar to the one that he developed in his teaching in Bordeaux, on which see Ziggelaar 1971, 177–9.

The thesis booklet that André Duhamel and Étienne Roux defended in Bordeaux in 1669 ended with a section on pneumatics presenting the waves of sound caused by the vibrations of the air.[96] In Milon's Parisian thesis booklet, the notion of waves is extended to the analysis of light. This extension is introduced by the image of sound as the monkey of light, coming from Kircher's *Ars magna lucis et umbrae*, which, as was noted, Pardies read when he was in Bordeaux.[97] The analogy between sound and light being however frequent in the seventeenth century, one must compare Pardies with other authors to catch the twist that he gave to this analogy. [98] Mersenne uses it at the beginning of the *Harmonie universelle* (1636–1637), but he adopts the opposite intellectual approach to that of Pardies by assuming that light and its properties are known, to infer from this knowledge several statements about sound.[99] Like Mersenne, Bourdin goes from light to sound, not the other way around.[100] Last, but not least, Kircher's image means that sound imitates light, not the reverse, and it emerges in a part of the *Ars magna lucis et umbra* that intends to derive the radiation of sound from the radiation of light and to establish the new discipline of *echologia* or *photocamptica* that would extend the laws of reflection and refraction from light to sound.[101] In short, whereas Pardies' predecessors explained sound by light, Pardies' intention was to explain light by sound.

His key concept for that was the wave that, in all likelihood, he took up from Francesco-Maria Grimaldi, whom he mentioned in several occasions.[102] In his *Physico-mathesis de lumine, coloribus et iride* (*Physico-mathematics on light, colors and rainbow*, 1665), Grimaldi explained the phenomenon of diffraction, which he was the first to observe and to describe systematically, by the undulation (*undulatio*) of a fluid matter. For him, light is a fluid animated by a very fast motion, the undulation of light, an accidental phenomenon that occurs occasionally.[103] When light encounters the borders of a small aperture, it refracts and moves with a certain fluctuation (*flutuatio*); similarly, the modification of light that makes both real and permanent colors appear is determined by "a very subtly wrinkled undulation, like a tremor of diffusion, with a very subtle fluctuation" (*undulatio minutissime*

[96] **Roux 1669**, 9–10 § 53–60.

[97] This image first appears in Kircher 1671, lib. II, pars I, chap, Exp. 3, p. 102. Kircher 1673, lib. I, sect. I, cap. 1 begins with a whole section entitled "Praelusio I. Sonus lucis simia est."

[98] On the analogy between light and sound in general, see Darrigol 2010a and idem 2010b.

[99] Mersenne 1636, book I prop. 1, p. 1–5. The main limitation of this analogy is that, for Mersenne, intentional species are necessary in the case of light, but not in the case of sound. See as well book I, prop. 25, p. 44–8.

[100] **Potier 1640**, 11 § 1, **Villette 1651**, 12 § 36.

[101] Kircher 1671, lib. II, pars I, cap. 6–7, p. 96–102.

[102] Pardies to Payen, 27 October 1666, in *Journal des sçavans*, 7 February 1667, p. 45; Bibliothèque municipale de Bordeaux, mss 934, f. 236v, Pardies to Newton, 21 May 1672, in Newton 1958, p. 97. Moreover, while it is known that Pardies, if he is not the author, at least got his hands on Rochon 1672; this work mentions Grimaldi 1665, 181.

[103] Grimaldi, 1665, prop. 2, p. 12.

crispata, et quidam velut tremor diffusionis, cum certa fluitatione subtilissima).[104] In the lengthy proposition that follows his explanation of colors, Grimaldi goes like Pardies from sound to light since he wishes to confirm his explanation by an analogy with sound.[105] But what was an accidental phenomenon for Grimaldi became in its elaboration by Pardies the unique hypothesis which explains all optical phenomena, sound constituting an operative analogy to understand light and the existence of subtle matter being established by the reinterpretation of barometric experiments that had been proposed in some of Bourdin's theses twenty years beforehand. Insofar as our aim is to stick to the 1671 thesis booklet, we will not reconstruct the wave optics of Pardies.[106] Rather, we would like to conclude our analysis of the 1671 thesis booklet with two remarks.

First, Pardies did not envision disputations only as occasions in which to make his students perform in front of an audience some exercises on elementary questions, but rather he saw them as opportunities to make known his own theories and to claim his identity as a member of the Republic of Letters. The scientific sophistication of his thesis booklets attests to this novel mode of employing theses. This transpires also from Pardies' references to his thesis booklets as if they were what we would today call his scientific publications, where the new concepts, results, or methods that he formulated were for the first time made public.[107] Such use of thesis booklets as an exchange currency between scholars became even more pronounced when the *collège de Clermont* passed under royal patronage and became the *collège Louis-Le-Grand*: the mathematical thesis booklet of 1686 opens with a dedication of the mathematicians of *the collège de Clermont* to the members of the *Académie des Sciences*.

Secondly, this new way of putting thesis booklets to use makes it all the more surprising that, after having advanced his theory of light, Pardies took up an Aristotelian thread and tried to confirm this new conception of light by invoking Aristotle's authority.[108] Pardies recalled the two approaches to light that can be found in the Aristotelian corpus. On the one hand, *De Caelo* gives a physical explanation of the generation of light by making it the effect of the friction of the moving sun and the moving stars with a subtle air.[109] On the other hand, *De anima* presents qualitatively light as the *energeia*, that is to say, the effective act, of the transparent medium as such.[110] Without seeking to resolve the tensions that exist between these

[104] Grimaldi 1665, prop. 22 and 43, p. 187–90 and p. 341–70.

[105] Grimaldi 1665, prop. 44, p. 370–95.

[106] Shapiro 1973, 209–44.

[107] **Duhamel 1669** was sent to the *Académie des sciences* and examined by Huygens (Ziggelaar 1971, 17). In Pardies 1670, Preface, Sig. A4v, Pardies claimed: "... il y a déja trois ans que j'ai donné publiquement tout ce que je mets dans ce discours"; in his letter to Oldenburg, 13 January 1672, in Oldenburg 1965–1973, vol. VIII, 455, he wrote: "Il y a deja plusieurs années que j'ay enseigné publiquement tout cecy et mesme je l'ay imprimé dans des theses."

[108] **Milon 1671**, 6 § 7.

[109] *De Caelo* II, chap. 7, 289 a19.

[110] *De anima* II, chap. 7, 418b9-10.

two approaches, Pardies sees in the first one a confirmation that the ether is the vehicle of light and in the second one the affirmation that light is not fire, but a fire endowed with efficient energy. Presumably, this amounts to identifying Aristotle's ether with subtle matter and the fire endowed with efficient energy with its undulating motion. This interpretation of Aristotle fully accepts the fundamental notions of Cartesian mechanics. Thus, if Pardies defends, against the Epicureans and against Descartes, the thesis that light is a quality, he understands by "quality" a motion of matter. If this interpretation is true, the objections that Pardies addresses to the Cartesian explanation of light assume a different meaning: for Pardies, Descartes failed to explain light according to the principles of his own mechanics because he limited himself to making it an *endeavor* to motion; only an undulating motion of subtle matter can give a proper mechanical explanation of light, that is, an explanation of light in terms of *motion*. Pardies criticizes Descartes for not having been Cartesian enough.

3. Now it is clear why Pardies was accused of Cartesianism by his fellows Jesuits. Recall that this accusation, which began in 1670 while Pardies was still in Bordeaux, led him successively, and each time in vain, to protest of his innocence, that is, that he was not a Cartesian, first by adding some remarks to his *Discours du mouvement local* (1670), then by publishing a *Discours de la connaissance des bêtes* (1672), and, finally, by helping Antoine Rochon, another Jesuit who was also suspected of holding Cartesian opinions, to write the *Lettre d'un philosophe à un cartésien de ses amis* (1672).[111] In the biography they give of Pardies, the *Mémoires de Trévoux* summarize the situation very aptly: "He used to say, when he was accused of being Cartesian, that no one had refuted Descartes more than he had; he was indeed telling the truth; but he was nevertheless suspected, and the suspicion was no less well-founded; for he was in fact following what everyone today takes from Cartesian physics…."[112]

The suspicion was well-founded from a social perspective in terms of how Pardies communicated his doctrines: just as Cartesians like Jean-Baptiste Denis or Jacques Rohault held open conferences for the public, so he invited his readers to attend his classes at the *collège de Clermont* to clarify the points of his books that remained obscure.[113] The suspicion was also well-founded from an intellectual per-

---

[111] See Ziggelaar 1971, 79–85 about the remarks that were added to the *Discours du mouvement local*, idem (95–7) about the contemporary perception of *Discours de la connaissance des bêtes* as a Cartesian work, idem (113–4) about the letters concerning Rochon and Pardies that the General of Jesuits Giovanni Paolo Oliva sent, and idem 118–20 about the contribution of Pardies to the *Lettre d'un philosophe*. Beaude 1976 criticized Ziggelaar 1971 for not having confronted more directly the problem of Pardies' crypto-cartesianism.

[112] Anonymous 1726, 670.

[113] Le Gallois 1672, 59 and 66. Pardies 1670, "Avis à ceux qui veulent apprendre la geometrie," 1st ed. (1671), Sig. a13v: "Si l'on veut se donner la peine de venir au College de Clairmont, l'Auteur de ces Elemens les y expliquera publiquement après la S. Remy," 2nd ed. (1673), Sig. A18: "Si l'on veut se donner la peine de venir au College de Clermont, l'Auteur de ces Elemens continuera de les y expliquer publiquement les Lundis et les Vendredis." Denis 1682, 81: "ceux qui y trouveront

spective with regard to Pardies' doctrines. From his first published book on comets to his book on the souls of beasts, he had an ambivalent attitude: on the one hand, he was indeed criticizing some of Descartes' theses; on the other hand, he had so totally assimilated the fundamental lessons of the new philosophers that he was moving in a space of knowledge structured by their principles.[114] As we have shown, this was also the case in the thesis booklet of 1671. He underlines certain weaknesses or inadequacies of the Cartesian explanation of light, but he does not question the Cartesian enterprise as such: optics being part of mechanics, to explain light is to show what motion of what matter accounts for it. While Bourdin preserved the structure of Aristotelian knowledge, Pardies overturned this structure by considering mechanics as the science of motion in general and this science of motion as the key to all physics.

There is a fundamental point here. As early as the theses of 1669, Pardies defined mechanics as "the science, which demonstrates by the first principles the production, conservation, communication, pressure, impulse, acceleration and extinction of local motion."[115] The definition of mechanics as a science of motion in general also opens the Preface of the *Discours du mouvement local*: "All the productions that come either from the industry of men or from the causes of nature, are made only by motion. Therefore, it is not possible to penetrate into the secrets of physics, nor to succeed in the invention and practice of the arts, without the help of mechanics, that is, without knowing the laws of motion."[116] But the most significant passage is the praise of mechanics that opens *La statique*:

> This treatise is a continuation of a *Discours du mouvement local*, which had already been published, with the intention of writing a whole mechanics and of putting in order all the science of motion. Those who know the way in which we proceed today in the consideration of nature, and in the practice of the arts, also know the advantages to be found in the knowledge of the laws of motion. And since it is certain that nothing is done in the arts without using mechanics, it must also be recognized that nothing can be explained in the particular effects of nature unless the demonstrations of this science are used. It is mechanics that prescribes the rules of … architecture … both civil and military. It is mechanics that builds ships and governs them. It constructs machines to lift the heaviest loads with ease. It regulates the conduct of the waters…. It animates the organs…. It makes rocks talk in arti-

---

quelque difficulté, pourront s'en éclaircir faciclement, en allant au College de Clermont; car le R.P. Pardies, qui y professe les mathematiques, prend la peine d'en expliquer toutes les Propositions, et de répondre aux objections qu'on luy forme sur cette Matiere [motion and the rules of percussion]." Pardies 1673a, Sig. A5r: "Ceux qui auront la curiosité d'apprendre à fond la Théorie et l'usage de ces Machines, pourront le faire aisément, s'ils veulent se donner la peine de venir au College de Clermont, où l'on en fera des entretiens publics, une fois la semaine, pendant quelque temps."

[114] Pardies 1672 analyzed in Roux 2013a and Pardies 1665a analyzed in Roux 2017, 117–8.

[115] **Duhamel 1669**, 3: "Mechanica vocetur scientia, quae motûs localis productionem, conservationem, communicationem, contensionem, impulsionem, accelerationem, extinctionem per prima principia demonstrat." Gabbey 1992, 311–4; and idem 2004, 22–3 emphasized this new definition of mechanics, that they found in Wallis 1670-71; in France, Poisson first made it explicit in his preface to Descartes 1668, Sig. A2.

[116] Pardies 1670, Sig. A2.

> ficial grottos where it imitates the singing of birds.... This is part of what mechanics does when it is employed by the artifice of men: but what does it not do, when it is employed by the industry of nature itself? Is it not mechanics which steadfastly consolidates the earth beneath our feet? Yes, it is it that rounds the surface of the sea and filters its waters through underground pipes, to bring out fountains and rivers; it is it that suspends the clouds in the air, that pushes them in various places by the wind and expresses the rain...; it is it that brings down heavy bodies, with that doubling of speed and that proportion which philosophers cannot sufficiently admire; it is it that gives the heavens a jolt and maintains them in such a regulated motion, it is it again that makes birds fly in the air, that makes fish swim in the water, and animals walk on the earth; it is by its means that the heartbeat, the circulation of blood, the distribution of spirits and breathing are performed, it is it that carries light and sound in circles on all sides.... In a word, nothing is done without it, neither in art nor in nature.[117]

To paraphrase what Leibniz said of Honoré Fabri, with such a definition of mechanics, Pardies abandoned the heart of the Aristotelian citadel to the enemy even if he defended a few bastions.[118] The most direct proof of this is the return to the tradition carried out in the *Optique divisée en trois livres* that Pierre Ango (1640–1694), one of Pardies' successors at the *collège de Clermont*, published in 1682, the year when the *collège de Clermont* received the patronage of Louis XIV. While the body of *Optique divisée* relies heavily on the manuscripts that Pardies had left to develop an optical theory, Ango's introductory words are for re-establishing the traditional structure of knowledge. According to the conventional order of disciplines, Ango recalls a theory of light like that of Pardies is the work of a geometrician, not of a philosopher.[119] This reminder goes hand in hand with a criticism of the Cartesians that implies a recapture of the citadel: whereas Aristotle had distinguished between physics and mechanics, things and principles of things, "the so-called philosophers who have recently written about physics" have wrongly called "physics" these writings, whereas they are only "a heap of various treatises on natural mechanics." Moreover, they have been completely mistaken in reducing the real qualities to "a heap of different dusts variously stirred, depending on how their whim made them think they should be stirred."[120] Also in 1682, Antoine-Louis de la Forest defended a thesis which, forgetting Pardies' new definition of mechanics, opened instead with the traditional definition of mechanics as the science of machines that make us lift great weights with small forces.[121]

---

[117] Pardies 1673b, Sig. a2r-a3v. This preface then continues to offer a history of mechanics which evokes Descartes' letter to Plempius for Fromondus, 3 October 1637, in Descartes 1996, 420-1, and finally ends with a presentation of the mechanics in six books that Pardies had projected. While in 1669 he divided mechanics into kinetics, statics, barytonics, ballistics, hydraulics and pneumatics (**Roux 1669**), here he merged barytonics and ballistics on the one hand, hydraulics and pneumatics on the other, but added a part on librations and vibrations of pendulums and another part on waves of light and sound (Pardies 1673b, Sig. a6– 11r which is rehearsed in his letter to Oldenburg, 13 January 1672, in Oldenburg 1965–1973, vol. VIII, 452–5).

[118] Leibniz to Des Bosses, 2 February 1706, in Leibniz 1875–1890, vol. II, p294.

[119] Ango 1682, 2.

[120] Ango 1682, 3–4.

[121] **La Forest 1682**, 3 § 1.

5.5 Conclusion

We have chosen to focus on optics in order to study as systematically as possible the thesis booklets defended at the *collège de Clermont* from 1637 to 1682. The general conclusions we have reached are the following.

1 Mathematics disputations, which were far from being supervised as strictly as philosophical disputations, vary greatly, whether in their form, in the subjects treated, in the capacity of the professors, or even in the doctrines that they defended. They constitute a richer material for social and intellectual history than had been assumed until now.

2 As far as the Jesuits' relationship to the Cartesian natural philosophy is concerned, the mid-1660s brought about a rupture: before 1665, Descartes' natural philosophy could indeed be criticized, but this criticism remained moderate; after 1665, the Cartesian natural philosophers became the main enemies of the Jesuits, those with whom one should not make a pact under any circumstances. This rupture can be explained in various ways. From a political perspective, one cannot ignore the fact that Descartes's works were put on the Index of Prohibited Books in 1663, but there were also more local factors, like the confrontation that comets occasioned in 1665. From an intellectual perspective, the Jesuits blamed the Cartesians for not respecting the traditional division of disciplines: mathematical hypotheses and fictions should not be confused with physical theses that explain the true nature of things. Ironically enough, about twenty years later, this is what *mutatis mutandis* the Cartesians would blame the Newtonians for. The division of disciplines is obviously an important part of what we have called the Aristotelian structure of knowledge: it is not something that can be changed lightly in an educational or research institution.

3 It was inevitable that some Jesuits would find themselves in a singular conundrum, as happened to Pardies. The paradox of Jesuit science was indeed that it aimed at defending tradition—the uniformity and solidity of doctrine—, while placing itself at the forefront of the new sciences. Seen from a little distance, there were two comfortable positions: the first one, illustrated by the Dominicans, was to reject completely the new sciences and to stick to the scholastic tradition; the second one, exemplified by the Royal Society and the *Académie des sciences*, was to reject the scholastic tradition and to move forward with the new sciences. Jesuits wanted to have their cake and eat it too: they placed themselves at the forefront of the new sciences, but they did so to control the sciences according to the requirements of tradition. As the case of Pardies shows, this might have turned into something like an impossible mission: to be on the forefront of scientific research implies, at one time or another, in one way or another, for one reason or another to stop being a scholastic.

Bibliography

Manuscripts

Cours de mathématiques, précédé d'un cours de théologie, 2 vols. [BnF (Paris), Latin 17861-17862].[122]

Cursus mathematicus: breviter, et clare figuris explicatus in gratiam eorum, qui mathesim volunt addiscere cito, ac facile [Bibliothek des Max-Planck-Instituts für Wissenschaftsgeschichte (Berlin), Rara B769c].[123]

Scientiæ contemplatricis thomus unus, quae est physiologia, sive de natura rerum corporearum... sub reverendo admodum Patre Pardies, illustrissimi Palensis senatus consiliarii filio, [Bibliothèque municipale (Bordeaux) Ms. 934].[124]

Mathematics theses defended at the collège de Clermont (1637–1682)

1634–1653: Pierre Bourdin (1595–1653)[125]

Pallu and Touchelée 1635

4–5 June 1635, Jacques Pallu and Jacques Touchelée, *Deo hominique Jesu Christo eiusque matri Virgini Mariæ deiparæ Encyclopædiam mathematicam D.D.V. [sic] Jacobus Pallu. Jacobus Touchelée Turonenses. Iidem sedebunt propugnatores sua illius Encyclopedia pro annua celebritate Academiæ Regii Collegii Flexiensis Societatis Jesu,* La Flèche: Georges Griveau [Bibliothèque municipale (Lyon), Bibliothèque Mazarine (Paris)].[126]

Cornouaille and Manchon 1638

19–20 June 1638, Pierre de Cornouaille and Jacques Manchon, *Deo hominique Jesu Christo eiusque matri Virgini Mariæ deiparæ Musæum mathematicum, D.D.C. [Dicant, dedicant, consecrant] Petrus de Cornouaille, Sylvanectensis,*

[122] Lecture notes by Paul Mercier of Bourdin's mathematics course at the collège de Clermont in 1636–1638.

[123] Lecture notes by an anonymous hand of Bourdin's course in mathematics.

[124] Lecture notes by doctor Raymond Dufour of a philosophy course given by Ignace–Gaston Pardies in Bordeaux in 1669.

[125] The list is organized by the presiding professor who wrote the theses, and under each professor organized chronologically. The names of the professors of mathematics who wrote the theses are not mentioned in the thesis booklets, but are known through the Jesuit archives. We follow the list given in Dainville 1954, 110–111.

[126] According to the dates we know, Bourdin had already left La Flèche for Paris when **Pallu and Touchelée 1635** was defended, we include it as the earliest existing thesis booklet written by Bourdin.

Jacobus Manchon, Parisinus, pro annua celebritate literaria collegii Claromontani Societatis Jesu, s.l.s.n. [BnF (Paris), Bibliothèque municipale (Lyon), Bibliothèque Mazarine (Paris), BL (London)].

Bussey 1638
20 June 1638, Claude Bussey, *Encyclopædia mathematica Collegii Claromontani Parisiensis Societatis Jesu. Propugnator, Claudius Bussey* [Public Library (New York)]

Culant 1639
27 February 1639, Jacques de Culant, *Deiparæ Virgini Mariæ Propositiones ex optica et geometria militari, D.D.C. [Dicat, dedicat, consecrat] Jacobus de Cullant Molinensis.... Disputabuntur in aula mathematica collegii Claromontani Societatis Jesu*, s.l.s.n. [BnF (Paris), Bibliothèque municipale (Lyon), Bibliothèque centrale du conservatoire national des arts et métiers (Paris)].

Petit 1639
22 May 1639, Antoine Petit, *Deo Optimo Maximo mathematicæ Positiones D.D.C. [Dicat, dedicat, consecrat] Antonius Petit, Meldensis.... Disputabuntur in aula mathematica collegii Claromontani Societatis Jesu*, s.l.s.n. [BnF (Paris)].

Henri 1639
9–10 July 1639, Yves Henri, *Augustæ cœli reginæ Mariæ Virgini Deiparæ ejusque integerrimo sponso D. Josepho Palatium mathematicum D.D.C. [Dicat, dedicat, consecrat] Yvo Henri Briocensis, pro annua celebritate literaria collegii Claromontani Societatis Jesu*, s.l.s.n [BnF (Paris), Bibliothèque municipale (Lyon), Bibliothèque Mazarine (Paris)].[127]

Potier 1640
30 June–1 July 1640, Charles Potier, *D.O.M. [Deo Optimo Maximo] Encyclopædia mathematica, ad agones panegyricos in Claromontano Parisiensis Societatis Jesu collegio. Agonista Carolus Potier Castrotheodoricus*, s.l.s.n. [Bibliothèque municipale (Auxerre), Bibliothèque municipale (Lyon), Bibliothèque Sainte Geneviève [BSG] (Paris), Bibliothèque centrale du conservatoire national des arts et métiers (Paris), Biblioteca nazionale centrale (Roma), BL (London)].

Gaillard 1640
25 November [1640], Pierre Gaillard, Soli iustitiæ omnia intuenti oculo Theses mathematicæ de optica deque mirabili oculi oeconomia D.D.C. [Dicat, dedicat, consecrat] Petrus Gaillard disputabuntur in aula mathematica collegii Claromontani

[127] Yves Henri is mentioned only in the BNF copy, which begins with a frontispiece in which the rectangles and circles are printed while the images inside them are drawn by hand. The Lyon copy includes printed figures of the three cosmological systems at the section "Stadium cosmographiæ," of optics and the physiology of the eye at the section "Gymnasium opticæ," and of military fortifications with the manuscript annotation "propugnaculum Le Bastion" at the section "Ænigma geometricomilitare."

Societatis Jesu, s.l.s.n. [BnF (Paris), Bibliothèque centrale du conservatoire national des arts et métiers (Paris)].[128]

Despont 1642
5–6 July 1642, Philippe Despont, *A.M.D.G. [Ad maiorem Dei gloriam] Encyclopædia mathematica, ad agones panegyricos in Claromontano Parisiensis Societatis Jesu collegio. Agonista Philippus Despont Parisinus*, s.l.s.n. [BnF (Paris), Bibliothèque centrale du conservatoire national des arts et métiers (Paris)].[129]

de Vic 1643
20–21 June 1643, Dominique de Vic, *Agones mathematici in Claromontano Parisiensis Societatis Jesu collegio celebrandi. Dissertationes singulis agonibus præmittentur contra Galilæum. Agonista Dominicus de Vic Parisinus*, s.l.s.n. [Bibliothèque municipale (Lyon), BnF (Paris), BSG (Paris), Bibliothèque centrale du conservatoire national des arts et métiers (Paris)].[130]

Gedoin 1644
2–3 July 1644, Louis Gedoin, *Agones mathematici in Claromontano Parisiensis Societatis Jesu collegio celebrandi cum Deo et B[eata] Virgini. Physico–mathematicæ prolusiones habebuntur…. Agonista Ludovicus Gedoin Parisinus*, s.l.s.n. [Bibliothèque centrale du conservatoire national des arts et métiers (Paris), Bibliothèque Mazarine (Paris)].

Foucquet 1646
9 and 11 June 1646, Yves Foucquet, *Exercitatio mathematica ad agones panegyricos. Cum Deo et Beata Virgini propugnator Yvo Foucquet, Parisinus, in Claromontano Parisiensis Societatis Jesu collegio*, s.l.s.n. [Bibliothèque centrale du conservatoire national des arts et métiers (Paris), Bibliothèque municipale (Lyon)].

Bouthier 1647
20–21 July 1647, Blaise Bouthier, *Æternæ memorie serenissimo Principis Henrici Bordoni Condæi Agones annuas D.D.C. [Dicat, dedicat, consecrat] mathematica Parisiensis collegii Claromontani Societatis Jesu. Propugnabit Deo duce et Virgine Blasius Bouthier, Parisinus*, s.l.s.n. [Bibliothèque centrale du conservatoire national des arts et métiers (Paris)].

Thierry 1648
20–21 June 1648, Pierre Thierry, *Exercitatio mathematica ad agones panegyricos. Cum Deo et Virgine propugnator Petrus Thierry Parisinus, in Claromontano*

[128] The BNF copy does not mention the year of the defense, the CNAM copy mentions neither the year nor the day. We argued that this exercise dates from 1640 in Collacciani and Roux 2017a, 55–6.

[129] There are two copies at the Bibliothèque centrale du conservatoire national des arts et métiers, one of which does not mention the name of the defendant.

[130] The BSG copy does not mention the name of the defendant.

Parisiensis Societatis Jesu collegio, s.l.s.n. [BnF (Paris), Bibliothèque centrale du conservatoire national des arts et métiers (Paris), Bibliothèque Mazarine (Paris)].[131]

Truel de Cohon 1648
[Unknown] July 1648, Jacques Truel de Cohon, *Positiones ex universa mathematica selectæ. Has mathematicæ positiones cum Deo et B[eata] Virgine propugnabit Jacobus Truel de Cohon, Alenconius in aula Claromontana Parisiensis*, s.l.s.n. [Bibliothèque de l'Arsenal (Paris), BnF (Paris)].

de Bourneuf 1650
26–27 June 1650, Jean de Bourneuf, *Agones mathematici in Claromontano Parisiensis Societatis Jesu collegio celebrandi. Cum Deo et B[eata] Virgini propugnabit Johannes de Bourneuf, Juliodunensis*, s.l.s.n. [Bibliothèque centrale du conservatoire national des arts et métiers (Paris)].

Villette 1651
1–2 July 1651, Pierre de la Villette, *Agones mathematici in Claromontano Parisiensis Societatis Jesu collegio celebrandi. Cum Deo et B[eata] Virgini propugnabit Petrus de la Villette, Mondiderinus*, s.l.s.n. [BSG (Paris), BL (London)].[132]

d'Herbelot 1651
1651, Edmond d'Herbelot, *Agones mathematici in Claromontano Collegio... propugnabit Edmundus d'Herbelot, Parisinus*, s.l.s.n. [Bibliothèque municipale (Grenoble)].[133]

1658–1661: Georges Faultrel (Unknown–1687)

de Ruffec and Richomme 1655
3 July 1655, Hyacinthe de Ruffec and Jean Richomme, *Exercitatio mathematica ad agones panegyricos, Cum Deo et B[eata] Virgine propugnabunt Hyacinthus de Ruffec, Parisinus, Ioannes Richomme, Rothomagæus, in Collegio Claromontano Societatis Iesu*, s.l.s.n. [Bibliothèque municipale (Toulouse), Bibliothèque Mazarine (Paris), Bodleian Library (Oxford)].

---

[131] According to the title-page, the disputation took place on 20 June only, but, according to the program, it took place on 20 and 21 June.

[132] **De Bourneuf 1650** and **Villette 1651** have an identical introductory page introduced by the title "Dissertatio generalis."

[133] Due to the Covid-19 pandemic we were not able to compare **Villette 1651** and **d'Herbelot 1651**, but our suspicion is that they are identical..

1661–1664: Nicolas d'Harouys (1622–1698)

Lamoignon 1663
15 June 1663, Chrétien François de Lamoignon, *Agones mathematici ad arcem copernicani systematis, Expugnatam in Collegio Claromontano Societatis Iesu,* Paris, Antoine Vitré [Bibliothèque interuniversitaire de Santé (Paris), BnF (Paris), Bibliothèque municipale (Coutances), Bibliothèque municipale (Lyon), BSG (Paris), Bibliothèque Mazarine (Paris), BM Library (London), ETH Bibliothek (Zürich), HAB (Wolfenbüttel), Legion of Honor Museum (San Francisco), Stanford University Library (Stanford)].[134]

Le Meilleur 1663
30 June 1663, Joseph Le Meilleur, *Positiones mathematicæ de triplici sphæra, armillari, terrestri, cælesti, propugnabuntur in Collegio Claromontano societatis Iesu a Iosepho le Meilleur Venetensi,* Paris, Edmond Martin [Bibliothèque municipale (Amiens), Bibliothèque municipale (Lyon), Bibliothèque municipale (Toulouse)].

Anonymous 1664
4 July and 6 July 1664, *Positiones physico–mathematicæ, de magnetis subiecto, effectu, et usu, propugnabuntur in Collegio Claromontano Societatis Iesu,* Paris, Edmond Martin, [Bibliothèque municipale (Amiens), Bibliothèque municipale (Lyon), Bibliothèque municipale (Toulouse), Bibliothèque Mazarine (Paris), Biblioteca Nacional de Portugal (Lisboa), Houghton Library (Harvard)].

1665–1669: Michel Beaussier (Unknown–Unknown)

Tarteron 1665
29 January 1665, Jérôme Tarteron, *De cometa ann. 1664 et 1665 observationes mathematicæ propugnabuntur ab Hyeronymo Tarteron, Parisino, in aula Collegi Claromontani Societatis Iesu,* Paris: Edmond Martin [BnF (Paris), Universiteitsbibliotheek (Ghent)].

Prou 1665a[135]
29 January 1665, Louis Prou, *De cometa ann. 1664 et 1665 observationes mathematicæ propugnabuntur a Ludovico Prou, Parisino, in aula Collegi Claromontani Societatis Iesu*, s.l.s.n. [Bibliothèque d'Amiens métropole (Amiens), Bibliothèque de l'observatoire (Paris)].

[134] According to the title-page, the disputation took place on 8 June 1663, but according to handwritten annotations in some copies, it took place on 15 June.

[135] **Tarteron 1665** and **Prou 1665a** are identical.

Ragayne de la Picottiere 1665a
12 June 1665, Louis Ragayne de la Picottiere, *De duplici cometa vero et ficto positiones mathematicæ, propugnabuntur a Ludovico Ragayne de la Picottière, Sagiensi, in Collegio Claromontano Societatis Iesu*, s.l.s.n. [BnF (Paris), Bibliothèque municipale (Lyon), Biblioteca nazionale centrale (Roma), BL (London), Herzog–August–Bibliothek (Wolfenbüttel)].

Prou 1665b
12 June 1665, Louis Prou, *De duplici cometa vero et ficto positiones mathematicæ, propugnabuntur a Ludovico Prou, Parisino, in Collegio Claromontano Societatis Iesu*, s.l.s.n. [Bibliothèque municipale (Toulouse), BSG (Paris), Bibliothèque de l'observatoire (Paris)].[136]

Ragayne de la Picottiere 1665b
13–14 June 1665, Louis Ragayne de la Picottiere, *De hypothesi cartesiana positiones physicomathematicæ, propugnabuntur a Ludovico Ragayne de la Picottiere, Sagiensi, in Collegio Claromontano Societatis Jesu*, s.l.s.n. [BnF (Paris), Bibliothèque municipale (Lyon), Bibliothèque du Centre Sèvres (Paris), Det Kongelige Bibliotek (Copenhagen)].

Prou 1665c
13–14 June 1665, Louis Prou, *De hypothesi cartesiana positiones physicomathematicæ, propugnabuntur a Ludovico Prou, Parisino, in Collegio Claromontano Societatis Jesu*, s.l.s.n. [BnF (Paris), BL (London), Biblioteca nazionale universitaria (Torino)].

Decombes 1665
[13–14 June] 1665, Jean Decombes, *De hypothesi cartesiana positiones physicomathematicæ, propugnabuntur a Joanne Decombes, in Collegio Claromontano Societatis Jesu*, s.l.s.n. [HAB (Wolfenbüttel)].

de la Bletonniere 1665
13–14 June 1665, Louis de la Bletonniere, *De hypothesi cartesiana positiones physicomathematicæ, propugnabuntur a Ludovico de La Bletonniere Burgundo, in Collegio Claromontano Societatis Jesu*, s.l.s.n. [BSG (Paris)].[137]

Anonymous 1666
2, 3 and 4 July 1666, *Ex optica selecta mathemata, cum Deo et B. Virgine explicabuntur ac propugnabuntur in Collegio Claromontano Societ. Iesu*, s.l.s.n. [Bibliothèque municipale (Lyon)].

Lamoignon de Basville 1666
10 July 1666, Nicolas de Lamoignon de Basville, *Ex optica et astronomia selecta mathemata, cum Deo et B. Virgine, explicabit ac propugnabit Nicolaus* de

[136] **Ragayne de la Picottiere 1665a** and **Prou 1665b** are identical.

[137] **Ragayne de la Picottiere 1665b**, **Prou 1665c**, **Decombes 1665**, and **de la Bletonniere 1665** are identical.

Lamoignon de Basville, Parisinus in Collegio Claromont. Societat. Iesu, s.l.s.n. [Bibliothèque d'Amiens métropole (Amiens), Bibliothèque municipale (Toulouse), BSG (Paris)].

de Maupeou 1666
23 July 1666, Pierre de Maupeou, Ex optica et astronomia selecta mathemata cum Deo et B. Virgine, explicabit ac propugnabit Petrus de Maupeou parisinus, in Collegio Claromont. Societ. Jesu [BnF (Paris), Bibliothèque d'Amiens métropole (Amiens), Bibliothèque Mazarine (Paris)].[138]

Benard de Rezay 1666
29 July 1666, Guillaume Benard de Rezay, *Ex optica et astronomia selecta mathemata, cum Deo et B. Virgine, explicabit ac propugnabit Guillelmus Benard de Rezay Parisinus, in Collegio Claromont. Societat. Iesu,* s.l.s.n. [Bibliothèque interuniversitaire de Santé (Paris), Bibliothèque centrale du conservatoire national des arts et métiers (Paris)].[139]

Castagnere de Chasteauneuf 1667
13 July 1667, Jean–Baptiste and Pierre–Antoine de Castagnere de Chasteauneuf, *De corporum cœlestium motibus ac phænomenis iuxta varias hypotheses astronomicas et de architectura militari propositiones mathematicæ, Deo duce, favente Deipara, propugnabunt Ioannes Baptista, et Petrus Antonius de Castagnere de Chasteauneuf, Camberienses, in Collegio Claromontano Societatis Iesu,* s.l.s.n. [Bibliothèque municipale (Lyon), BSG (Paris)].

Anonymous 1667
July 1667, *De corporum cœlestium motibus ac phænomenis iuxta varias hypotheses astronomicas et de architectura militari Propositiones mathematicæ,... propugnabuntur in Collegio Claromontano Societatis Jesu* [Bibliothèque municipale (Toulouse)]

Bolé de Champlay 1668
20 March 1668, Jules–Louis Bolé de Champlay, *De corporum cœlestium motibus ac phænomenis iuxta varias hypotheses astronomicas et de architectura militari Propositiones mathematicæ, Deo duce, favente Deipara, propugnabit, Iulius*

[138] Due to the Covid-19 pandemic we were not able to compare **Anonymous 1666** and **de Maupeou 1666**, but our suspicion is that they are identical..

[139] The section on optics is identical in **Benard de Rezay 1666** and **Lamoignon de Basville 1666**, but the latter adds § 8. The section on optics in **Anomymous 1666** and **Lamoignon de Basville 1666** are identical, except that the latter omits § 8 and § 22, but adds § 29. The section on astronomy which is mentioned in the titles (*ex astronomica*) of **Benard de Rezay 1666** and **Lamoignon de Basville 1666** includes fairly similar comparisons of the systems of the world, but the later adds a section on eclipses [or: "on the eclipse"]. **Anomymous 1666** does not include the comparison of the systems of the world present in **Benard de Rezay 1666** and **Lamoignon de Basville 1666**, but includes the section on eclipses of **Lamoignon de Basville 1666**.

Ludovicus Bolé de Champlay, Parisinus, in Collegio Claromontano Societatis Iesu, s.l.s.n. [BnF (Paris), Bibliothèque municipale (Lyon)].[140]

Anonymous 1668
16–17 June 1668, *Propositiones mathematicæ ex statica mechanica et architectura militari, Deo duce, favente Deipara, propugnabuntur in Collegio Claromontano Societatis Iesu*, s.l.s.n. [Bibliothèque municipale (Lyon)].

Colbert de Seignelay 1668
30 August 1668, Jean–Baptiste Colbert de Seignelay, Regi armis omnia expugnanti architecturam militarem sapientia omnia constituenti totius mundi constitutionem bellis pacisque arbitro bellatricem pacificamque mathesin consecrat Joannes Bapt[iste] Colbert de Seignelay, Parisinus, s.l.s.n. [Sorbonne, BnF (Paris), Bibliothèque municipale (Lyon), Bibliothèque Mazarine (Paris)].[141]

Anonymous 1669
1–2 June 1669, *Positiones mathematicæ de calendario Romano, mundi systemate, antiquis ac novis sideribus et architectura militari, Deo duce, favente Deipara, propugnabuntur in Collegio Claromontano Societatis Iesu*, s.l.s.n. [Bibliothèque municipale (Lyon)].

de Mesgrigny 1670
14 June 1670, Jean–Baptiste Joseph–Ignace de Mesgrigny, *Positiones ex variis matheseos partibus, Deo duce, favente deipara propugnabit Johannes Baptista Josephius Ignatius de Mesgrigny, Aquensis, in Collegio Claromontano Societ. Jesu*, s.l.s.n. [BnF (Paris), Bibliothèque municipale (Lyon)].[142]

Hervé 1670
27 June 1670, Charles–Bénigne Hervé, *Positiones ex variis matheseos partibus propugnabit Car. Benig. Hervé*, s.l.s.n. [Bibliothèque d'Amiens métropole (Amiens)].[143]

---

[140] **Castagnere de Chasteauneuf 1667**, **Bole de Champlay 1668**, and perhaps **Anonymous 1667** are identical. **Bole de Champlay** originally bore the date "20 January" but this is corrected by hand to "20 March."

[141] The last page indicates that the thesis booklet was defended in August 1668: Renaudot, "Gazette du 1 Septembre 1668," in Renaudot 1669, 914, gives the precision that it was on 30 August, Colbert having defended his philosophy thesis on 29 August.

[142] The section on optics in **Mesgrigny 1670**, 7–14 is borrowed from Anonymous 1666, its section on military architecture is borrowed from **Anonymous 1668**, 10–11, § 21–4, and from **Anonymous 1669**,14–15, § 3–6. The problems to be solved at the end of each paragraph are sometimes, but not always, different. A few developments of Anonymous 1666 do not figure in **Mesgrigny 1670**, for example the condemnation of Descartes' subtle matter expressed in **Anonymous 1666**, 4, § 6.

[143] Due to the Covid-19 pandemic we were not able to compare **De Mesgrigny** 1670 and **Hervé 1670**, but our suspicion is that they are identical.

1670–1673: Ignace–Gaston Pardies (1636–1673)

Roux 1669
1669, Étienne Roux, *Theses mathematicæ et mechanicæ*, Bordeaux. [Universitaire Bibliotheken (Leiden)].[144]

Duhamel 1669
August 1669, André Duhamel, *Clarissimmo Viro Domino D.[omino] Carolo Duhamel [...] Burdegalensis Societatis Jesu primitias offert et consecrat Andreas Duhamel, Burdegalensis. Theses mathematicæ. Has theses Deo duce atque auspice Maria propugnabit Andreas Duhamel in aula Collegii Burdegalis*, Bordeaux: s.n. [Bibliothèque municipale (Lyon)].[145]

Anonymous 1671a
19–20 June 1671, *Theses mathematicæ de optica propugnabuntur in Collegio Claromontano Societatis Iesu*, s.l.s.n. [BnF (Paris), Bibliothèque municipale (Lyon)].

Milon 1671
6 July 1671, Alexandre Milon, *Theses mathematicæ de optica, propugnabuntur in Collegio Claromontano Societatis Iesu ab Alexandro Milon, Turonensi,* s.l.s.n. [BnF (Paris), Bibliothèque municipale (Lyon), Bibliothèque Mazarine (Paris), King's College Library (London)].[146]

Anonymous 1671b
[1671–Unknown] *Theses de re militari*, s.l.s.n. [Bibliothèque municipale (Lyon), King's College Library (London)].[147]

Le Mazier 1672
24–26 June 1672, Louis le Mazier and al., *Theses mathematicæ de vario mundi systemate, Propugnabuntur in Collegio Claromonatno Societatis Jesu, Propugnatores Die XXIV Ludovicus le Mazier, Die XXV, Due e Societate Jesu, Die XXVI, unus a Societate Jesu*, s.l.s.n. [BnF (Paris), Bibliothèque municipale (Lyon)].

[144] As a manuscript note indicates, this copy was sent by Pardies to Huygens, through the intermediary of Galois.

[145] **Roux 1669** and **Duhamel 1669** are identical. According to the dates we know, Pardies was still in Bordeaux when these theses were defended. We include them here as the earliest known theses written by Pardies.

[146] **Anonymous 1671a** and **Milon 1671** are identical.

[147] **Anonymous 1671b** has no title-page. The King's College copy is bound with **Milon 1671**. The same printing matrices were used.

1673–1676: Claude-François Millet de Chales (1621–1678)

Godley 1674
15–17 June 1674, Jean Godley and al., *Theses mathematicæ de hydrostatica, architectura militari, et astronomia, Has theses, Deo duce, et auspice Deipara, propugnabuntur in Collegio Claromontano Societatis Jesu.* (*Propugnatores : diebus Veneris XV et Sabbati XVI, duo e Societate Jesu, Die Dominica XVII Joannes Godley, Londinensis*) s.l.s.n. [BnF (Paris), Bibliothèque municipale (Lyon), Gottfried Wilhelm Leibniz Bibliothek (Hannover)].

de Geldrop 1674
20 July 1674, Embert de Geldrop, *Theses mathematicæ de hydrostatica, architectura militari, et astronomia, Has theses, Deo duce, et auspice Deipara, propugnabit Embertus de Geldrop, Mechliniensis, in Collegio Claromontano Societatis Jesu*, s.l.s.n. [Bibliothèque municipale (Lyon)].[148]

Anonymous 1675
5, 6 and 7 July 1675, *Theses mathematicæ de geometria practica, mechanica, statica, geographia, optica, Deo duce, et auspice Deipara, propugnabuntur In Collegio Claromontano Societatis Jesu*, Paris: Veuve Edmond Martin [BnF (Paris)].

Anonymous 1676
19, 20 and 21 June 1676, *Theses mathematicæ de hydrostatica, architectura militari, et astronomia propugnabuntur in Collegio Claromontano.* [Bibliothèque interuniversitaire Sorbonne (Paris), Bodleian Library (Oxford), BL (London)].[149]

1677–1682: Jean de Fontaney (1643–1710)

de La Hillière 1680
1680, François–Joseph de La Hillière, *Theses mathematicæ de architectura militari. Has theses tuebitur in Collegio Claromontano La Hillière.* [Public Library (New York)].

La Forest 1682
21 June 1682, Antoine–Louis de la Forest, *Theses mathematicæ de mechanica et hydrostatica, Propugnabuntur in Collegio Claromontano Societatis Jesu,* (*Has theses, Deo duce, et auspice Virgine, tuebitur Antonius Ludovicus de La Forest Constantinopolitanus*), Paris: Gabriel Martin [BnF (Paris), Public Library (New York)].

---

[148] **Godley 1674** and **de Geldrop 1674** are identical.

[149] **Anonymous 1676** and **de Geldrop 1674** are identical except for one page added in the former (p. 6, "de meteororum æquilibrium").

References

Ango, Pierre. 1682. *L'optique divisée en trois livres*. Paris: Michallet.

Anonymous. 1726. Review of Pardies. *Œvres complètes. Journal de Trévoux ou Mémoires pour l'histoire des sciences et des beaux-arts. Avril* 1726: 664–693.

Ariew, Roger. 1995. Pierre Bourdin and the Seventh Objections. In *Descartes and his Contemporaries: Meditations, Objections, and Replies*, ed. Roger Ariew and Marjorie Greene, 208–225. Chicago: University of Chicago Press.

———. 1997. Critiques scolastiques de Descartes: le *cogito*. *Laval théologique et philosophique* 53 (3): 587–603.

———. 1999. *Descartes and the Last Scholastics*. Ithaca: Cornell University Press.

———. 2013. Censorship, Condemnations and the Spread of Cartesianism. In *Cartesian Empiricism*, ed. Mihnea Dobre and Tammy Nyden, 25–46. Dordrecht/Heidelberg/New York/London: Springer.

———. 2014. *Descartes and the First Cartesians*. Oxford: Oxford University Press.

Armogathe, Jean-Robert. 1969. Dom Desgabets et Port–Royal. *Chroniques de Port–Royal* 17 (18): 68–87.

———. 1977. *Theologia cartesiana: L'explication physique de l'eucharistie chez Descartes et Dom Robert Desgabets*. The Hague: Martinus Nijhoff.

Baker, Tawrin. 2016. Why All This Jelly? Jacopo Zabarella and Hieronymus Fabricius ab Aquapendente on the Usefulness of the Vitreous Humor. In *Early Modern Medicine and Natural Philosophy*, ed. Peter Distelzweig, Benjamin Goldberg, and Evan Ragland, 59–90. Dordrecht: Springer.

Beaude, Joseph. 1976. Sur le cartésianisme d'Ignace-Gaston Pardies. A propos de l'ouvrage du Père A. Ziggelaar. *Revue d'histoire des sciences* 29–33: 261–267.

Bessot, Didier. 2005. Synthèse et développement de techniques d'anamorphoses au XVIIe siècle: les traités du père Jean-François Niceron. *Mélanges de l'École Française de Rome. Italie et Méditerranée* 117 (1): 91–129.

Bigourdan, Guillaume. 1918. La station astronomique du collège de Clermont (première période) et la mission astronomique de Siam. *Comptes rendus hebdomadaires des séances de l'Académie des sciences* 166: 833–840 and 871–875.

Bonelli, Maria Luisa. 1959. Una lettera di Evangelista Torricelli a Jean-François Niceron. In *Convegno di studi Torricelliani in occasione del 350o anniversario della nascita di Evangelista Torricelli (19–20 ottobre 1958)*, 37–41. Faenza: Fratelli Lega.

Bourdin, Pierre. 1641. *Le Cours de mathématique, representé par figures, et cartes, et clairement expliqué dans toutes ses parties, avec quantité de connoissances et pratiques nouvelles, Le tout en faveur de ceux qui veulent apprendre les mathematiques en peu de temps*. Paris: François Pelican.

———. 1645. *Le Cours de mathématique, representé par figures, et clairement expliqué, avec quantité de connoissances et pratiques nouvelles. Le tout en faveur de la Noblesse, et de ceux qui veulent apprendre en bref la Mathematique du temps rapportée à la Milice*. Paris: François Pelican.

———. 1661. *Le Cours de mathématique contenant en cent figures une idée generale de toutes les parties de cette Science*. Paris: Sébastien Bernard.

Brice, Germain. 1685. *Description nouvelle de ce qu'il y a de plus remarquable dans la ville de Paris*. 2 vols. La Haye: Abraham Arondeus.

Brockliss, Laurence. 1981. Aristotle, Descartes and the New Science: Natural Philosophy at the University of Paris 1600–1740. *Annals of Science* 38: 33–69.

———. 2002. Harvey, Torricelli and the institutionalization of new ideas in seventeenth–century France. In *Wissensideale und Wissenskulturen in der Frühen Neuzeit Ideals and Cultures of Knowledge in Early Modern Europe*, ed. Wolfgang Detel and Claus Zittel, 115–134. Berlin: Walter de Gruyter.

———. 2006. The Moment of No Return: The University of Paris and the Death of Aristotelianism. *Science and Education* 15: 259–278.

Cavalieri, Bonaventura. 1635. *Geometria indivisibilibus continuorum nova quadam ratione promota*. Bologna: de Ducis.

———. 1647. *Exercitationes Geometricæ Sex*. Bologna: J. Monti.

Chabaud, Gilles. 1994. *Sciences en jeux: les récréations mathématiques et physiques en France du XVIIe au XVIIIe siècle*. Unpublished PhD, École des hautes études en sciences sociales (Paris).

Colie, Rosalie L. 1966. *Paradoxia epidemica. The Renaissance tradition of Paradox*. Princeton: Princeton University Press.

Collacciani, Domenico, and Sophie Roux. 2017a. La querelle optique de Bourdin et de Descartes à la lumière des thèses mathématiques soutenues au collège de Clermont. In *Chemins du cartésianisme*, ed. Antonella Del Prete and Raffaele Carbone, 51–84. Paris: Garnier.

———. 2017b. Pierre Bourdin, anti–cartésien ou jésuite ordinaire? *Historia Philosophica* 15: 89–113.

Compère, Marie-Madeleine and Dominique Julia. 1984–2002. *Les collèges français (xvie—xviiie siècles)*, 3 vols. Paris: INRP/Éditions du CNRS.

Coret, Jacques. 1662. *Theses propugnatæ a jesuitis, die 12. decembris an. 1661. In Collegio Claromontano Parisiens. In quarum decimâ nonâ appendice manifesta continetur hæresis qua fidei catholicæ fundamentum & Ecclesiæ gallicanæ doctrina concutiuntur*. [Paris]: s.n.

Cosentino, Giuseppe. 1970. Le matematiche nella *Ratio studiorum* della Compagnia di Gesù. *Miscellanea storica ligure* 2: 169–213.

D'Orgeix, Émilie. 2005. Fortification et perspective militaire au XVIIe siècle en France. *Les Cahiers de la recherche architecturale et urbaine* 17: 91–105.

Darrigol, Olivier. 2010a. The Analogy between Light and Sound in the History of Optics from the ancient Greeks to Isaac Newton. *Centaurus* 52 (2): 117–155.

———. 2010b. The Analogy between Light and Sound in the History of Optics from the ancient Greeks to Isaac Newton. Part 2. *Centaurus* 52 (3): 206–257.

De Bruycker, Angelo. 2009. To the adornment and honour of the city': the mathematics course of the Flemish Jesuits in the seventeenth century. *British Society for the History of Mathematics Bulletin* 24: 135–146.

de Dainville, François. 1954. L'enseignement des mathématiques dans les Collèges jésuites de France du XVIe au XVIIIe siècle. *Revue d'histoire des sciences* 7–1: 6–21 and 7–2: 109–123.

De Lucca, Denis. 2012. *Jesuits and Fortifications: The Contribution of the Jesuits to Military Architecture in the Baroque Age*. Leiden-Boston: Brill.

de Rochemonteix, Camille. 1889. *Un collège de Jésuites aux XVIIe et XVIIIe siècles: Le Collège Henri IV de La Flèche*, 4 vols. Le Mans: Leguicheux.

Denis, Jean–Baptiste. 1682. *Mémoires, conférences et observations sur les arts et les sciences présentées à Monseigneur Le Dauphin*. 2nd ed. Paris: Laurent d'Houry.

Descartes, René. 1668. *Traité de la Mécanique, composé par Monsieur Descartes: de plus l'Abrégé de Musique du même auteur, mis en français avec les éclaircissements nécessaires par N. P. P. D. L.* [Nicolas Poisson], Paris: Charles Angot.

———. 1996. *Œuvres de Descartes*, ed. Charles Adam and Paul Tannery, revised ed. ed., 11 vols. Paris: Vrin.

Dhombres, Jean, and Patricia Radelet-de Grave. 2009. *Une mécanique donnée à voir. Les thèses illustrées défendues à Louvain en juillet 1624 par Grégoire de Saint-Vincent S.J.* Turnhout: Brepols.

Du Breuil, Jacques. 1649. *Troisième et dernière partie de la perspective pratique où se voient les beautez et les Raretez de cette Science [...] et les effets admirables des trois rayons. Droit, reflechi et brisé*. Paris: Vve F. Langlois.

Dupont-Ferrier, Gustave. 1921–25. *Du collège de Clermont au lycée Louis-le-Grand*, 2 vols. Paris: De Boccard.

Forest–Duchesne, Nicolas. 1650. *Florilegium universale liberalium artium et scientiarum, philologicum, mathematicum, philosophicum ac theologicum*. Paris: Lesselin.

Gabbey, Alan. 1992. Newton's Mathematical Principles of Natural Philosophy: A Treatise in Mechanics? In *The Investigation of Difficult Things: Essays on Newton and the History of Exact Sciences in Honour of D.T. Whiteside*, ed. Peter M. Harman and Alan E. Shapiro, 305–322. Cambridge: Cambridge University Press.

———. 2004. What was 'Mechanical' About 'The Mechanical Philosophy'? In *The Reception of the Galilean Science of Motion in Seventeenth-Century Europe*, ed. Carla Rita Palmerino and J.M.M.H. Thijssen, 11–24. Dordrecht: Springer.

Gal, Ofer, and Raz Chen-Morris. 2010. Baroque Optics and the Disappearance of the Observer: From Kepler's *Optics* to Descartes' Doubt. *Journal of the History of Ideas* 71-2: 191–217.

Galluzzi, Paolo. 1993. Gassendi e l'affaire Galilée delle leggi del moto. *Giornale Critico Della Filosofia Italiana* 72: 86–119.

Garnier, Jean. 1678. *Systema bibliothecae collegii parisiensis Societatis Jesu*. Paris: Sébastien Marbre-Cramoisy.

Gassendi, Pierre. 1642. *De apparente magnitudine solis humilis et sublimis epistolæ quatuor, in quibus complura physica, opticaque problemata proponuntur, et explicantur*. Paris: Louis de Heuqueville.

Gatto, Romano. 1994. *Tra scienza e immaginazione. Le matematiche presso il collegio gesuitico napoletano (1552–1670 ca.)*. Florence: Olschki.

Graindorge, André. 1942. Une page d'histoire des sciences. Vingt-deux lettres inédites d'André de Graindorge à P.-D. Huet. *Mémoires de l'Académie nationale des sciences, arts et belles-lettres de Caen*. N.S. 10: 245–337.

Grimaldi, Francesco Maria. 1665. *Physico-mathesis de lumine, coloribus et iride*. Bologna: V. Benatiis.

Guldin, Paul. 1635–1641. *De centro gravitatis trium specierum quantitatis continuae*. Vienna: Formis Gregorii Gelbhaar.

Hellyer, Marcus. 2005. *Catholic Physics: Jesuit Natural Philosophy in Early Modern Germany*. Notre-Dame: University of Notre Dame Press.

Hunt, James L., and John Sharp. 2011. The refractive anamorphic viewer of J.-F. Niceron: Reconstructing a 17th century optical toy. *American Journal of Physics* 79: 2023–2029.

Huygens, Christiaan. 1888–1950. *Œuvres complètes de Christian Huygens*, Société Hollandaise des Sciences, ed., 22 vols. The Hague: Martinus Nijhoff.

Jones, Phillip S. 1947. The Identity of the Author of a Hitherto Anonymous Work. *Scripta Mathematica* 13: 119–120.

Jourdain, Charles. 1862–1866. *Histoire de l'université de Paris aux XVIIe et XVIIIe siècles*. Paris: Hachette.

Kepler, Johannes. 1604. *Ad Vitellionem Paralipomena Quibus Astronomiae Pars Optica Traditur*. Frankfurt: Claudius Manius et Haeredes Joanni Aubri.

Kircher, Athanasius. 1671. *Ars Magna Lucis et Umbræ in decem libros digesta quibus admirandæ Lucis et Umbrae in mundo, atque adeo universa natura, vir effectusque uti nova, ita varia novorum reconditiorumque spcciminum exhibitione, ad varios mortalium usus, panduntur*. 2nd ed. Amsterdam: Janssonius a Waesberge Weyerstraet. (1st ed. 1645).

———. 1673. *Phonurgia nova sive Conjugium mechanico-physicum Artis et naturae Paranympha Phonosophia Concinnatum qua universa sonorum natura, proprietas, vires effectuumq prodigiosorum Causae.... Describitur*. Campidoniae: per Rudolphum Dreherr.

Klein, Boris. 2017. *Les chaires et l'esprit. Organisation et transmission des savoirs au sein d'une université germanique au XVIIe siècle*. Lyon: Presses universitaires de Lyon.

Le Dividich, Aude. 1996. *L'enseignement des* mathématiques *en France (1600–1670), Unpublished PhD*. École des Chartes (Paris).

———. 2003. Afficher, distribuer: usages éphémères de l'imprimé dans l'enseignement des sciences mathématiques en France au XVIIe siècle. *Revue de la Bibiothèque nationale de France* 14: 56–63.

Le Gallois, Pierre. 1672. *Conversations de l'Académie de Monsieur l'abbé Bourdelot, contenant diverses recherches, observations, expériences et raisonnements de physique, médecine, chymie et mathématique. Le tout recueilli par le Sr Le Gallois. Et le parallèle de la physique d'Aristote et de celle de Mons. Descartes, leu dans ladite Académie*. Paris: Thomas Moette.

Leibniz, Gottfried Wilhelm. 1875–1890. *Die philosophischen Schriften*. Carl Immanuel Gerhardt, ed. 7 vols. Berlin: A. Asher & Comp.

Leinsle, Ulrich. 2006. *Dilinganae Disputationes. Der Lehrinhalt der gedruckten Disputationen an der Philosophischen Fakultät der Universität Dillingen 1555–1648*. Regensburg: Schnell & Steiner.

Lerner, Michel-Pierre. 2001. La réception de la condamnation de Galilée en France au XVIIe siècle. In *Largo Campo di Filosofare. Eurosymposium Galileo 2001*, ed. José Montesinos and Carlos Solís Santos, 513–548. La Orotava: Fundación Canaria Orotava de Historia de la Ciencia.

Leurochon, Jean. 1622. *Selectæ propositiones in tota sparsim mathematica pulcherrimæ. Quas in solemni festo sanctorum Ignatii et Xaverii et anniversaria colegii Mussipontani celebritate literaria propugnabunt Mathematicarum auditores*. Pont-à-Moussson: S. Cramoisy.

———. 1629. *Selectæ propositiones in tota sparsim mathematica pulcherrimæ ad usum et exercitationem cerebrium academiarum. Propositiones arithmeticæ, Geometricae, Mechanicae, Cosmographicae, Musicae, Opticae*. Pont-à-Moussson: G. Bernard.

Lindberg, David C. 1976. *Theories of Vision from Al-Kindi to Kepler*. Chicago: University of Chicago Press.

Lukács, Ladislao, ed. 1986. *Ratio atque institutio sudiorum Societatis Iesu (1586 1591 1599), Monumenta paedagogica Societatis Iesu*. Vol. 5. Rome: Institutum historicum Societatis Iesu.

Lüthy, Christoph. 2006. Where Logical Necessity Becomes Visual Persuasion: Descartes' Clear and Distinct Illustrations. In *Transmitting Knowledge: Words, Images and Instruments in Early Modern Europe*, ed. Ian Maclean and Sachiko Kusukawa, 97–133. Oxford: Oxford University Press.

Malcolm, Noel. 1998. The Title Page of *Leviathan* Seen in a Curious Perspective. *The Seventeenth Century* 13: 124–155.

Matton, Sylvain. 2018. De Pierre Bourdin à Olivier Heroult: Descartes dans les collèges au XVIIe siècle. *Kritikon Litterarum* 45: 194–223.

Mersenne, Marin. 1634. *Questions inouyes, ou, Récréation des scavans: qui contiennent beaucoup de choses concernant la théologie, la philosophie, & les mathématiques*. Paris: Chez Iaques Villery.

Mersenne, Marin. 1636. *Harmonie universelle contenant la théorie et la pratique de la musique, ou il est traité de la nature des sons, et des mouvemens, des consonances, des dissonances, des genres, des modes, de la composition, de la voix, des chants, et de toutes sortes d'instruments harmoniques*. Paris: Sébastien Cramoisy.

Meyer, Véronique. 1993. Les thèses, leurs soutenances et leurs illustrations dans les universités françaises sous l'Ancien Régime. *Revue de la Sorbonne* 12: 43–111.

———. 2002. *L'illustration des thèses à Paris dans la seconde moitié du XVIIe siècle*. Paris: Paris-Musées.

———. 2017. *Pour la plus grande gloire du roi. Louis XIV en thèse*. Rennes: Presse Universitaires de Rennes.

Newton, Isaac. 1958. *Isaac Newton's papers & letters on natural philosophy and related documents*. Cambridge: Cambridge University Press.

Niceron, Jean-François. 1638. *Perspective curieuse ou Magie artificielle des effets merveilleux de l'Optique par la vision directe, de la Catoptrique, par la réflexion des miroirs plats, Cylindriques et Coniques, de la Dioptrique, par la refraction des Crystaux*. Paris: P. Billaine.

Oldenburg, Henry. 1965–1973. *The Correspondence of Henry Oldenburg*, ed. and trans. Alfred Rupert Hall and Marie Boas Hall, 9 vols. Madison, Wisconsin: University of Wisconsin Press.

Pantin, Isabelle. 2008. Simulacrum, Species, Forma, Imago: What Was Transported by Light into the Camera Obscura? *Early Science and Medicine* 13: 245–269.

Pardies, Ignace-Gaston. 1665a. *Dissertatio de motu et natura cometarum*. Bordeaux: Pierre du Coq.

———. 1665b. *Remarques sur les comètes et autres phaenomènes extraordinaires de ce temps*. Bordeaux: G. de la Court.

———. 1670. *Discours sur le mouvement local*. Paris: Edme Martin.
———. 1671. *Éléments de géométrie ou par une méthode courte et aisée l'on peut apprendre ce qu'il faut sçavoir d'Euclide, d'Archimede, d'Appollonius, et les plus belles inventions des anciens et des nouveaux géomètres*. Paris: Sebastien Mabre-Cramoisy.
———. 1672. *Discours de la connaissance des bêtes*. Paris: Sebastien Mabre-Cramoisy.
———. 1673a. *Deux machines propres à faire les cadrans avec très grande facilité decrites et expliquées par le P. Ignace Gaston Pardies de la Compagnie de Jésus*. Paris: Sebastien Mabre-Cramoisy.
———. 1673b. *La Statique ou la science des forces mouvantes*. Paris: Sebastien Mabre-Cramoisy.
Pascal, Blaise. 1647. *Expériences nouvelles touchant le vuide: faites dans des tuyaux, syringues, soufflets & siphons de plusieurs longueurs & figures, avec diverses liqueurs, comme vif–argent, eau, vin, huyle, air, etc. avec un discours sur le mesme sujet*. Paris: Margat.
Petit, Pierre and Valeriano Magni. 1647. *Observation touchant le vuide, faite pour la première fois en France, contenue en une lettre écrite à Monsieur Chanut [...]Par Monsieur Petit intendant des fortifications, le 10 novembre 1646. Avec le discours qui en a esté imprimé Pologne sur le mesme sujet, en juillet 1647*. Paris: Sebastien Cramoisy et Gabriel Cramoisy.
Pierius, Jacques. 1648. *An detur vacuum in rerum natura*. Paris: n.d.
Platter, Felix. 1583. *De corporis humani structura et usu*. [Basel]: Frober.
Plempius, Fortunatus Vospicus. 1632. *Ophthalmographia, sive Tractatio De Oculi fabrica, actione, et usu praeter vulgatas hactenus Philosophorum ac Medicorum Opiniones*. Amsterdam: Henri Laurent.
Ramelli, Agostino. 1588. *Le diverse et artificiose machine del Capitano Agostino Ramelli*. Paris: in casa de l'autore.
Raphael, Renée. 2017. *Reading Galileo: Scribal technologies and the Two New Sciences*. Baltimore: Johns Hopkins University Press.
Renaudot, Théophraste. 1669. *Recüeil des gazettes nouvelles ordinaires et extraordinaires. Relations et recits des choses avenues tant en ce Royaume qu'ailleurs, pendant l'année mil six cent soixante–huit*. Paris: Bureau d'Adresse.
Rice, Louise. 1999. Jesuit Thesis Prints and the Festive Academic Defence at the Collegio Romano. In *The Jesuits: Cultures, Sciences, and the Arts, 1540–1773*, ed. John W. O'Malley, Gauvin Alexander Bailey, Steven J. Harris, and T. Frank Kennedy, 148–169. Toronto: University of Toronto Press.
———. 2007a. *Apes philosophicae*: Bees and the Divine Design in Barberini Thesis Prints. In *I Barberini e la cultura europea del Seicento. Atti del convegno internazionale (Roma 7–11 dicembre 2004)*, ed. Lorenza Mochi Onori et al., 181–194. Rome: De Luca.
———. 2007b. POMIS SUA NOMINA SERVANT: The Emblematic Thesis Prints of the Roman Seminary. *Journal of the Warburg and Courtauld Institutes* 70: 195–246.
Righini-Bonelli, M.L. 1976. *Il Museo di Storia della scienza a Firenze*. Milano: Electa.
Rochon, Antoine. 1672. *Lettre d'un philosophe à un cartésien de ses amis*. Paris: Thomas Jolly.
Romano, Antonella. 1999. *La Contre–réforme mathématique*. Rome: École Française de Rome.
———. 2006. Teaching Mathematics in Jesuits Schools. Programs, Course, Content, and Classroom Practices. In *The Jesuits II: Cultures, Sciences and the Arts, 1540–1773*, ed. John W. O'Malley, Gauvin Alexander Bailey, Steven J. Harris, and T. Frank Kennedy, 355–370. Toronto: University of Toronto Press.
Roux, Sophie. 2013a. A French Partition of the Empire of Natural Philosophy (1670–1690). In *The Mechanization of Natural Philosophy*, ed. Daniel Garber and Sophie Roux, 55–98. New York/ Dordrecht/Boston/London: Kluwer.
———. 2017. The Two Comets of 1664–1665: A Dispersive Prism for French Natural Philosophy Principles. In *The Idea of Principles in Early Modern Thought*, ed. Peter Anstey, 98–146. New York/London: Routledge.
———. 2019. The Condemnations of Cartesian Natural Philosophy under Louis XIV (1661–1691). In *Oxford Handbook to Descartes and Cartesianism*, ed. Delphine Antoine-Mahut, Steven Nadler, and Tad Schmaltz, 755–779. Oxford: Oxford University Press.

Sabra, Abdulhamid I. 1981. *Theories of Light from Descartes to Newton*. Cambridge: Cambridge University Press.

Scheiner, Christoph. 1619. *Oculus, hoc est: Fundamentum opticum: in quo ex accurata oculi anatome, abstrusarum experientiarum sedula pervestigatione, ex invisis specierum visibilium tam everso quam erecto situ spectaculis, nec non solidis rationum momentis, radius visualis eruitur*. Innsbruck: apud Danielem Agricolam.

———. 1630. *Rosa ursina, sive Sol ex admirando focularum et macularum suarum phænomenon varius*. Brescia: A. Phaeum.

Schott, Gaspar. 1657. *Magia universalis naturae et artis, sive recondita naturalium & artificalium rerum scientia*. Würzburg: Henricus Pigrin.

Shapiro, Alan E. 1973. Kinematic Optics: A Study of the Wave Theory of Light in the Seventeenth Century. *Archive for History of Exact Sciences* 11 (2–3): 134–266.

———. 2008. Images: Real and Virtual, Projected and Perceived, from Kepler to Dechales. *Early Science and Medicine* 13 (3): 270–312.

Simon, Gérard. 1988. *Le Regard, l'être, l'apparence dans l'optique de l'Antiquité*. Paris: Seuil.

———. 2019. In *Kepler, rénovateur de l'optique*, ed. Delphine Bellis and Nicolas Roudet. Paris: Classiques Garnier.

Smith, A. Mark. 2015. *From Sight to Light. The Passage from Ancient to Modern Optics*. Chicago: University of Chicago Press.

Sommervogel, Carlos. 1890–1932. *Bibliothèque de la Compagnie de Jésus*, nouvelle édition, 12 vols. Paris: Picard.

Truci, Isabella. 1976. Le Anamorfosi di Jean François Niceron all'Istituto e Muso di Storia della Scienza di Firenze. *Annali dell'Istituto e Museo di storia della scienza di Firenze* 1: 57–64.

Vanpaemel, Geert. 2008. Mechanics and Mechanical Philosophy in Some Jesuit Mathematical Textbooks of the Early 17th Century. In *Mechanics and Natural Philosophy before the Scientific Revolution*, ed. Walter Roy Laird and Sophie Roux, 259–274. New York: Kluwer.

Ziggelaar, August. 1971. *Le physicien Ignace-Gaston Pardies S.J. (1636–1673)*. Odensee: Odensee University Press.

Zittel, Claus. 2009. *Theatrum philosophicum. Descartes und die Rolle ästhetischer Formen in der Wissenschaft*. Berlin: Akademie-Verlag.

———. 2011. Conflicting Pictures: Illustrating Descartes' *Traité de l'homme*. In *Silent Messengers. The Circulation of Material Objects of Knowledge in the Early Modern Low Countries*, ed. Sven Dupré and Christoph Lüthy, 217–260. Berlin: Lit-Verlag.

Chapter 6
Subtilis, Inutilis: The Jesuit Pedagogy of Ingenuity at La Flèche in the Seventeenth Century

Raphaële Garrod

> *la philosophie n'est qu'une poesie sophistiquée.*
>
> *Philosophy is merely sophistical poetry (Montaigne, "Apologie de Raimond Sebond," (II,12) in Montaigne 1595, 305. Edition owned by the Jesuit college of La Flèche, catalogue of the old library (fonds ancien or fonds historique) accessible online at http://www.bibliotheque-du-prytanee.fr/catalogues. Most of the texts used in this chapter are found in editions listed in this catalogue. The date of entry in the collection, when known, is indicated in individual notices. All translations are mine unless otherwise stated. I thank Paul J. Smith and Karl Ennenkel for their invitation to the 'Emblems and the Natural World' conference held at the University of Münster in December 2015, where I presented an early version of this chapter. The research leading to this chapter has received funding from the European Research Council under the European Union's Seventh Framework Programme (FP7/2007-2013)/ERC grant agreement no 617391. I am grateful to Tim Chesters for his careful editing of my prose.*

Abstract Using the spider emblem of the ingenium from Antoine de Bourgogne's *Mundi lapis lydius* (*Compass of the World,* 1639) as its pretext, this chapter explores the Jesuit pedagogy of ingenuity at La Flèche in the seventeenth century. It makes sense of this image and its motto by exploring the ways in which ingenuity was represented, assessed, and managed in the manuals of rhetoric, poetics, logic, natural philosophy and mixed mathematics which feature in the La Flèche historical library.

R. Garrod (✉)
Magdalen College, University of Oxford, Oxford, UK
e-mail: raphaele.garrod@mod-langs.ox.ac.uk

S. Berger, D. Garber (eds.), *Teaching Philosophy in Early Modern Europe*, Archimedes 61, https://doi.org/10.1007/978-3-030-84621-3_6

Keywords Ingenuity · Spider · Clocks and clockworks · Emblems · Subtlety · Cognition · Jesuit pedagogy · Early modern dialectic · Early modern rhetoric · Early modern poetics · Early modern astronomy · Early modern mixed mathematics · Classical reception

6.1 The *Ingenium* Emblem. Commentary

Vanity: The fine blade of wit
Truth: Fine, therefore useless

Commentary

Quam aptae sint humanis usibus Cameracenae, Hollandicae, ac Cortacenses telae, manifestus docet quotidiana necessitas, quam ut quempiam lateat. Eam vero, quam sibi ipsa et colus, & fusus, & lanipendia, & textrix aranea conficit, non minus aperte constat tam inutilem esse, quam subtilem; imò verò ideò maxime inutilem, quia maxime subtilem. Sic nihil verae sapientiae magis inutile, quam Sophistae subtilitas. Ideo invisam Palladi finxere veteres araneam. Ut mucronis ergo, sic ingenii acies sit, non ut fetae; subsistat, ut penetret. Quorundam acumen tenui cuspide frangitur, primoque molimine deficit. Validissima etiam, si ad ultimum extenuentur, infirma sunt: sic robur omne tenuitas atterit. Ingenium, ut eae quas diximus telae, si bonis artibus applicabile pretiosa supellex, & utilis; sin minus, ut araneae textum, laboriosa, & futilis.

Quis Sophistarum bella dinumeret? quis inenodabiles nodos? quis veri professionem? quis patrocinium falsi? quis ignorantem scientiam? dum hoc unum certo scire se fatentur quod nihil sciant, ac ne viam quidem scientiae adhuc invenerint? Et verissime quidem. Si enim, juxta unanimem ipsorum consensum, nihil recte scitur, quod non per causas, easque proximas, & efficientes quidem aut finales, rite demonstratur; tales vero attributorum nulli hactenus assignari potuerunt, nisi solis immutabilibus et propriis accidentibus, tamque exiguo numero, ut fere an numerus dici possit ambigam; quis non videt, eos inter ignaros potius quam sapientes esse censendos?

Eorum porro, quae scire perhibentur, nihil non rationibus inter se pugnantibus problematice varium, nihil tam solidum, ut non mox aeque stolidum esse comprobetur. Exemplis, quod loquor, evincerem, nisi propositum brevitatis studium, vel ex nunc alio me vocaret.

Quapropter, ut paucis expediam, genus hoc hominum horologiis conferes, quae dum veri regulam profitentur, ipsamet a vero deviant; & hoc uno semper conveniunt, quod inter se numquam conveniant. De hora certe nemo crebrius ambigit, aut interrogat, quam qui ejusmodi loquaces rotas circumferunt: sic nulli minus sapiunt, quam qui maxime. Et e diverso ignorans non dubitat.

(continued)

Pejus est, quod subjungo. Raro magni errores, nisi ex magnis ingeniis prodiere. Nullum magnum ingenium sine mixtura dementiae: quidni & nequitiae? Da sane, inquit vulgus, ingenii maxime subtilis artificem, dabo plerumque moribus quam maxime perditis nebulonem.

Quam perniciosus Reipublicae Catilina fuerit, saepius Cicero prodidit, quam Consularis modestia postulabat: at Crispus illum magna vi animi fuisse ait, sed ingenio malo. Tu an bono sis, nescio: quod ob sensuum fallaciam, caeco; ob rerum omnium incertudinem, erroneo; ob humanae fragilitatis conditionem, periculoso; hoc certo scio.

Habe et hoc corodinis loco. Ingenium quo velocius, eo proprius ruinae, utinam non eternae! Quam id vellent tot illustres animae, quarum nimia subtilitas, caeli claustris infracta, sibi aliisque barathrum referavit!

Daily necessity teaches clearly the usefulness of the sturdy cloth made in Holland, Cambrai, and Brabant, which addresses human needs. The cloth that she weaves from her own self—to itself its own warp, spindle, thread and puff of raw wool—is obviously as useless as it is fine—really, all the more useless because it fine. In a similar fashion, nothing is more useless to true wisdom than sophisticated subtlety. Hence the ancients have invented the fable according to which Athena hated the spider. Wit should be acute like the edge of a sword, not like a new mother—it must hold its position to pierce through. The very blade of any sharp wit, overly sharpened, will chip; even the best arguments become faulty when refined to their limits—fineness ultimately diminishes strength. Wit, like the cloths just mentioned, when applied to good arts, is a tool both precious and useful, when not, like the spider's woven stuff, it is both laborious and useless.

Who will keep count of the wars of sophists, of their knotty problems that no one can solve? Who shall reckon professions of truth? Defenses of falsehood? Who, the ignorant science? So they deemed that they knew that one thing for sure: that they knew nothing; and yet from there they could not find any path to certain knowledge? This is all very true indeed. If, in accordance with the unanimous agreement of that same lot, nothing is rightfully known that has not been properly demonstrated by its causes: proximate ones, and efficient ones at that, or even final ones, truly then none of these can ever be ascribed to any attributes so far, except immutable and proper accidents; and they are in such small number, that I shall even debate whether it can be called a number. Who would not see, that these should be deemed to belong with the ignorant rather than the learned?

In turn, of those things supposedly known, nothing that is not problematically variable, upheld by arguments that contradict each other, nothing so solid that is not soon proven to be stupid. I would adduce examples of what I say, were it not for the fact that my explicit attempt at brevity calls me away from here to there.

(continued)

For this reason, to put it in a few words, this type of men you shall compare with clocks, which as they utter the rule of truth, they deviate from that very rule; they always keep time individually, while they are never in sync with each other. For sure, no one has more doubt or asks more questions about the hour than those who have made those babbling wheels rotate in this way; thus no one knows less, than those who know most. And the ignorant man is not thrown into doubt by the experience of opposites.

Worst still, what I am adding here. Great errors rarely happen, except when brought forth by great minds. There is no great mind without a touch of madness, why not without wickedness too? Granted for sure, the plebs says, the artifice of the subtle mind; granted even more, most often, the sorry wretch with corrupt habits.

Cicero shows more often than his consular modesty demanded how pernicious Catilina was for the Republic: yet Sallust says he had a formidable strength of soul, but a wicked wit. You, whether your wit is good, I know not: that it is blind because of the falsity of the senses, wrong because of the uncertainty of all things, threatened because of the frailty of the human condition, that for sure I know.

Now, here comes the final flourish: the faster the wit, the fitter to its own ruin, if only not an eternal one! How they hope as much, these many brilliant souls, whose excessive subtlety, broken by the bolted doors of heaven, had them, and others with them, sent back to the abyss!

6.2 The Ambivalences of *Ingenium*

6.2.1 *Subtle Wit, Useless Wit*

This emblem dedicated to *ingenium* opens Antoine de Bourgogne's 1639 *Mundi lapis lydius* (*Compass of the World*), an emblem book held by the library of the Jesuit College of La Flèche, where René Descartes studied (Fig. 6.1).[1] It features in the catalogue of the *fonds ancien* or *historique* of the college. In the foreground of an elegant urban perspective, a spider has spun a fine web within the alcove of a classical building, which stands out against the backdrop of the domes and spires of the city churches behind it. The spider's craft exemplifies the ingenious products of one's wit. The first meaning of the term *ingenium* in early modern Latin and polyglot dictionaries is one's inborn nature and temperament; its second definition, the one explicitly targeted in this emblem, identifies *ingenium* with the cognitive

[1] Little is known about Antoine de Bourgogne (1593/4–1657); he was canon and archdeacon of the cathedral of Bruges. His *Mundi lapis lydius, sive vanitas per veritatem falsi accusata et convicta* (*Compass of the World, or Vanity Accused of Falsehood and Condemned by Truth*) was first published in Latin in 1639, translated into Dutch in 1643, and re-edited in 1654.

Fig. 6.1 "Ingenium emblem," Antoine de Bourgogne, *Mundi lapis lydius, sive vanitas per veritatem falsi accusata et convicta* (Antwerp: Vve Cnobbaert, 1639), 2–5. (Courtesy of The Rare Book & Manuscript Library, University of Illinois at Urbana-Champaign)

expressions of one's such inborn nature, namely, wit.[2] The motto of this emblem warns the reader that a subtle or fine wit is a useless one, while being itself a rhetorical display of such witty subtlety: the paronomastic pairs *vanitas* and *veritas*, the rich rhyme between *subtilis* and *inutilis*, finally the verbal ellipsis that Latin allows, all contribute to the lapidary wit of this epigrammatic maxim. The commentary also demonstrates a playful awareness of the natural-philosophical meanings of subtlety as material tenuousness hardly perceptible by the senses but also as cognitive complexity: these qualities initially defined subtlety for Gerolamo Cardano (1501–1576) in *De subtilitate* (*On Subtlety*).[3] This emblem defines subtlety as an extreme performance of ingenuity that ultimately renders it inane. Yet as this chapter will show, this emblem dramatizes the tension between the condemnation of such wit in rhetorical invention, epistemology, and the mechanical arts on the one hand, and its stylistic display on the other; this tension is central to Jesuit pedagogy. This emblem is indeed an ironic, virtuosic firework of satirical wit, which denounces the very means it employs to catch the attention of its readers. In so doing it promotes a pedagogical definition of the *ingenium* as one's malleable nature, that is, as a site of pedagogical, social and theological *docilitas*.

[2] See the entry *ingenium* under *gigno* in Estienne 1531, 411^{v}-12^{r}.

[3] Cardano 1642, 1. The first edition of the *De subtilitate* dates back to 1550. It was followed by two subsequent editions incorporating amendments by Cardano himself in 1554 and 1560. I am using the French translation listed in the *fonds historique* of the La Flèche library.

6.2.2 *Thinking with Images, Reforming One's Nature: The* Ingenium *in Jesuit Pedagogy*

That de Bourgogne's witty emblems should feature in the library of the Jesuits of La Flèche is no surprise: ingenuity was central to Jesuit schooling, intent at exercising the wits of their pupils in order to discipline their nature, thus tying together the two period meanings of *ingenium*. As one's inborn nature and temperament, the *ingenium* was the very site of pedagogical exercise and reform for the Jesuits.[4]

Following a long scholastic tradition arching back to Hugh of Saint Victor's *Didadascalion*, the Jesuits assumed a pedagogical understanding of the *ingenium* as one's embodied cognitive abilities, that is, one's mental power to discover and combine data stored in the memory.[5] In his European best-seller, *Examen de los ingenios para las ciencias* (*The Examination of Men's Wits*, first edition in 1575, second expurgated and posthumous edition in 1594), the Spanish physician Juan Huarte de San Juan describes the *ingenium* as the mental power to manipulate cerebral images stored in the memory—these images were the archetypes of specific arts and disciplines. The peculiarities of a person's temperament and the nature of their brain determined their aptitude to learn and excel in specific arts and disciplines: Hippocrates and Galen were good physicians, because they had neatly imprinted in their memory a model of the human body and its workings.[6] Half political treatise on good governance, half medical pamphlet on eugenics, the *Examen* provides a curious medical typology of *ingenia* intended to help parents beget the "right" children for a harmonious civil society, and to advise the Prince on how to engineer such a harmonious civil society. Its pedagogical potential struck the Jesuits at once; Antonio Possevino provided the first Latin translation of the *Examen* geared toward pedagogy under the title *Cultura ingeniorum* (1603). Three copies of the text—a 1652 Spanish edition, a 1663 Latin translation, and a 1675 French one—feature in the La Flèche library.[7] At the heart of Jesuit pedagogy, one therefore finds the *ingenium* understood as the ability to generate knowledge out of the manipulation of cerebral representations—an ability deriving from one's inborn, embodied nature.

Indeed for the Jesuits, one's wit had to be disciplined through teaching, and the very site of that pedagogical disciplining (*docilitas*) was the *ingenium* as one's inborn nature and temperament. While one's *ingenium* was indeed, one's inborn nature, such nature, like good soil, had to be cultivated to bear fruits and was

---

[4] See Marr et al. 2018.

[5] Hugh of Saint Victor 1961, 91: "Hoc ad naturam ingenio" ("Concerning aptitude as related to natural endowment"), book 3, ch. 7.

[6] Huarte de San Juan 1655, 7–8. The fonds historique of the La Flèche library owns the 1675 edition. For a detailed account of the scholastic sources of the notion of pedagogical *ingenium* and its importance in Descartes' early *Regulae*, see Garrod and Marr, eds., 2021.

[7] Possevino 1603, vol. 1, 1–49. The second Latin translation was the work of the German humanist Joachim Caesar. A revised *Scrutinium* was printed in 1637, which took into account the expurgated Spanish edition of the *Examen* in 1594. The La Flèche library owns a 1663 edition of that revised version.

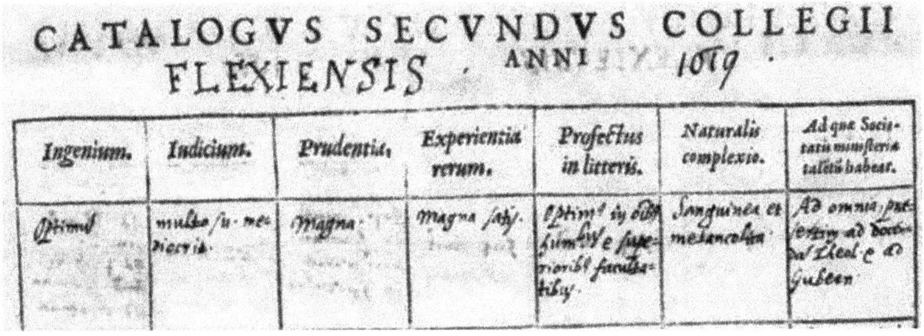

CATALOGVS SECVNDVS COLLEGII FLEXIENSIS . ANNI 1619 .

Ingenium.	Iudicium.	Prudentia.	Experientia rerum.	Profectus in litteris.	Naturalis complexio.	Ad quæ Societatis ministeria talentum habeat.
Optimum	[illegible]	Magna	Magna satis.	[illegible]	Sanguinea et melancolica	Ad omnia, [illegible]

Fig. 6.2 Pedagogical profiling of the teacher Natalis Deniau at La Flèche in 1619: "Catalogus secundus of the Collège de La Flèche (1619), ARSI, FRANC. 11, fol.70ʳ, © Archivum Romanum Societati Iesu"

therefore malleable. In this respect, the *ingenium* was also the site of teachability or "docility" that the Jesuit targeted in their students, just as it defined the pedagogical expertise of their teachers. Thus the Jesuits attributed the posts in their colleges on the basis of a fine-grained cognitive and temperamental profiling of their teachers, whose *ingenia* were assessed alongside their judgement, erudition, prudence, experience, and complexion, in order to determine what discipline they were likely to be best suited for (Fig. 6.2).

De Bourgogne's emblem condemns exceptional or paradoxical expressions of subtle wit, and promotes a view of the *ingenium* as one's teachable, docile, and commonsensical nature: the emblem is therefore a performance of what the good pedagogy of the *ingenium* should be. In this respect, the *ingenium* emblem and the whole book from which it is excerpted are paradigmatic of the Jesuit pedagogy of ingenuity, which intended to sharpen one's wit in order to educate one's nature. The *Compass* exercises the former in order to reform the latter.

The whole book emblematizes a much older Petrarchan allegorical dialogue in two parts, *De remediis utriusque fortunae* (*On Remedies Against Good or Bad Fortune*).[8] The *Remedies* pits frivolous Joy (*Gaudium*) against vitriolic Wisdom (*Sapientia*). Petrarch's *Gaudium* is translated into de Bourgogne's *vanitas* motto and its worldly values, and *Sapientia* into the *veritas* motto. In the first part of the dialogue, *Sapientia* mercilessly undermines *Gaudium*'s positive assessment of what constitute good fortune: ingenuity, glory, wealth, a noble birth etc. Not only does de Bourgogne systematize the dialogic form of the *Remedies* into the striking double-motto structure of the *Compass*, his commentaries for each emblem are creative paraphrases of Petrarchan highlights carefully excerpted from the *Remedies*. Thus, most conceits and epigrams in the text of our emblem are lifted verbatim from Petrarch's dialogue on the same topic in the *Remedies*, entitled "On Wit" (*De ingenio*).[9] De Bourgogne retains from his source the punch of epigrammatic wit put

[8] The La Flèche library owns the 1581 Basel edition of Petrarch's complete works, in the first volume of which this dialogue features. This extensive Petrarchan borrowing remains a silent one: Petrarch's name is not mentioned once.

[9] Petrarch 1581, vol. 1, 5. "De ingenio," book I, ch.7 in *De remediis utriusque fortunae*.

to ethical use, and tightens the blows towards disciplining the "inborn nature" of his reader. One has to make sense of the image and its commentary, to unravel its web of ironic analogies and erudite allusions in order to reassess the true value of those criteria which socially defined a good nature: the constitutive parts of eloquence—wit (*ingenium*), memory (*memoria*), and verbal invention (*copia*)—but also physical strength (*robur*) and beauty (*forma*), or social status and its signs, such as a good name (*bonum nomen*), noble blood (*clarus sanguis*), free birth (*ingenita libertas*) all demand such scathing reassessment; indeed they are all the topics of emblems in the *Compass*.[10] Thus the *Compass* repeatedly points to another scale of "true" goods ordered according to Christian eschatology, and in doing so, redefines the norms of "a good nature."

This chapter maps the redefinition of ingenuity in de Bourgogne's spider emblem in poetics and rhetoric first, then in logic, epistemology, and mixed mathematics; it studies how this redefinition is carried out by referring to emblematic, rhetorical, logical, natural-philosophical and mechanical sources available at La Flèche in the seventeenth century. In so doing, it unearths the contradiction of a stylistically ingenious condemnation of subversive wit and promotion of docility and commonsense which underpins the Jesuit pedagogy of ingenuity: once you have tied your students in the intellectually alluring knots of a witty emblem, how do you instill the lesson that ingenuity should *not* be about potentially dangerous displays of wit, but about commonsensical and docile social performance? This is the challenge that this emblem takes up.

6.3 Rhetorical Ingenuity: Invention—Judgement

The first target of de Bourgogne's emblem is rhetorical and poetic ingenuity. The first structural contradiction of this emblem, however, is to encapsulate the condemnation of ingenuity as subtle invention in emblematic form—the emblem as a genre epitomizes rhetorical and poetic ingenuity in the Jesuit curriculum. Getting the students to make sense of this emblem is one way of sharpening their wit in order to reform their nature; this disciplining first amounts to an exercise in textual and visual interpretation intended to foster sound judgement rather than subtle invention.

6.3.1 Jesuit Ingenuity and Emblematics

From its inception in the sixteenth century, the genre of the emblem has been defined as witty "thinking with images". The twelve emblem books recorded in the La Flèche's *fonds historique* bear testament to the pedagogical vogue the genre enjoyed

[10] See de Bourgogne 1639, 2–5, 6–9, 10–13, 17–19, 20–22, 26–28, 30–33, and 34–36.

in the order. Its classics are represented, such as a sixteenth-century edition of Joachim Camerarius's *Centuria* (*Centuries*) and a heavily annotated seventeenth-century edition of Andrea Alciato's *Emblemata* (*Emblems*).[11] Some Jesuit texts are also present, such as René Rapin's 1660 *Pax Themidis cum Musis* (*Peace of Themis with the Muses*) and an instance of the neighboring genre of device collections, Pierre Le Moyne's 1666 *Art des devises* (*Art of Devices*).[12]

As early as Guillaume de la Perrière's 1544 *Théâtre des bons engins* (*Theater of Fine Devices*), one of the most influential emblem books of Renaissance France, emblematics have been associated with both the display and exercise of one's *ingenium* understood as natural wit. La Perrière's title means just that: spectacle—*théâtre*—of the products of his wit—*engins*.[13] As a rhetorical and poetic exercise, the emblem discloses its maker's ability to weave together pithy conceits in both visual and verbal forms; yet it also demands that its audience exercise their own wit by summoning at once their attention and erudition, in order to make sense of the moral lesson encapsulated in such visual and textual conceits. The definition of the emblem provided by Nicolas Caussin (1583–1651), a professor of rhetoric at La Flèche and a well-regarded preacher in Paris in the 1630s, bears testament to this relationship between witty rhetorical invention and interpretation by locating the emblem alongside the symbol and the apologue within an ancient tradition of witty wordplay, such as the pun and the enigma: "the witty invention of enigmas, and the subtle games of puns were held in great repute in ancient times."[14] Caussin defines the emblem as "some ingenious symbol, both sweet and suited to a moral topic, consisting of a picture and epigram, usually pointing towards some more eminent doctrine."[15] For Caussin an emblem was a pleasing wit sharpener which led the students from the learned elucidation of conceits to moral reflection. Making and solving puns, enigmas and emblems therefore featured in Jesuit programs of studies—it is within this context that Descartes's reference to "childlike" anagrammatic games in the *Regulae ad directionem ingenii* (*Rules for the Direction of the Native Intelligence*) makes sense.[16] Seventy years after Caussin, Joseph de Jouvancy, the author of a detailed program of studies for the first classes of grammar, humanities,

---

[11] Camerarius 1590; Alciato 1602.

[12] Rapin 1660; Le Moyne 1649–1650 and idem 1666.

[13] La Perrière 1544. Available at http://www.emblems.arts.gla.ac.uk/french/books.php?id=FLPa. On Jesuit emblematics, see Dekoninck 2016, consulted online on 30 August 2018 https://doi.org/10.1163/2468-7723_jho_COM_192540. On emblematics and wit, see Praz 2001; Montagu 1968.

[14] Caussin 1623, 2: "Aenigmatum vero solers inventio, griphorumque lusus aculeati, magnam habent antiquitatem commendationem." *De symbolica aegyptorum sapientia* (*On the Symbolic Wisdom of the Egyptians*) is Caussin's own textbook on enigmas and emblems; it consists of an eclectic compilation drawing from the hermetic *Hieroglyphica* of Horapollo, the patristic *Stromata* of Clement of Alexandria, and Epiphanus' *Physiologus*, one of the most common bestiaries in the Middle Ages.

[15] Caussin 1623, 4: "Emblema [...] est enim proprie symbolum aliquod ingeniosum, suave, & moratum, ex pictura et lemmate constans, quo aliqua gravior sententia indicari solet."

[16] Descartes 1996, X, 391. "Regula 7" in *Regulae ad directionem ingenii*.

and rhetoric, the *Ratio discendi and docendi* (*Reasoned Programme for Learning and Teaching*) (1690) locates wordplay in general and emblematics in particular within "polymathy" (*polymathia*), which includes diplomatics, numismatics, and the art of making and deciphering puns, enigmas, and emblems.[17] Unlike Caussin, he warns against indulging in polymathy as a potentially sinful expression of curiosity. One can take ingenuity too far; the erudite wit of emblematics needs to be contained within the bounds of counsel and *doxa*.[18] Despite this warning, Jouvancy proceeds with a detailed typology of witty games, and gives precise rhetorical prescriptions for their performance: the student should use an entertaining variety of style, pepper his presentation with a solid dose of erudition, and, if possible, structure it in a way that would allow the quick wits in his audience to guess the answer to his enigma or the meaning of his emblem before he actually reveals it. Seventy years after Caussin, Jouvancy registers both a growing awareness of the potential dangers of witty excesses in the pedagogy of emblematics, and the solid grip the genre holds in the Jesuit classroom as a wit sharpener. De Bourgogne's spider emblem of the *ingenium* is characteristic of this ambivalence towards rhetorical and poetic ingenuity.

6.3.2 Redefining Rhetorical Wit

The commentary on the image and its motto makes it clear that the fable underpinning de Bourgogne's emblem is that of Arachne in Ovid's *Metamorphoses*. The Ovidian fable was a recurring motif in classical accounts of rhetorical invention and became integral to commonplaces on rhetorical ingenuity which all featured in the humanist editions and commentaries of the La Flèche library: Denis Lambin's 1566 edition of Cicero's complete works, the 1582–3 Aldine edition and commentary on Cicero's works—including the volume of rhetorical treatises, namely the *Brutus (De oratore)*, the *Orator*, the *Topica* (*Topics*), and the *Partitiones oratoriae* (*Partitions of Oratory*)—, Sebastian Gryphius's 1540 edition of Quintilian's *Institutiones oratoriae* (*Institutes of Oratory*), as well as Josse Bade 1528 commented edition of Quintilian.[19] They also fed into the textbooks of the Jesuit order itself, such as Nicolas Caussin 1619 *Eloquentiae sacrae et humanae parallela* (*Parallels of Sacred and Human Eloquence*). This set of rhetorical commonplaces is the matrix of de Bourgogne's commentary.

In Ovid's fable, Arachne typifies the bad *ingenium*, that is, a boastful, indocile inborn talent. The Lydian maid, an outstanding weaver, refuses to acknowledge that Athena taught her her craft. The outraged goddess summons her to a weaving contest; Arachne's tapestry, which depicts all sorts of Olympian deceits, is a mimetic

[17] Jouvancy 1690, 94–100: "De polymathia, seu philologia" art.IV.

[18] Jouvancy 1690, 94–5.

[19] Cicero 1566; idem 1583; Quintilian 1540; Bade 1528.

masterpiece of energetic variety and color which infuriates Athena so much that she tears it to pieces.[20] Arachne kills herself out of despair, but a repentant Athena turns her into a spider, suspended by a thread and bound to weave out of her own body. Arachne's tale was prime matter to discuss the role of one's own nature and wit in poetics taken in its etymological sense of the art of making first, and in the narrower, rhetorical sense of "writing a good speech, creating a good text": this was a tale about the ability to weave—*textus*—and to represent.[21] De Bourgogne's commentary fully engages in this discussion and makes its spider emblem a symbol of ongoing debates about the nature of rhetorical invention. The masterful exploitation of the etymology of *subtilis* (*sub-tela*, the warp) sustains throughout the simile between weaving and rhetorical invention. De Bourgogne's commentary first contrasts the useful, sturdy cloths of Brabant and Cambrai that answer human needs with the useless, subtle lace that the spider, the autarkic spinner and weaver, produces. A brief allusion to Arachne's tale ("Athena hated the spider") allows de Bourgogne to shift the focus allegorically from weaving threads to inventing text and, in particular, arguments: the spider web becomes, in a few sentences, the symbol of poor argumentative invention, or sophistry. Another classical commonplace underpins this shift, namely the epigrammatic rant against the sophist's insubstantial arguments in Diogenes Laertius' life of Ariston of Chios: "Dialectical reasonings, he said, are like spiders' webs, which, though they seem to display some artistic workmanship, are yet of no use."[22] The underlying warp (*subtela*) of the good cloth of Cambrai, the product of industry, is sturdy; whereas the spider lace, the useless product of ingenuity, is fine (*subtilis*) thread on which nothing else can be woven. In the seventeenth century, the commentary tradition turned Laertius's view on spider webs into a gloss of Ovid's fable; thus one reads in a La Flèche textbook: "Antoninus Pius used to say that the subtleties of sophists were similar to the spider webs, in which we see much artifice, but from which we derive nothing useful; and that the sophists, that is, those philosophers whose science consists of vain subtleties, were to the wise men what Arachne was to Pallas."[23]

In the emblem, the simile of subtle weaving expresses a conceptual foundation of rhetorical accounts of invention, namely the pair *ingenium* (inborn talent)/*industria* or *excertitatio* (repeated work and practice). Invention as the ability to find both the right subject-matter (*res*) and its right expression (*verba*) was the prerogative of the

[20] Ovidius Naso 1977, vol. 3:6, 1–145.

[21] Ballestra Puech 2006, "L'Abeille et l'araignée" in Perras 2016, 231–46.

[22] "Ariston of Chios" in "Life of Zeno," book 7, in Diogenes Laertius 1594, 537: "Similes autem esse orationes dialecticas dicebat aranearum telis, quae etsi quid artificiosum indicare videantur, inutiles sunt Laertius." This edition is in the La Flèche library. English translation from Diogenes Laertius 2005, vol. 2, 161 (book VII, ch.2).

[23] Ovidius Naso 1665, 133: "Antonius Pius avoit accoustumé de dire que les finesses des Sophistes estoient semblables aux toiles des Araignées où l'on void beaucoup d'artifice, & d'où l'on tire peu d'utilité; & que les Sophistes, c'est à dire ces Philosophes dont la science consiste en de vaines subtilitez, estoient au regard des sages ce qu'estoit Arachné au regard de Pallas." The 1655 print of this text is at La Flèche.

good *ingenium*. As far as the invention of the *res* of text was concerned, according to Bade glossing Quintilian, the *ingenium*, fueled by a good memory, involved the ability to perceive similarities easily.[24] Caussin for his part further defines quick-wittedness (*sollertia*, a quality of the *ingenium*) as innate topical mastery, that is, the ability to draw good arguments from the places of arguments, the *loci*: "The tenth source of invention: reason, and quick-wittedness in handling the places of arguments."[25] The simile of subtlety in de Bourgogne's commentary fits both Bade and Caussin's account of ingenious topical invention. When it comes to inventing the right expression (*verba*), the *ingenium* as the ability to perceive similarities fuels the making of metaphors. In the La Flèche textbooks, Aristotle discusses metaphor as an instance of diction in the *Poetics* and notes that the ability to coin metaphors is the prerogative of particularly supple *ingenia*, as it cannot be taught.[26] This view fostered pedagogical resistance: thus Beni sends his reader back to Aristotle's *Rhetoric*—where making metaphors is discussed within the context of urbane, witty uses of language alongside neologisms and jokes—and notes that the *novator* Francesco Patrizi faulted Aristotle for denying that industry (*industria*, *diligentia*) could allow one to learn how to coin metaphors.[27]

In de Bourgogne's text, subtle rhetorical invention first amounts to argumentative flimsiness and overall weakness of composition (*sub-tela* also means "design"): sophisms are not rhetorically or logically sturdy enough to underpin a truly persuasive speech. De Bourgogne consciously steers away from a standard interpretation of the rhetorical *ingenium* as ability to generate *copia*: he explicitly disregards the generative etymological root of *ingenium* (from *genere*) in demanding not a "pregnant" wit (that of mothers), but one as sharp as a blade—rather than a performance of one's power to invent pleasing analogies, wit in rhetorical invention should demonstrate one's ability to hold steadily one's argumentative grounds. In this process, the sword has replaced the spider as the *analogon* of the rhetorical *ingenium* and glosses an implicit, commonplace qualifier of one's wit as sharp (*acutus* or *acre*)—a qualifier which features, for example, in Caussin's definition of *ingenium* as the intellect, inasmuch as it is particularly sharp and fast (*acre et sollers*).[28] Subtlety results in overly sharpening and therefore chipping the blade of one's argument, similes or metaphors, which end up lacking substance. De Bourgogne's attack on subtle wit in rhetorical invention therefore condemns *la pointe*, the over-refined, obscure and ultimately inane conceit—a condemnation which belongs in a

---

[24] Bade 1528, fol.VIII^{v}.

[25] "Decimus fons inventionis: ratio, et solertia in locis," in Caussin 1619, 137 at La Flèche.

[26] Beni 1613, 465r (CXXII): "id quod unum non aliunde quaesitum sit opportet, versatilisque ingenii indolem prae se ferat."

[27] Beni 1613, 466v. The *novator* and critic of Aristotelianism Francesco Patrizi (1529–1597) published a *Discorso della diversità de' furori poetici* (*Discourse on the Variety of Poetic Furor*) as well as a full treatise on poetics. See Patrizi 1553 and idem 1586.

[28] Caussin 1619, 104.

longstanding rhetorical and poetic reflection on the danger of metaphorical obscurity and the need for proportion between the two elements that the metaphor collates.[29]

De Bourgogne thus reassesses rhetorical ingenuity in significant ways: rather than the effortless, brilliant source of bold arguments and conceits ultimately refined into inanity, he contends that the *ingenium* should generate argumentative sturdiness. Yet the architectonic, variegated similes of subtlety—the spider web, the blade—*are* the argumentative warp on which this whole passage is built, and their persuasive punch owes much to those ingenious stylistic processes that de Bourgogne condemns: if you ask for a sharp *ingenium*, expect the *pointe* of lapidary conceits. In this strongly prescriptive passage, however, the blade of wit cuts through, as it helps to choose the right Petrarchan commonplaces and the right order in which to string them: rather than a performance of copious rhetorical invention, the reader witnesses a display of rhetorical judgment and disposition. Rhetorical wit thus proves a *critical* tool rather than an inventive one, and such critical *ingenium* ought to be disciplined in order to work well. De Bourgogne finally substitutes a pedagogical definition of the *ingenium* as teachable to that of the *ingenium* as outstanding, inborn nature that underpins the *vanitas* of subtle, inventive wit in rhetoric. Talent *can* be taught, nature ought to be disciplined, art is *not useless*: after all, Arachne was merely an ungrateful student, and, in a concluding statement reminiscent of Huarte, "applied to the good arts, the *ingenium* can be a useful tool." In one final surprising reversal in this first section, de Bourgogne opposes industry and toil (*industria* and *labor*): industry helps cultivate one's nature, while the vain quest for subtlety means the labor of racking one's brain to produce sterile conceits. The spider web emblematizes both useless toil and conceited subtlety in the subsequent Jesuit emblematic tradition: the two devices featuring it in Daniel de la Feuille's *Devises et Emblèmes* (*Devices and Emblems*), a best seller of early modern emblematics, are characteristic in this respect.[30] Device 9 echoes Laertius and laconically stigmatizes the excessive subtlety of scholastic dialectic by mocking the disputational method: "A spider web. Yes and no (stated in Latin, French, Italian, Spanish, Dutch, English and German)" (Fig. 6.3).

As for device 12, it features a torn spider-web that the spider is busy repairing, that is, a perfect instance of endless, useless toil: "A spider toiling over repairs of its broken web, 'I mend my interrupted work'" (stated in Latin, French, Italian, Spanish, Dutch, English and German) (Fig. 6.4).

Following the condemnation of subtlety in rhetorical invention as argumentative flimsiness, logical, epistemological and mixed-mathematical ingenuity come under

[29] See Beni 1613, 457r-466v, and among the moderns, Gerard Vossius, "De metaphoris longe petitis," chapter 8 in Book 4, "De metaphora," in Vossius 1643, 102–3, at La Flèche.

[30] This book *does not feature* in the La Flèche library: de la Feuille (1640–1709) was a Huguenot who fled his native France to find refuge in Amsterdam. His *Emblèmes et devises* "can ... be seen and used as a touchstone of the sixteenth and seventeenth century emblematic tradition." (Daniel de la Feuille introduction page, Emblem Project Utrecht: https://emblems.hum.uu.nl/f1691_introduction.html)

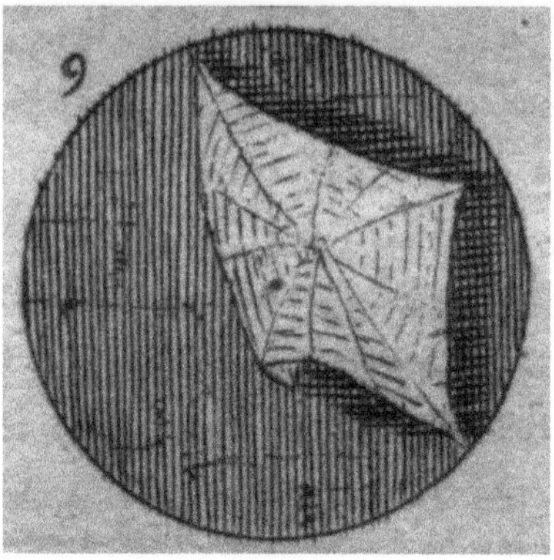

Fig. 6.3 Device 9: "une toile d'araignée, ita & non, Ouy & non, si e no, Si y no, Ia en Neen, Yes and no, Ya und nein," in Daniel de La Feuille, *Devises et Emblèmes* (Amsterdam: Daniel de la Feuille 1691), 33. (Courtesy of the Utrecht Emblem Project)

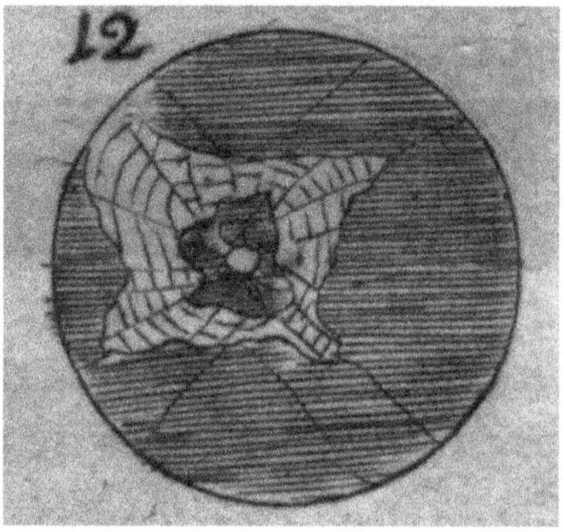

Fig. 6.4 Device 12: "Une Araigne qui travaille à refaire sa toile rompuë, *Interrupta retexam*, je racommode mon travail interrompu, *Raccommod le mie rotte trame,* confietto mi tela, Ick verstelle my gestaeckt werk, I repair my work that was left, Ich verbessere meine gelassene arbeit." In Daniel de la Feuille, *Devises et Emblèmes* (Amsterdam: Daniel de la Feuille, 1695), 33. (Courtesy of the Utrecht Emblem Project)

attack in the next section of de Bourgogne's commentary on the *Ingenium* emblem. While he has picked on rhetorical subtlety as tenuous, he satirizes logical, epistemological and mixed-mathematical subtlety as pedantic complexity leading to dissent in this second moment, thus focusing on the spider web as an emblem of sterile intellectual knottiness or complexity.

6.4 Philosophical and Mixed-Mathematical Ingenuity: Paradoxes of Ignorant Science, Pedantic Precision

De Bourgogne's next target is the very possibility of devising those "solid," or useful arguments in philosophy—arguments that would be non-contradictory, and yield certain knowledge. De Bourgogne thus tears into logical and scientific ingenuity in this second moment of his commentary. His first attack on rhetorical ingenuity as the tenuous fabrication of similes and metaphors has aptly ended in a text devoid of images, the conceit of the blade having vanished into gnomic statements. By contrast, a striking image of philosophical discord concludes his vivid, epigrammatic diagnosis of logical and epistemological failure: that of the loquacious wheels of asynchronic clocks. This classical commonplace taken from Seneca the Younger's Menippean satire, the *Apolocolyntosis* or *Gourdification of Claudius*, also features in the preface of Petrarch's *Remedies*, alongside the oxymora of the opening set of rhetorical questions: "who will count Sophists's wars? who, the knots of their unsolvable problems?"[31] Flares of sharp, satirical style that are borrowed from Petrarch, himself borrowing from Seneca, thus open and conclude this second attack.

6.4.1 Structure: Philosophical Commonplaces

In contrast with the composition of the preceding section, rhetorical invention and judgment do not mean excerpting dazzling Petrarchan conceits to set them carefully in analytical paraphrase; in this second moment of the commentary, Petrarchan excerpts instead frame this satirical and self-reflexive dramatization of skeptical, anti-scholastic commonplaces. In a long variation on the skeptical and patristic paradox of ignorant science, the text moves seamlessly from a condemnation of dialectical hair-splitting in disputation (logical ingenuity) to an attack against the possibility of causal, scientific demonstration in natural philosophy in particular, and the quest for apodictic principles in general (epistemological ingenuity). *Copia*—previously denounced as a product of the verbose ingenuity of a pregnant wit in rhetoric—take the form of impossible accountancy in this section: too many sophist wars, defenses of falsehood to be counted, too few proximate causes and proper accidents to deserve numbering. This underlying mathematical theme comes to the fore in the mixed-mathematical and mechanical simile of philosophical disputes as dissenting clocks: time computation—either based on celestial observations and reliant on gnomons and astrolabes, or, based on the mechanics of pendulum clocks and geared escapement clocks—epitomize mixed-mathematical subtlety for the early moderns. Failed logic, epistemology in need of reform, and a concluding mechanical simile of philosophical dissent connected by a running, mathematical metaphor: the condemnation of ingenuity in this second moment of de Bourgogne's

[31] Seneca 1632 606^{v}, 1913, 2: "Horam non possum certam tibi dicere, facilius inter philosophos quam inter horologias conveniet" (What hour it was I cannot certainly tell; philosophers will agree more often than clocks). The 1632 edition is at La Flèche. The rhetorical questions as well as the comparison between dissonant clocks and discordant philosophers can be found in Petrarch 1581, vol. 2, 105.

text highlights the concomitance between the emergence of the culture of ingenuity and that of the "New Science."[32]

6.4.2 From Skepticism to Satire: The Condemnation of Logical and Epistemological Ingenuity

Dialectic as the epitome of logical subtlety or ingenuity is de Bourgogne's first target. Dialectic assumes consent in that it is meant to allow for the confrontations of endoxical views in debate—views shared by all the people, or most of them, or the wise among the people, or most of them.[33] Such sifting through of opinions imperceptibly paves the way to science in that it often leads to the discovery of apodictic principles; dialectic was therefore commonly called "the road to science."[34] Yet in de Bourgogne's text, dialectic ends in dissent because its scholastic jargon fails to assume a shared, commonsensical use of language; nor can it be the road to truth because it is a sophistical practice whose knotty arguments are often absurd. These skeptical arguments end up in satire: pulverized by skepticism, the epistemological content of de Bourgogne's text—its *res*—dissolves into satirical laughter and its witty display of puns, that is, pure *verba*. Early modern skepticism was indeed well represented in the La Flèche library, which featured a 1557 Basel edition of Pico della Mirandola's complete works, a 1569 and a 1621 edition of Sextus Empiricus' *Against the Mathematicians* and *Pyrrhonian Hypotyposes,* as well as the 1595 edition by Marie de Gournay of Montaigne's *Essais*, whose "Apologie de Raymond Sebond" (Essai 2.12) popularized Pyrrhonism in the vernacular.

Echoed in Laertius' attack against the flimsy arguments of sophists mentioned above, Sextus Empiricus equates sophisms—*sophismata* or fallacies that only bear the appearance of truth—with dialectical reasoning themselves in his *Pyrrhonian Hypotyposes*, and mocks the *useless* absurdities they might lead to: water is black, snow is frozen water, therefore snow is black.[35] Skepticism thus allowed for the identification between sophistry and dialectic, and between false arguments and subtle ones. Such identification fed into Veronese's allegory of dialectic, depicted as

[32] For significant accounts of the importance of ingenuity in the emergence of the New Science, see Gal and Chen-Morris 2013. For specific examples, e.g. Francis Bacon, see Jalobeanu 2014; Lewis 2014, 113–163; on Galileo, see Reeves 2008.

[33] Aristotle, *Topica*, I.100b22–24.

[34] The definition of dialectic as either the road to science (through disputation) or the queen of science (because its procedures guarantee logical rigor) are scholastic commonplaces which feature in the scholastic textbooks mentioned by Descartes in his correspondence: Descartes to Mersenne, 30 September 1640, in Descartes 1996, III, 185. See for example Eustachius a Sancto Paulo 1609, 18, which mentions dialectic as the queen of sciences; among the Coimbra commentators, Fonseca 1578, 334–36: this chapter dedicated to the dialectical argument and the invention of its middle term points out the overlap between opinion and science, dialectical and demonstrative syllogisms; "probable" opinions may well be proven true through debate.

[35] On sophisms, see Sextus Empiricus 1569, 487–8.

Fig. 6.5 Paolo Veronese, *La Dialettica*, (oil on canvas 150 × 220 cm), 1575–77. Venice, Ducal Palace, Sala del Collegio. (Credit: Didier Descouens CC BY SA licence)

Arachne holding a spider web on the ceiling of the "sala dell collegio" in Florence (Fig. 6.5).[36]

After dialectic, the epistemological procedures of *scientia* proper, that is, demonstrative, causal knowledge of what is general—are De Bourgogne's second target in this passage, namely the apodictic discovery of principles in any given disciplines, and that of the middle term in the demonstrative syllogism. The discovery of the principles of a given discipline are a matter of intellectual intuition or apprehension in the Aristotelian theory of the demonstration that features in the *Posterior Analytics*;[37] yet the sectarian variety of such first principles and the dissent this variety fuels was a skeptical commonplace.[38] It is within that skeptical tradition that the intuition of such principles is defined not so much as the apprehension of truth, but as a witty exercise designed to generate pleasing fictions. Sextus Empiricus thus equates the most ingenious philosophers with good rhetoricians championing falsehood—echoed in de Bourgogne's "Defenses of falsehood" (*falsi patrocinium*)—because their main concern is not so much the discovery of truth, but the rhetorical ability to generate belief (*ad fidem faciendam*) by inventing credible lies or

[36] I am grateful to Astrid Zenkert for calling this painting to my attention.

[37] Aristotle, *Posterior Analytics*, II.19.100b5–17, repeated in Eustachius Sancto Paulo 1609, 226 and in Fonseca 1578, 223.

[38] The first part of Sextus Empiricus's *Adversus Mathematicos* amounts to a doxographical survey of the contradictory views on the nature of truth held by Greek philosophical sects from pre-Socratic philosophers to the Hellenistic Stoics. Sextus Empiricus 1569, 119–184.

fictions.[39] The principles of natural philosophy adopted by various philosophical sects—from Epicurean atoms to Stoic pneumas, Montaigne ironically reminds us in the "Apologie"—are just such credible fables.[40] Read through the skeptical lens, *scientia* and poetics coalesce: the Ovidian fable of Arachne therefore proves a suitable emblem for scientific discourse too.

(Academic) skepticism itself is swept in de Bourgogne's Pyrrhonian condemnation of philosophy: surely, the skeptical "I know nothing" offers some sort of certainty and is therefore a path to truth. It is difficult not to hear in this ironic suggestion a satire of the Cartesian method, which does indeed take the skeptical "I know nothing" as it revolutionary epistemological starting-point; the *Discourse on the Method* had been published the year before de Bourgogne's *Compass*. Yet, rather than the Cartesian, analytical "chains of reasons" modelled on mathematical reasoning, De Bourgogne goes on to denounce the epistemological standards of the Aristotelian demonstration. In a truly demonstrative syllogism, the middle term should express the proximate, efficient or even final cause. The ability to discover quickly the middle term in a demonstration is defined as quick-wittedness (ἀγχίνοια) in the *Posterior Analytics*, a term translated by the Latin *ingenium* in scholastic textbooks.[41] Yet such proximate final or efficient cause often amount to either a *proprium* (an essential attribute) or even the specific difference (the defining essential attribute) of the subject of the first premise in the syllogism. In other words, the inferential force of the Aristotelian, causal demonstration often derives from the fact that the middle term makes explicit either shared predicates in the definition (genus + specific difference) or, in the absence of definition, the description (which includes the list of essential attributes) of the major and minor terms in the first two premises; that is, the demonstration merely "unpacks" a definition or description. However, de Bourgogne takes stock of the humanist and skeptical critiques of such requirements, which also resounded in early modern scholastic textbooks: finding such specific differences or essential attributes is exceedingly difficult; we only know very few of them.[42] The

[39] Sextus Empiricus 1621, 197: "Fortasse ergo in philosophia quoque, ex ii qui quaesierunt veritatem, ii qui sunt quidem maxime intelligentes, eo quod valeant ingenio, videntur esse apti ad persuadendum fidemque faciendam, etiam si falsum defendant" ("it could therefore be that in Philosophy as well, among those who would have been investigating the truth, those who are indeed the most clever, because they have a good *ingenium*, seem to be the best at generating belief, even when they argue in favour of falsehood").

[40] See the ironic survey of the endless diversity of natural-philosophical first principles in Greek philosophy in Montaigne's "Apologie de Raimond Sebond," II,12 in Montaigne 1595, 352–53.

[41] Aristotle, *Posterior Analytics*, I.34, 89b10–11. See also du Laurens 1600, 535 which features in the La Flèche library: "Ingenium autem appellat ἀγχίνοια, id est, solertiam, quae definitur 'ευστοχια, id est, inveniendi et conjiciendi promptitudo" (Indeed he calls *ingenium* ἀγχίνοια, that is, quickness of mind, which is defined as 'ευστοχια, that is, speed in discovering and inferring). Du Laurens glosses Niccolò Leoniceno's translation of ἀγχίνοια as *ingenium* in his Latin text of Galen's *Ars parva*, which states that quick-wittedness (*ingenium*) is dependent on subtle brain matter. Galen 1508, 6^{v}.

[42] Occurrences of the (genus + specific difference) definition were far and few, as the humanist Rudolph Agricola pointed in his 1515 *De inventione dialectica libri tres*. The impact of such rarity on the true demonstration "propter quid" (by proximate causes) was noted by Eustachius a Sancto

stringent epistemological requirements defining a good scientific demonstration in scholastic terms mean that such demonstrations are very rare, and true science, very limited. De Bourgogne concludes his attack on epistemological forms of ingenuity with a satirical and highly reflexive flourish: epistemology takes itself as its object in the suggested disputation of "whether or not the very small numbers of known *propria* qualifies as a number," which mocks an existing, epistemological type of scholastic disputations.[43]

Their *res* corroded by skeptical laughter, subtle dialectic and ingenious *scientia* end up, once again, as a rhetorical matter of style. Dialectical subtlety equated with sophistical syllogisms turned jokes had a well-known precedent in Michel de Montaigne's "bacon and thirst" sophism: "Get rid of those thorny problems of dialectics—they are excessive: our lives are never amended by them [...]. What will he do when they harass him with some sophistical syllogistic subtlety: bacon make you drink; drinking quenches thirst: therefore bacon quenches your thirst? Let him simply laugh at it: it is cleverer to laugh at it than to answer it."[44] Skeptical laughter easily turns sophistic syllogisms and the jargon of dialecticians into jocular nonsense and puns: Pyrrhonian skeptical arguments land their reader into satire. The same skeptical and ultimately satirical vein instantiated in Montaigne takes the form of the anti-scholastic pun in the opening rhetorical questions borrowed from Petrarch's preface to the *Remedies*. The knotty problems that cannot be undone (*inenodabiles nodos*) thus mocks the *enodatio* (elucidation of meaning) section in a scholastic "question" (*quaestio*): *enodationes* were meant to clarify the terms of the *quaestio* before delving in the subject-matter proper.[45] In a similar fashion, the skeptical attacks on *scientia* end in pure wordplay. Stylistic brevity in philosophical argument is de Bourgogne's final advice in this section: there is no need to stoop even to the lowest, weakest, most rhetorical form of proof, the example, to bring the whole condemnation of demonstration home, as examples breed *copia*, whereas emblematic satire demands economy of means. Such economy of means finds its

---

Paulo: "Sunt autem rarissimae eiusmodi demonstrationes, quia accuratae rerum definitiones ut plurimum nos latent." (These sorts of demonstrations are indeed very rare, because precise definitions of things are usually hidden to us). Eustachius a Sancto Paulo 1609, 224. For the skeptical condemnation of the shortcomings of demonstration, see Sextus Empiricus 1621, 228–56.

[43] Eustachius a Sancto Paulo thus dedicates the first five "preliminary questions" (*quaestiones proemiales*) of the *Summa logica* to establishing the epistemological status of the discipline he is about to expose, such as "Quomodo, seu quot in partes dividenda sit dialectica." (How, or rather, into how many parts should dialectic be divided). Eustachius Sancto Paulo 1609, 1–9.

[44] Montaigne, "De l'institution des enfans," (I.26) in Montaigne 1595, 97: "Ostez toutes ces subtilitez espineuses de la Dialectique, dequoy nostre vie ne se peut amender [...]. que fera-il si on le presse de la subtilité sophistique de quelque syllogisme: le jambon fait boire, le boire desaltere, parquoy le jambon desaltere? Qu'il s'en mocque. Il est plus subtil de s'en moquer que d'y repondre."

[45] "Enodatio" is thus a well-defined section in the commentary structures of the Coimbra commentaries on Aristotle. See for example the *enodatio* of the controversy on the nature of the heavens in Aristotle 1600, 67, which starts by highlighting the polysemy of the words "immortal" and "incorruptible."

perfect expression in the oxymoronic pun "solid science/stupid science," which turns good science (*solidus*) into a crass, or crude one (*stolidus*), since *scientia* merely relies on the trite, obvious unfolding of known definitions. The pun thus offers a witty take on the longstanding commonplace of ignorant science, a requirement of true faith in the patristic tradition,[46] a condemnation of rationalist claims in the skeptical, philosophical one.[47]

6.4.3 Dissenting Clocks and the Defense of Commonsense: Against Mixed Mathematical Ingenuity

Another satirical commonplace leads de Bourgogne from deriding epistemological ingenuity to mocking mixed-mathematical subtlety: it is also at this point that the stylistic ingenuity of the text shifts from verbal to visual wordplay, and provides a witty because allusive commentary on the spider emblem. De Bourgogne's initial satirical simile is borrowed from Seneca the Younger's Menippean satire, *The Gourdification of Claudius*: it allows him to mock mixed-mathematical and mechanical ingenuity or subtlety, epitomized in clock-making. "Philosophers will agree more often than clocks" Seneca says, pointing out that precise time-keeping was even less likely to happen than philosophical consensus.[48] Such likelihood is altogether absent from de Bourgogne's take on the same simile: philosophers, like clockmakers, always stray from the absolute rule of truth they all individually claim to utter, yet all disagree about. A striking example of such disagreement about time-keeping features in the La Flèche library: *L'usage de l'un et l'autre astrolabe, particulier et universel...corrigez, augmentez, et remis en meilleur ordre* (*The Use of This and That Astrolabes, The Particular One and the Universal One...Amended, Augmented, and Put Back into Better Order*). Amended by the printer Jacques Moreau, this is the fourth, 1625 edition of the 1545 *L'usaige de l'astrolabe, avec un traicté de la sphere* (*The Use of the Astrolabe, With a Treatise on the Sphere*) by Dominique Jacquinot, one of the very first treatises in the vernacular to describe the constitutive parts of the astrolabe and its use. Every single page in the 1625 edition is heavily annotated by Moreau, who relentlessly pinpoints both mistakes made by his predecessor David Robert, who amended the third, 1617 edition of the book, and

---

[46] The paradox thus structures one of Chrysologus' sermons: "there the wise men of the time he made stupid, the observers of the world he taught no to see a thing, the teachers of science he made devoid of science, the enquirers in all things he sent back ignorant." (Hinc sapientes saeculi fecit stultos, speculatores mundi docuit nihil videre, professores scientiae fecit scientiam non habere, inquisitores rerum omnium dimisit ignaros), Chrysologus 1845, 0470D.

[47] On this point, Sextus thus turns a paradox into an attack *ad hominem* against Chrysippus the Stoic who stated that "the fool is ignorant of all things": since Chrysippus was, like all dogmatists and human beings, a fool, his statement does not necessary hold and ascertaining a criterion of truth is not possible. See Sextus 1621, 182.

[48] See above n. 32 for the full text and translation.

discrepancies between the 1545 celestial computations of Jacquinot and his own.[49] Time-keeping reliant on the mechanical instantiation of celestial motions had been a paradigm of rationalism since Cicero: in de Bourgogne's commentary, it becomes yet another instance of fruitless pedantic dissent.[50]

In the early modern period, the mechanical arts became the touchstone of ingenuity.[51] Nicolas Caussin thus noted in his rhetoric textbook that good *ingenia* excelled in mechanical arts and techniques, and went on to illustrate his point by listing all sorts of marvels of mechanics and craft—the automaton turtledove of Architas, the fall of Phaeton engraved on the gemstone of a signet ring, the whole story of Christ's passion finely delineated on one's nails.[52] Chief among witty engineers was Archimedes: his fabled glass copy of all celestial motions was the model against which Angelo Poliziano assessed the ingenious wonder of Lorenzo della Volpaia's clock which prompted much curiosity in Florence at the end of the fifteenth century.[53] Indeed, the joint early modern development of mixed mathematics and mechanics contributed to making the clockmaker one paradigmatic figure of mechanical wit, as Cardano highlighted in his *On Subtlety*.[54] Cardano thus lists among the wondrous artificial inventions demonstrating subtlety and improving on those of the ancients that of the mechanical clock with its cogwheels.[55] Cardano also reminds us that, as a mixed-mathematical art, the making of time-keeping devices belongs with architecture: the ninth book of the *De architectura* touches on sundials. In the emblem, poised in her niche at the center of a rectilinear net in the architectural display of this grand urban perspective, the spider evokes a sundial: her motif finds a structural echo on the two clock towers in the background. This development on mechanical and mixed-mathematical subtlety in the commentary thus invites the reader to exercise his own wit by reading in the emblematic spider a visual metaphor of ingenious time-keeping—a metaphorical reading which also puns astronomical, technical jargon: the *rete*, the last movable disc on top of the astrolabe which represents the main constellations of the zodiac and especially their most radiant stars, was also called the spider (Fig. 6.6).

---

[49] Jacquinot's original text was a luxury manuscript made for Catherine de Medicis (Condé library, Château de Chantilly, Ms. 323). See Jacquinot 1545, idem 1617 and idem 1625.

[50] Cicero, *De natura deorum* II.38, in Cicero 1566, IV, 101–7.

[51] See Verrin 1993.

[52] Caussin 1619, 104. Caussin lifts verbatim this list of ingenious marvels of mechanics and crafts from the book of his fellow Jesuit writer Jacopo Pontano, See Pontano 1620, vol. 3, 72.

[53] Angelo Poliziano, "Letter to Francesco della Casa, 4 August 1484" *Epistolae* IV.8 in Poliziano 1498, f.I^{r}-^{v}. For a detailed account of the reception of this mechanical wonder in the Ficinian context, with a transcription and translation of the letter, see Toussaint 2002.

[54] In the *De subtilitate*, the clock's pendulum features as an example of double motion (by density and rarity), a clock which marks the hour both with its arm and with a strike is noted as an example of outstanding craftsmanship, and clocks reliant on wheels rather than suspended pendulum are listed as newly discovered instances of subtle arts: Cardano 1642, 10a, 185a–b, 390b.

[55] Cardano, 1642, 387a.

Fig. 6.6 "L'araigne du Zodiaque" in Dominique Jacquinot, *L'usaige de l'astrolabe avec un traicté de la sphere* (Paris: Jean Barbé, 1545), 19. (Courtesy of the History of Science Museum, University of Oxford)

The clock simile of philosophical disagreement sweeps in the same satirical condemnation philosophical and mixed-mathematical ingenuity for their pedantic, rationalist demands, be they impossible criteria of demonstrative truth or the quest for numerical accuracy. The clocks (sundials, astrolabes, but also pendula) of mixed-mathematics, thus provide the emblematic spider of the *ingenium* with yet another meaning, that of pedantic precision in the mixed-mathematical and mechanical endeavors which characterized the emergence of the "New Science."

6.5 Style and the Commonsensical Discipline of Wit

De Bourgogne's emblem relentlessly castigates ingenuity as subtlety under all its guises old and new: in rhetorical invention, in scholastic disputations and demonstrations, or in mixed-mathematical computations.

The castigation itself, however, promotes other forms of ingenuity. Ingenuity as a stylistic, witty practice is the most apparent. In rhetoric, against the flimsy *copia* of metaphors, favor the sharp, aphoristic conceits of critical judgment: the former is useless brain-racking, the latter is useful discipline of one's *ingenium*. In dialectic and philosophy, dissolve the complexity of subtle arguments with skeptical laughter: satire as a witty literary genre and style upholding brevity shapes this moment of the commentary which defines the ingenious philosopher as a good story-teller, and turns claims to *scientia* into puns. In mixed mathematics, give up on the pedantic quest for precision, and revel in the polysemous visual pun of an emblematic spider denoting a variety of time-keeping devices.

These stylistic expressions of wit are meant to bring home the lesson that one's *ingenium* as wit ought to be disciplined because it is integral to one's *ingenium* as temperament and moral nature. The appropriate expressions of wit which indicate a good temperament are pedagogical docility, sound judgment, and commonsense. Thus in rhetoric, subtle wit (*ingenium*) as the rhetorical ability to craft metaphors is useless and ought to be disciplined by one's nature (*ingenium*) as docile site of good learning. Similarly, logical, epistemological, and mixed-mathematical ingenuity only foster sterile dissent, and ought to be reined in by a collective, doxical form of knowledge uncluttered by pedantic attempts at precision: the common man demonstrates that he is not puzzled by the experience of contradiction. Good sense, or commonsense—*ingenium* as inborn nature, but also collective identity—should silence the excesses of logical, epistemological and mixed-mathematical subtlety—the *ingenium* of scholastics or *novatores* alike.

That this discipline should happen through the ingenious means of witty genres and their related stylistic expressions (the emblem, satire, the conceit, puns) in de Bourgogne's text is a stark reminder that the *ingenium* relates, par excellence, to the invention and manipulation of mental representations. De Bourgogne's emblem is replete with striking images, be they the architectural vista and spider web of the picture itself, or the conceits and similes which punctuate the text: wit as a chipped blade; philosophers as dissenting clocks; the speed of a mind crashing onto the gates of heaven. The *ingenium* in the Huartian and Cartesian sense of the embodied site where the making and manipulation of images takes place, either towards generating knowledge (wit), or towards moral and religious disciplining (inborn nature) is therefore not only the topic of de Bourgogne's lesson, but its very medium. In this respect, the culture of ingenuity and its ambiguities encapsulated in De Bourgogne's emblem offers a new take on a very old, Aristotelian philosophical view: that there is no thinking without images.[56] That lesson was still very much central to the first methodical attempt of one of La Flèche's pupil, the *Regulae ad directionem ingenii* (*Rules for the Direction of One's Native Intelligence*) of René Descartes.[57]

[56] Aristotle, *De anima*, III.7, 431a16.

[57] The *Regulae*, left unfinished, were in all likelihood written in the 1620s. There Descartes defines the *ingenium* as follows "propie autem ingenium appellatur [vis] cum modo ideas in phantasia novas format, modo jam factis incumbit" (this power is properly called ingenium either when it generates new ideas in the phantasy, or when it applies itself to those already made there). Descartes 1996, X, 416. See Sepper 1996.

Bibliography

Agricola, Rudolf. 1515. *De inventione dialectica libri tres*. Louvain: Thierry Martin.

Alciato Andrea. 1602. *Emblemata cum commentariis*, ed. Claude Mignault. Paris: Jean Richet.

Aristotle. 1600 [1592]. *Commentarii in quatuor libros De caelo, meteorologicos, parva naturalia*, ed. collegium Conimbricenses. Cologne: Lazarus Zetner.

Bade, Josse. 1528 [1516]. *Commentarii Io. Badii Ascensii in M. Fabii Quintiliani institutiones oratorias*. Lyon: Josse Bade.

Ballestra Puech, Sylvie. 2006. *Métamorphoses d'Arachné: l'artiste en araignée dans la littérature occidentale*. Geneva: Droz.

Beni, Paolo. 1613. *In Aristotelis poeticam commentarii*. Padua: Francisco Bolzetta.

Camerarius, Joachim. 1590. *Symbolorum et emblematum ex re herbaria desumptorum centuria una collecta*. Leipzig: Philipp Vögelin.

Cardano, Gerolamo. 1642 [1556]. *Les Livres de Hierome Cardanus médecin milannois intitulez. De la Subtilité, et subtiles inventions*. Trans. Richard Le Blanc. Rouen: Vve du Bosc.

Caussin, Nicolas. 1619. *Sacrae et humanae eloquentiae parallela*. 2nd ed. Paris: Sébastien Chappelet.

———. 1623 [1618]. *De Symbolica aegyptorum sapientia*. Cologne: Johann Kinck.

Chrysologus, Petrus. 1845. *Sermones*, ed. Jacques-Paul Migne (Patrologia Latina 52). Paris: Migne.

Cicero, Marcus Tullius. 1566. *M. Tulli Ciceronis Opera omnia*, ed. Denis Lambin, 4 vols. Paris: J. Du Puys.

———. 1583. *In M. Tulii Ciceronis De Rhetorica volumen primum Aldii Mannucii commentarius*, ed. Aldus Manutius. Venice: Aldus Manutius.

de Bourgogne, Antoine. 1639. *Mundi lapis lydius, sive vanitas per veritatem falsi accusata et convicta*. Antwerp: Vve Cnobbaert.

———. 1643. *Des wereldts proef-steen of te de ydelheydt door de waerhyed beschuldight ende overluyght van valscheydt*, ed. and Trans. P. Geschier. Antwerp: Cnobbaert.

de la Feuille, Daniel. 1691. *Devises et Emblèmes anciennes et modernes, tirées des plus célèbres auteurs*. Amsterdam: Daniel de Feuille. https://emblems.hum.uu.nl/f1691_introduction.html

Dekoninck, Ralph. 2016. Jesuit Emblematics Between Theory and Practice. *Jesuit Historiography Online*. https://doi.org/10.1163/2468-7723_jho_COM_192540.

Descartes, René. 1996. *Œuvres de Descartes*, ed. Charles Adam and Paul Tannery, revised 16, 11 vols. Paris: Vrin.

Diogenes Laertius. 1594. *De Vitis, dogmatibus et apophthegmatibus clarorum philosophorum libri X*. Trans. Thomas Albrandino, comm. Isaac Casaubon. Paris: Henri Estienne.

———. 2005. *Lives of Eminent Philosophers*. Trans. R.D. Hicks, 2 vols. Cambridge, MA: Harvard University Press.

Du Laurens, André. 1600. *Historia anatomica humani corporis*. Paris: M. Orry.

Estienne, Robert. 1531. *Thesaurus linguae latinae*. Paris: Robert Estienne.

Eustachius a Sancto Paulo. 1609. *Summa philosophica quadripartita—summa logica*. Paris: Claude Chastellain.

Fonseca, Pedro da. 1578. *Institutionum dialecticarum libri octo*, 2nd ed. Cologne: Maternus Cholinus.

Gal, Ofer, and Raz Chen-Morris. 2013. *Baroque Science*. Chicago: University of Chicago Press.

Galen. 1508. *Galeni medicinalis ars Nicolai Leoniceno interprete quae et ars parva dicitur*. Trans. Niccolò Leoniceno. Venice: Jacob Pentius de Zeucho.

Garrod, Raphaële with Alexander Marr, ed. 2021. *Descartes and the Ingenium: The Embodied Soul in Cartesianism* (Brill's Studies in Intellectual History 323). Leiden: Brill.

Huarte de San Juan, Juan. 1622. *Scrutinium ingeniorum, pro iis qui excellere cupiunt*. Trans. Joachim Caesar. Leipzig: Kote.

———. 1655 [1650] *L'Examen des esprits pour les sciences, ou se montrent les differences d'esprit qui se trouvent parmi les hommes…augmenté suivant la dernière impression d'Espagne*. Trans. Charles Vion D'Alibray. Paris: Guignard et fils.

———. 1663. *Scrutinium ingeniorum, pro iis qui excellere cupiunt*. Trans. Joachim Caesar. Revised edn. Lena: L. Levenhaus, impr. Samuel Krebs.
Hugh of Saint Victor. 1961. *The Didascalicon of Hugh of Saint Victor: A Medieval Guide to the Arts*. Trans. J. Taylor. New York: Columbia University Press.
Jacquinot, Dominique. 1545. *L'usaige de l'astrolabe, avec un traicté de la sphere*. Paris: Jean Barbé.
———. 1617. *L'usaige de l'astrolabe avec un petit traicté de la sphere...corrigé esclarcy et augmenté ...suivant la reformation du calendrier par David Robert*, ed. David Robert. Paris: Jerome de Marnef.
———. 1625. *L'usage de l'un et l'autre astrolabe, particulier et universel...corrigez, augmentez, et remis en meilleur ordre*, ed. Jean Moreau. Paris: Jean Moreau.
Jalobeanu, Dana. 2014. *The Hunt of Pan: Francis Bacon's Art of Experimentation and the Invention of Science*. Bucharest: Zeta Books.
Jouvancy, Joseph de. 1690. *Ratio discendi et docendi*. Paris: Barbou frères.
La Perrière, Guillaume de. 1544. *Théâtre des bons engins*. Paris: Denis Janot. http://www.emblems.arts.gla.ac.uk/french/books.php?id=FLPa
Le Moyne, Pierre. 1649–50. *Les Poésies du P. Pierre Le Moyne de la Compagnie de Jesus: suivi de Devises héroïques et morales*. Paris: Augustin Courbé.
———. 1666. *De l'art des devises, avec diverses recueils de devises du mesme autheur*. Paris: Sebastien Cramoisy.
Lewis, Rhodri. 2014. Francis Bacon and Ingenuity. *Renaissance Quarterly* 67-1: 113–163.
Marr, Alexander, Raphaële Garrod, José Ramón Marcaída, and Richard J. Oosterhoof. 2018. *Logodaedalus: Word Histories of Ingenuity in Early Modern Europe*. Pittsburg: Pittsburg University Press.
Montagu, Jennifer. 1968. The Painted Enigma and French Seventeenth-Century Art. *Journal of the Warburg and Courtauld Institutes* 31: 307–335.
Montaigne, Michel de. 1595. *Les Essais*, ed. Marie de Gournay. Paris: Abel L'Angelier.
———. 2003. *Essays*. Trans. Michael Screech. Dover: Penguin.
———. 2004. *Les Essais*, ed. Pierre Villey and Verdun Saulnier, intr. Marcel Conche, 3rd edn. Paris: Presses Universitaires de France.
Ovidius Naso, Publius. 1665. *Les Métamorphoses d'Ovide, traduites en françois par P. Du Ryer avec des explications historiques, morales et politiques sur toutes les fables*. Trans. P. Du Ryer, 2nd edn. Paris: De Sommaville.
———. 1977. *Metamorphoses*. Trans. F.J. Miller, 3rd ed., 6 vols. London: Heinemann.
Patrizi, Francesco. 1553. *Discorso della diversità de' furori poetici*. Venice: Giovanni Griffio.
———. 1586. *Della poetica*. Ferrara: V. Baldini.
Perras, Jean-Alexandre. 2016. *L'Exception exemplaire: inventions et usages du génie (XVIe–XVIIIe siècles)*. Paris: Garnier.
Petrarch, Francesco. 1581. *De remediis utriusque fortunae* in *Operae quae extant omnia*, 4 vols. Basel: Heinrich Peter.
Poliziano, Angelo. 1498. *Epistolae* in *Opera omnia*, vol. IV. Venice: Aldus Manutius.
Pontano, Jacopo. 1620 [1617]. *Attica bellaria, seu literatorum secundæ mensæ, ad animos ex contentione et lassitudine studiorum lectiunculis exquisitis, jucundis, ac honestis relaxandos*, 3 vols. Augsburg: Andréa Asperger, Johann Hertzroy.
Possevino, Antonio. 1603. *Cultura ingeniorum* in *Bibliotheca selecta rationis studiorum*, 2 vols. Venice: Altobellum Salicatium.
Praz, Mario. 1964 [1939]. Emblem, Device, Epigram, Conceit. In *Studies in Seventeenth-Century Imagery*, 2 vols, II, 11–54. Rome: Edizione di storia e litteratura.
Quintilian, Marcus Fabius. 1540. In *Institutiones oratoriae*, ed. Sebastien Gryphius. Lyon: Sebastien Gryphius.
Rapin, René. 1660. *Pax Themidis cum musis*. Paris: Sebastien Cramoisy.
Reeves, Eileen. 2008. *Galileo's Glassworks: The Telescope and the Mirror*. Cambridge, MA: Harvard University Press.

Seneca, Lucius Aenneas. 1632 [1605]. *Annaei Senecae philosophi opera quae exstant omnia a Justo Lipso emendata et scholiis illustrata*. Antwerp: Balthassar Moreti for Plantin.

———. 1913. In *Apocolocyntosis (The Gourdification of Claudius)*, ed. W.H. Rouse. London: Heinemann.

Sepper, Dennis. L. 1996. *Descartes's Imagination: Proportions, Images, and the Activity of Thinking*. Berkeley: University of California Press.

Sextus Empiricus. 1569. *Adversus mathematicos ...Pyrrhoniarum hypotyposeωn*. Trans. Gentian Hervet and Henri Estienne. Paris: Martin Jeune.

———. 1621. *Adversus mathematicos*. Geneva: Pierre et Jacob Chouet.

Toussaint, Stéphane. 2002. Ficino, Archimedes and the Celestial Arts. In *Marsilio Ficino: His Theology, His Philosophy, His Legacy*, ed. M.J.B. Allen, M. Davies, and V. Rees, 307–326. Brill: Leiden.

Verrin, Hélène. 1993. *La gloire des ingénieurs: l'intelligence technique du XVIè au XVIIIè siècles*. Paris: Albin Michel.

Vossius, Gerard. 1643. *Commentariorum rhetoricorum, sive, Oratoriarum institutionum libri sex*. Leiden: Jean Maire.

Chapter 7
Manuscripts as Pedagogical Tools in the Philosophy Teaching of Jean-Robert Chouet (1642–1731)

Ann Blair and Anja-Silvia Goeing

Abstract This chapter analyzes how the physics courses of Jean-Robert Chouet (1642–1731) changed across the twenty years of his career as a professor of philosophy, first at the Academy of Saumur (starting in 1664) then at the Academy of Geneva (1669–86). We compare eight surviving student manuscripts, noting much continuity but also some changes in organization, presentation, and content (in particular a greater attention to the topics of place and extension important to Cartesianism). Teaching by dictating a coursebook to students allowed the professor to adjust his course at every iteration. The students also exercised individual choice in the format, layout, and trappings of their manuscript coursebook, which could include an alphabetical index or decorative elements. The most famous of Chouet's students whose coursebooks survive is Nicolas Fatio de Duillier (1664–1753), who was later a friend of Newton's. He studied philosophy with Chouet in 1678–80 and his coursebook, which unfortunately does not include the section on physics, is exceptionally beautifully kept and illustrated.

Keywords Student manuscripts · Physics courses · Academy of Geneva · Academy of Saumur · Jean-Robert Chouet · Nicolas Fatio de Duillier

---

Warm thanks to the librarians who helped us on our visits to Harvard, Yale, and the Salle Sénebier of the Bibliothèque de Genève. We are grateful to the editors of the volume for inviting Ann to contribute to the conference and both of us to the volume, and for their helpful comments.

A. Blair
Harvard University, Cambridge, MA, USA
e-mail: amblair@fas.harvard.edu

A.-S. Goeing (✉)
Harvard University, Cambridge, MA, USA

University of Zurich, Zürich, Switzerland
e-mail: agoeing@fas.harvard.edu

S. Berger, D. Garber (eds.), *Teaching Philosophy in Early Modern Europe*, Archimedes 61, https://doi.org/10.1007/978-3-030-84621-3_7

In the late seventeenth and early eighteenth centuries, manuscript academic lecture scripts were an important part of a learned written culture that included printed books, pamphlets and broadsides (e.g. theses printed for disputations or as teaching aids), and periodicals. Recent attention to the uses of manuscripts in the handpress era has shown that handwriting was preferred over print as a mode of dissemination in a variety of circumstances, notably to produce a small number of copies at lesser expense or to attempt to limit or control circulation.[1] Manuscripts played an especially important role in academic contexts, since handwriting was (and is still) valued as an aid to retention and gave flexibility to both teachers and students in creating a record of a course. Transcripts of courses from early modern European universities survive in considerable numbers but are hard to study systematically since they are often scattered across many libraries as a result of the movement of students and teachers, and the impact of the modern market in rare materials.[2] We most often become aware of these scripts when they have been published, either at the time by the professor or his students (often just after the latter's death), or in modern editions. Student scripts have been the basis for many published works, from the teachings of Aristotle to seventeenth-century Paris professors, to modern works by Kant, Hegel or Ferdinand de Saussure, although that millennial practice may become superseded by the use of audio- and video recordings of lectures.[3] For the historian student manuscripts offer unique insight into not only the content of courses that were never published, but also the methods of teaching and learning that can be inferred from the surviving manuscripts. Of course those manuscripts cannot capture all the personal and oral interactions between teacher and students in the classroom or outside it, but they give us valuable access to the student experience that complements other sources such as official regulations and curricula and printed theses or treatises.

In this article we use eight surviving student manuscripts from the physics courses of the Calvinist Geneva professor Jean-Robert Chouet (1642–1731) to examine the continuities and changes across nearly twenty years of his teaching and the various decisions of the students who made the scripts. Chouet is best known to

[1] For recent work on manuscripts in the age of print see Hall 2008; Yale 2016; Blair 2015; for functions of manuscripts in the context of higher education see Goeing 2017, 79–89 ("School regulations as scribal publication").

[2] Jacob Schmutz has undertaken the first bibliography dedicated to student manuscripts in early modern France which is currently in progress. See also his study of 176 manuscripts surviving from the teaching at the Convent of the Cordeliers in Schmutz 2008, especially 395ff and appendix 1.

[3] On student course books see Leonhardt 2008; Blair 2008; and the literature cited there. For a case study of lecture scripts put into print see Goeing 2007. There is also a growing literature on student notes entered as commentary on printed texts, notably in the collèges of the University of Paris in the late sixteenth century, which we will not address here. For an entry into this field see Compère et al. 2009; Couzinet and Mandosio 2004; Blair 1991; Grafton 1981.

historians for his "philosophia novantiqua" which sought to bridge Aristotelianism and Cartesianism during his career as professor of philosophy, first at the Huguenot academy of Saumur in France from 1664, then at the Academy of Geneva where he was appointed in 1669.[4] Chouet focused on philosophy throughout his career and declined to take a public stance in the theological debates of the day, which divided Calvinists between liberal and orthodox factions. After exposure as a student to the more liberal environments of Nîmes and Saumur, once back in Geneva he protected his teaching by insisting on the separation and independence of philosophy from theology and rejected a rationalist conception of divine action in favor of a voluntarist one in which God is never constrained by human reason.[5]

Chouet was well connected in Geneva: his father Pierre was a printer-bookseller and his mother Renée Tronchin came from a family of noted intellectuals including two professors of theology—Théodore and Louis Tronchin, Chouet's grandfather and uncle respectively. Chouet studied philosophy in Geneva and Nîmes (to 1662) and theology briefly in Geneva (1662–1663) before focusing on philosophy.[6] On his return from teaching at Saumur, Chouet held positions of responsibility in Geneva first within the Academy (including the post of rector from 1679 to 1681) then in the city government. In 1677 he was elected to the "Council of the 200" and in 1686 to the "Council of the 25" at which point he left the Academy. He was Secrétaire d'Etat and Garde des Archives from 1689 to 1698, Syndic (Council leader) in 1699, 1703 and 1707, and First Syndic in 1711, 1715, and 1719.[7] From 1701 to 1727, he also held the office of Scholarch, with the duty of representing the interests of the Academy and educational program to the city government.[8] In these roles Chouet worked for reform and increased secular control of the Academy by the city but in cooperation rather than confrontation with the Company of Pastors, which remained an ecclesiastical partner in the city government.[9]

---

[4] We are very grateful for the existing scholarship on Chouet notably the crucial work of Heyd 1973, 1982, Sina 2008, 2010, each of which cites further bibliography (see especially Heyd 1983, 19). On Heyd's career see Heyd 2020. On Chouet's nomination in Saumur see Fatio 2015, 130; and on Chouet's nomination in Geneva see Fatio 2015, 202, 205–208. For literature on the Genevan Academy see Goeing 2021.

[5] Heyd 1982, ch. 2, e.g. 86; for this general strategy in Geneva see 165ff. On Chouet's evolution from rationalism to voluntarism see Heyd 1979, 532, 535, 541. On the greater liberalism of the environment in Saumur, see Fatio 2015, 48ff. But Chouet discusses theological questions in his correspondence; see Sina 2008.

[6] From his studies with David Derodon in Nîmes and Kaspar Wyss in Geneva we have a set of theses explictly authored by Chouet, see Chouet 1662.

[7] Budé 1899, 238ff.; Borgeaud 1900, appendix, 636; Santschi 2005; Sina 2010, vol. 1, XIII-L; Fatio 2015, 130, 765, 799, 816.

[8] On the role of scholarch see Borgeaud 1900, 148–152.

[9] Heyd 1982, 178–89.

7.1 Chouet's Teaching

Chouet published little—mainly theses defended by students[10]—but we have multiple manuscripts from his courses on three parts of philosophy—logic, metaphysics, physics. Chouet's courses were delivered orally, probably by dictation, and recorded in student manuscripts which could also circulate to others who were not present.[11] Chouet was reportedly much appreciated by his students,[12] and his favorable reputation may have favored the circulation of these manuscripts. Teaching by oral lecture and manuscript circulation was the norm at French universities as well as in Geneva, as we can observe from the student manuscripts that survive from the courses of the philosophy professors who were near contemporary with and succeeded Chouet.[13] In Geneva Antoine Léger the younger (1652–1719), who studied with Chouet, taught philosophy from 1686 to 1713 then theology until 1719, and was succeeded by his student Jean-Antoine Gautier (1674–1729) who taught philosophy at the Geneva Academy from 1696 to 1723—for each of them we have a number of student scripts too (for an external view of some of these see Fig. 7.7 below).[14] Chouet typically merits a mention in general histories only for the experiments that he carried out for a general public on Wednesdays, reportedly to great acclaim, but he evidently kept these quite separate from his physics courses where he made no mention of them.[15]

From Chouet's teacher and precedessor in Geneva, Kaspar Wyss, originally from Zurich, we have Wyss's copy of his teacher Derodon's course and the autograph manuscript of his own courses in logic and physics.[16] These latter were also published shortly after his death in 1668 by his widow who addressed dedications of the volumes to officials of Zurich and of Geneva respectively, in order to solicit

[10] A student from Saumur, Timotheus Rovierus, published *Theses ex universa philosophia selectae* (Chouet and Rovierus 1667). Heyd lists ten theses presided by Chouet in Geneva, Heyd 1982, 292. Sina also notes a thesis defended by Chouet in 1661 under the direction of Wyss in Geneva, *Disputationum physicarum nona*: Chouet 2010, vol. 1, XXXV and Sina 2008, XX note 26 (Wyss and Chouet 1661). See also Stelling-Michaud 1959, vol. 2, 506 for a *De motu* (Geneva, 1659).

[11] Chouet's courses were in one case at least explicated in a private setting by a recent convert (François Deschamps) who was accused of tendentious conclusions—the fear of which motivated regulations forbidding unauthorized private teaching; Heyd 1982, 167–69.

[12] Heyd 1982, 51.

[13] On student note-taking at Geneva see Blair 2012 (including two images from Bibliothèque de Genève (BGE) Ms Lat. 322, 50–51). For various French examples see Schmutz (note 3); Blair 1993; on Adrien Geffroy's course in 1731–33 in Paris: Firode 2008.

[14] For detailed descriptions of course scripts by these professors (among others) at BGE with "Ms Lat." call numbers see the invaluable Jeger 2016. On Chouet's successors see Dufour 2014, "Gautier, Jean-Antoine;" Fatio 2014, "Léger, Antoine"; and Galiffe 1884, vol. 6, 311–12 (Gautier), vol. 2, 390 (on Léger's marriage). On the succession to Léger and Gautier, see Heyd 1982, 189–97.

[15] See for example Lacki 2007, 233; for the constraints on his ability to create a bridge between the school and the larger intellectual context, see Heyd 1982, 104, 114.

[16] For Wyss's autograph manuscripts see Zurich Zentralbibliothek Ms D184–185 and Car X 162 as reported in Gagliardi and Forrer 1982, 430, no. 1629. On Derodon's metaphysics, see Ragni 2020.

financial help in view of the calamity of his death for herself and her children.[17] Interestingly by contrast we have no autographs by Chouet and none of Chouet's courses were published (until 2010). A number of courses by other professors in philosophy were published in the seventeenth century, either during the teacher's lifetime or after the teacher's death by grateful students or (as in the case of Wyss apparently) needy heirs.[18] One factor which likely worked against posthumous publication in Chouet's case was his exceptionally long life after he ceased teaching. By the time Chouet died in 1731 his last students were already 45 years past their studies with him—indeed Chouet even outlived the two who succeeded him as professors of physics. Chouet could also have published his courses himself although he never did.

Autographs can also survive from professors' teaching –sometimes annotated in preparation for publication (either during the author's lifetime or afterward),[19] but in Chouet's case we have none. We therefore have no additional clues about how Chouet prepared and delivered his courses, whether by speaking from memory or from notes or by reading from a full text. From our modern vantage point it seems difficult to dictate and repeat sentences to students without reading from a complete text, but we know that others taught without a full text. In particular a near-contemporary biographer reported of Petrus Ramus (1515–72) teaching at Paris about one century before Chouet that he relied on "brief notes on a piece of paper which he held in his hand and glanced down at as he was speaking," and "when he returned home he used to jot down in shorthand what he had lectured and commented on," and from those his assistant would write up a full clean copy. By that method there never was a full text autograph preceding the lecture but a text might be generated after the fact by the professor and his assistant (or a student).[20]

Teaching by dictation and manuscript coursebook offered several advantages: the teacher could change his presentation at will from one iteration to the next and expended no extra effort toward publication, and at the same time the students

[17] On Wyss's teaching see Sina 2010, XXVI-LVII. Among the sources are Kaspar Wyss, *Logica quae est cursus pars prima* (Geneva: de Tournes, 1668) and *Physica quae est cursus pars secunda* (Geneva: de Tournes, 1669). Wyss published a number of theses with students in Geneva–the Bayerische Staatsbibliothek owns eighteen such ranging from 1656 to 1668, but also before his move to Geneva a *Dialectologia Sacra* (Zurich: Johann Jakob Bodmer, 1650). See Marti-Weissenbach 2012.

[18] Jean-Cécile Frey published some of his courses during his lifetime; see Blair (1993). The courses of the French Franciscan Claude Frassen were published in his lifetime as part of a concerted effort by his order to promote teaching of their favorite Doctor of the Church, John Duns Scotus, as opposed to the Jesuits and Dominicans who favored Thomas Aquinas. Schmutz 2008, 388–89.

[19] For surviving autographs see the cases of Wyss (note 17), Frey (Blair 1993, 96, 158), Frassen (Schmutz 2008, 432–33)—in the latter two of which the autographs were annotated for publication although Frey's were never published and Frassen's were.

[20] "Hoc autem scire juvabit, hunc Ramo nostro morem fuisse, ut de re qualibet, vel Gorgiae in modum thesi posita dicturo sufficerent notae breves, quas in chartula gerebat, et obliquis oculis dicendo inspectabat." "Reversus domum, ex more, praelecta sibi aut enuntiata Tachugraphois exscribebat." Sharratt 1975, 194–95, 190–91.

would necessarily attend to every word as they wrote the lectures down. The students had great flexibility in making their course books which they could customize according to the expense and time each devoted to it—some course books were small and functional, others more or less decorated. Illustration for explanatory purposes played a minor role in Chouet's physics courses which were heavily textual, but the manuscripts typically show careful attention to their presentation. Many carry trappings we associate with a printed book such as title page, table of contents, alphabetical index, and layout on the page to highlight divisions in the text, including changes in size of script at the beginnings of sections or paragraphs, and in some cases lovely decorative elements. It is well known that early printed books imitated the look and format of late medieval manuscripts, but the influence also flowed in the other direction after a time, in that manuscripts of the early modern period imitated the look and many features of printed books.[21] Some features, like headings, pagination, tables of contents and indexes, facilitated later consultation of the coursebook, whether during the student years or beyond. Others, like the title page and stylistic elements marking the beginnings and endings of sections, may have served to convey the authoritative nature of the manuscript as a finished product worthy of "publication." Even if these coursebooks were not printed they were clearly finished, sometimes with a decorative flourish or a colophon, and designed to be saved and made available to others—at least to one's later self but possibly also to friends or descendants.[22]

The surviving manuscripts we have found span Chouet's career from 1667 (when he taught at Saumur) to 1685, the year before he left his professorship in Geneva. Four of the eight manuscripts belong to the Bibliothèque de Genève (BGE). They are titled "Syntagma Physicum," dated 1678–79, and they are similar though not identical in content. Two other scripts of Chouet's "Syntagma Physicum" from Geneva 1684 and 1685 are extant in libraries in Karlsruhe and at Harvard respectively. In addition two manuscripts with different titles survive—a "Cursus physicae" from Chouet's time in Saumur (1667), now at the Wellcome Institute in London, which is closely related to the later "Syntagma" manuscripts, and a "Compendium physicae" of 1684 now at Yale which offers a summary form of the 1678–79 Syntagma possibly prepared for a publication which was never carried out.

In this period physics usually featured in a multipartite philosophy course including logic, ethics, then metaphysics and physics which were taught over two years at the University of Paris for example.[23] Although he may have taught ethics at Saumur

[21] For examples of manuscripts copying commonplace headings from printed books see Blair 2010, 246.

[22] On efforts to preserve manuscripts, both of the "finished" and less finished variety, see Keller et al. 2018.

[23] Brockliss 1987, 187. The quadripartite course typical of French universities was not always the norm elsewhere; ethics for example was not taught in Italian and Spanish Scotist curricula, see Schmutz 2008, 399–400.

Chouet did not in Geneva, leaving that field to a colleague (Fig. 7.1).[24] Two of the Geneva manuscripts are part of a multi-part course along with logic and metaphysics in one case and just metaphysics in the other, ending in both cases with the physics; the other scripts include only the physics—maybe some students came to study only physics with Chouet or the other parts of their philosophy course have simply been lost in the interim. Physics formed part of the teaching of philosophy from the origins of the Academy in 1559 when Jean Calvin and Theodore Beza designed it to provide higher education to Calvinist students and future ministers. The Academy in Geneva served this purpose along with similar Reformed institutions in Lausanne, Berne, Basel, and Zurich in the Swiss Confederation.[25] Though it likely started as a minor subject at the Academy, physics gained special importance in the seventeenth century as the site of debates about recent scientific ideas (e.g. Copernicanism, experiments about the void, and Cartesianism) but also about the role of God in nature and specifically the argument for God from the design of the physical world.[26] Certainly physics is the subject in which the most Chouet course manuscripts survive, including not only the full course called "cursus" then "syntagma" and the summary "compendium physicae," but also (in the case of two of the manuscripts in Geneva) short texts on more specific topics such as living creatures, the armillary sphere, and the magnet.[27]

7.2 The Geneva Manuscripts and the Role of Dictation

The manuscripts currently in Geneva offer the best-studied point of departure into the Chouet course manuscripts and the use of dictation to create them. Fortunately, Mario Sina and a team of scholars have published a comparative critical edition of

[24] Two ethical theses appear in Chouet's *Theses ex Vniversa Philosophia selectae* (Saumur, 1667), reproduced in fascimile in Sina 2010, vol. 1, XCIX- C. Another Geneva manuscript comprises a metaphysics by Chouet and an ethics taught by a different professor: "Ethica dictata a Clarissimo viro Domino Vincentio Minutolio in inclyta Geneuensium Academia Professore 1685"; Archives Turrettini, Ms. L. 2. (Pars II, 1–93) as discussed in Sina 2010, vol. I, CIV. By contrast Chouet's successor Antoine Léger did teach ethics; see BGE Ms. Lat 225 for a surviving script from his ethics course, which features a lovely floral decoration of the course title reproduced in Fig. 7.1. For some discussion of Chouet's metaphysics, see Ragni 2020, 71–77.

[25] Geneva was allied with the Swiss Confederation as early as the fifteenth century but did not join it until 1815. On the history of the Academy, see Borgeaud 1900; Magnin and Muller 2012; Lacki 2007 provides some images of the buildings there that date back to the sixteenth century.

[26] See for example Feldhay and Heyd 1989; Klauber 1998.

[27] BGE Ms Lat. 322 (Morel de la Pize) starts with logic and metaphysics and BGE Ms Lat. 323 (Abraham de Livron) with metaphysics. Then both include, after the Compendium physicae and the Syntagma physicum these two shorter texts: Brevis introductio ad doctrinam de corporibus viventibus sive animatis, Sphaerae artificialis seu armillaris brevis explicatio; in addition Ms Lat. 323 further adds a De magnete. We have not examined the Compendium, except in the unusual Yale manuscript; the Compendium in the Geneva manuscripts is reprinted in Sina 2008, 463–75; followed by the Brevis introductio ad doctrinam de corporibus viventibus sive animatis, 476–84.

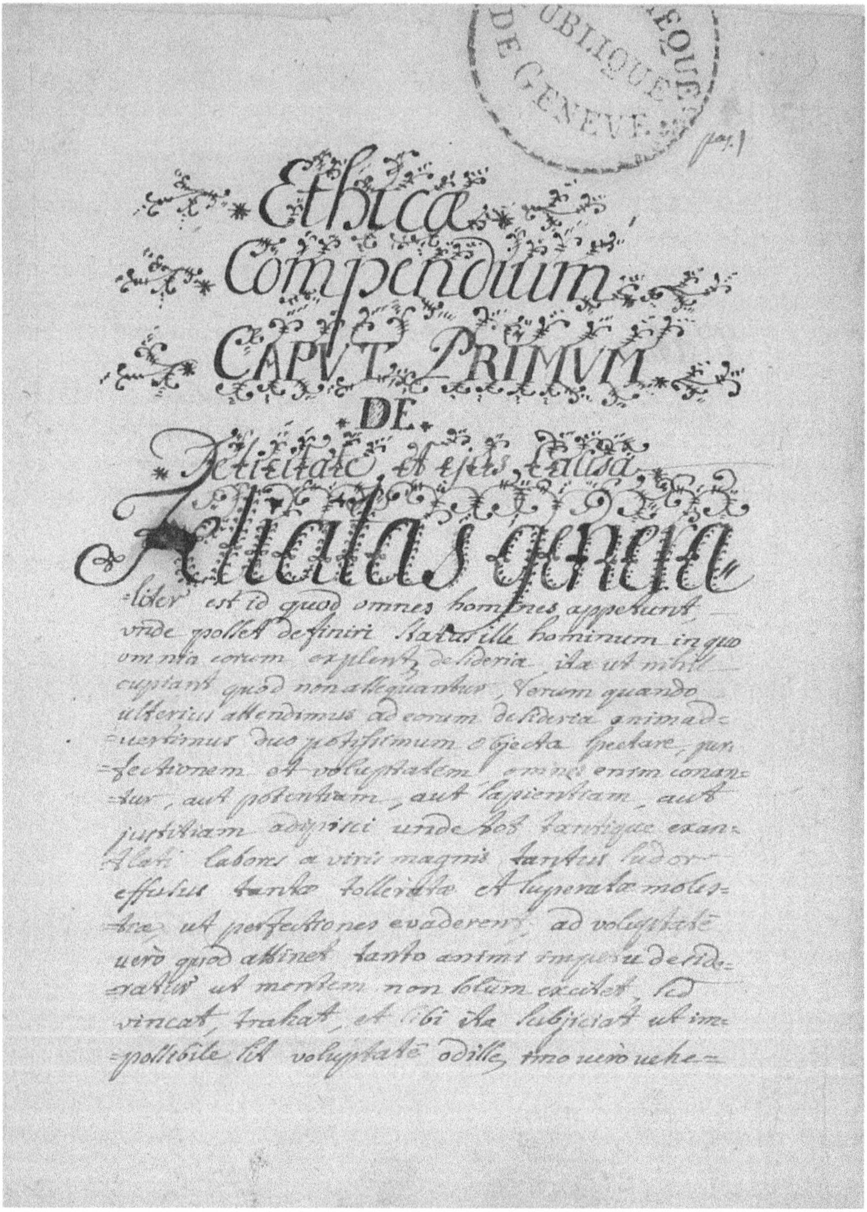

Fig. 7.1 First page of a student manuscript of an ethics course by Chouet's successor Antoine Léger. The student note-taker is unidentified. The title reads: "Ethicae compendium. Caput Primum, De Felicitate et eius causa" (Chapter One, On Happiness and its cause). The flowery decoration calls attention to the title by framing it on all sides. The decorative scheme does not recur within the text. BGE, Ms. Lat 225, f. 1r (see Jeger 2016, pp. 947–48)

these four Geneva manuscripts.[28] They have added footnotes to mark where the scripts differ, e.g. because of words and paragraphs omitted, or passages that differed between manuscripts or were ordered differently. Overall the manuscripts are very similar (with a few orthographical differences, or words missing from time to time), but the few rearrangements or changes in the sizes of paragraphs presumably represented Chouet's decision on how to present the material in different iterations of his teaching.

Of the four manuscripts at the BGE two are anonymous and undated, and contain only the "Syntagma physicae" or part of it. But the other two appear in multi-part philosophy courses and are inscribed— by Paul Morel de la Pize, a French refugee who went on to become a minister,[29] and Abraham de Livron (1660–1730), from Gex in nearby France, who matriculated at the academy in 1678.[30] The manuscripts by de Livron and Morel de la Pize can be compared directly because their inscriptions indicate that the sections on physics stem from the same year (1679) and they are almost identical. From them we can form a hypothesis about how the students wrote their coursebooks —under dictation from Chouet. Indeed de Livron's inscription reads "dictated (*dictata*) by Chouet ... and diligently caught with the ear (*excepta*) by Abraham de Livron."[31] (See Fig. 7.2 for the title page of this manuscript.) The differences between the two students' scripts are small: a few words here and there are missing or differently transcribed, such as "arbitrio" (Ms Lat. 323-de Livron) vs "attributo" (Ms Lat. 322-Morel de la Pize) —probably an aural misunderstanding— or "potentia" (Ms Lat. 323) vs "sapientia" (Ms Lat 322) —not an aural error but perhaps an inadvertent or intentional substitution.[32] The lacunae of single words make clear that the two did not copy from each other—both wrote from a third source, i.e. what the professor was dictating. Chouet evidently dictated slowly and comprehensibly: both students were able to write almost every word that

[28] Chouet 2010, vol. 2. The edition is based on BGE Ms Lat. 322 and 323 described above, along with BGE Ms Lat. 292, containing a complete Syntagma Physicum, and BGE Ms Jallabert 37 containing part of the Syntagma Physicum starting at the chapter "De calore et frigore" (for comparison that chapter is located at folio 177 of 505 folios in BGE Ms Lat. 292). The three Ms. Lat. call numbers are described in detail in Jeger 2016, 1256–60 (Ms Lat. 292), 1390–98 (Ms Lat. 322), 1399–1406 (Ms Lat. 323); the library in Geneva has no detailed description of Ms Jallabert 37.

[29] Paul Morel de la Pize was a French refugee from Cheylard, in Ardèche, who was around 19 years old in 1678, though he does not figure in the lists of students at the Academy (which were not rigorously kept). From 1706 until 1712, he was minister without a parish at La Neuveville in the Jura. See Anonymous 1870–71, 312, and Arnaud 1979, I, 615. We are grateful to Max Engammare for these references. See also Germiquet 1889, 43.

[30] Stelling-Michaud 1959, vol. 4, 348.

[31] "A D.º Chouëto, Philosophiae Professore in Publico Geneuensi Lycaeo Dictata. et ab Abrahamo De Liuron Diligenter Excepta." On the meanings of "excipio" see Lewis and Short 1879, sense B.2.a. For comparison "excepit" also appears in the Karlsruhe manuscript discussed below, while Morel de la Pize and the writers of the Houghton and Yale manuscripts used forms of "scribere" (to write) or "perscribere" (to write in full) to describe their work.

[32] Chouet 2010, vol 2, 21, 133.

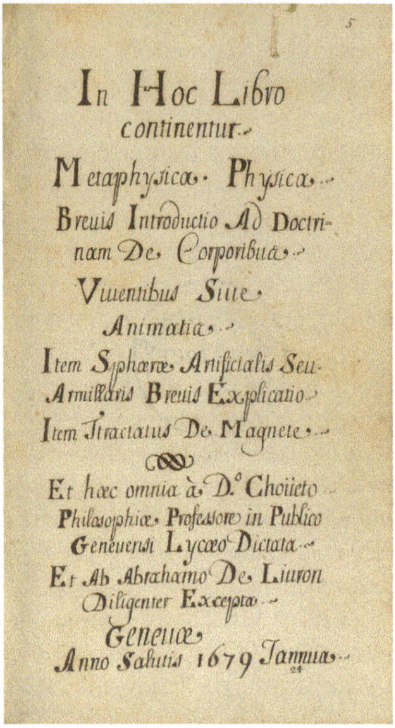

Fig. 7.2 Title page of a coursebook in metaphysics and physics, written by Abraham de Livron under dictation of Jean-Robert Chouet, January–December 1679. This title page provides the information that one would expect of a printed title page: title first, followed after a decorative element by the author, scriptor, place and date. But the layout in even lines of similar size differs from printed title pages which often deployed a range of fonts and font sizes. BGE, Ms. Lat 323, f. 5r (see Jeger 2016, pp. 1399–1406)

he said. Dictation was forbidden at the university of Paris in the middle ages, but the explicit and repeated regulations on that point indicate that there were indeed pressures on professors to dictate their courses. By the late fifteenth century the regulations against dictation were lifted and dictation became the dominant mode of teaching philosophy at the University of Paris. Dictation has remained a central element of French education for many generations—it was still in use as a means of transmitting information through the 1960s and remains a classic exercise for teaching spelling, though by the twentieth century dictating was characteristic of primary and secondary schools rather than universities.[33]

Nevertheless two larger discrepancies between Ms. Lat 322 and 323 warrant attention. Missing arguments and a sequence of two missing subchapters in Ms. Lat. 322 (compared with MS Lat 323) suggest that Morel de la Pize was absent from the course at those junctures.[34] But Morel de la Pize also changed the order of paragraphs within chapters occasionally, which would be highly unusual in the ordinary practice of dictation.[35] Did Chouet lecture on physics twice in that same year, maybe

[33] On medieval Paris see Hajnal 1959, 117–23; on the long history of the practice of dictation see Waquet 2003, 75.

[34] Chouet 2010, vol. 2, 110.

[35] Chouet 2010, vol. 2, 37 (inversion of subchapters), 66, 147.

in the morning and in the afternoon, as professors did in large universities such as Padua? If so he might have changed the order of presentation between the two iterations in the same year.[36] Or did the manuscripts sometimes or regularly result from copying over later the notes that the student or a friend had taken in class, during which process a student might re-arrange the material? Second-order copying of that kind strikes us as perhaps necessary for the most elaborately presented scripts, but in most cases these student manuscripts appear functional rather than beautifully executed and probably did not require recopying from notes taken in class. On the other hand copying was viewed as a pedagogical virtue in itself and some pedagogues recommended copying over one's notes a second time in order to remember them better, so students may have regularly or sometimes copied over their notes, whether or not they were told to by their teacher.[37] The main conclusion we draw is that these manuscripts are a finished product that cannot tell us everything we might like to know about the processes by which they were made in the classroom and outside it.[38] Nevertheless they offer an opportunity to assess the evolution of Chouet's teaching of physics across his career.

7.3 Trends in Chouet's Courses Across His Career—From the Wellcome Manuscript of 1667 to the Karlsruhe Manuscript of 1684

We can trace evolution over time in Chouet's teaching thanks to manuscripts surviving from before and after the Geneva scripts of 1678–79. Given our focus on methods of teaching and learning we have only sampled a few passages across the different manuscripts and cannot offer a proper philosophical analysis.[39] The earliest of the Chouet manuscripts we have found, now at the Wellcome Institute in London, dates from 1667 when Jacob de Normandie (Normandius) wrote that his

[36] Belloni 1986, 63–106. See also Goeing 2016.

[37] Francesco Sacchini's manual on note-taking which included the advice to copy notes over twice originated in a Jesuit context, but its translation into French in 1786 was dedicated to the Geneva pastor and professor of theology David Claparède (1727–1801), which suggests that the book was expected to be well received there; see Sacchini 1786. For some discussion of Sacchini see Blair 2010, ch. 2, 70.

[38] In one Chouet manuscript containing his logic and metaphysics three addenda appear at the end to be inserted in specified places in the text, which suggests that Chouet himself issued them. Chouet, Syntagma logicum, BGE Ms Lat. 220. See Jeger 2016, 924. This manuscript from 1673 to 1674 is attributed to the student Jean Le Clerc (1657–1736) who became professor of philosophy in Rotterdam then Amsterdam. On Chouet's later correspondence with Le Clerc see Heyd 1982, 78ff.

[39] For background on the topics we mention, see Garber and Ayers 1998, especially Ariew and Gabbey 1998; Garber et al. 1998; Gabbey 1998.

"Cursus physicae" was dictated by Chouet at the Huguenot academy in Saumur.[40] De Normandie, a Geneva citizen and half-brother of Chouet (by the second marriage of Chouet's mother with Michel de Normandie in 1662) had matriculated at the Academy in Geneva in 1663, after graduating from Latin school there, and followed Chouet to Saumur in 1665 when Chouet taught there from 1664 to 1669.[41] De Normandie went on to advocate for radical reform in Geneva and to serve as legal adviser to William of Orange before and after his accession as king of England, then to the Prussian crown.[42]

The latest surviving manuscripts from 1684 to 1685 enable us to trace changes in Chouet's teaching in physics across almost twenty years. Barthélémy Franconis (Franconius) inscribed the manuscript now in Karlsruhe in 1684, two years after he matriculated in Geneva.[43] It comprises the first part of Chouet's Syntagma physicum.[44] Harvard's Houghton Library owns a manuscript entitled "Opera physica" containing the Syntagma physicum and a Compendium physicae.[45] Although a later

[40] "Dictati A Domino Domino Johanne Roberto Choueto in Academia Salmuriensi Philosophia Professore et a me Jacobo Normandio conscripsit Anno Domini 1667 Soli Deo gloria." Chouet, Cursus Physicus, Wellcome Library (London), Ms 1633, 333.

[41] Jacob's father Michel de Normandie was a member of the Petit Conseil. Michel had married in 1662 as his second wife Théodora Tronchin, the sister of Louis Tronchin and mother of Jean-Robert Chouet, so Jacob was a half-brother of Chouet. See Fatio 2015, 164n1.

[42] Heyd 1982 discusses de Normandie's calls for reform without noting that he was a student of Chouet, 178–80, 211. Stelling-Michaud 1959, vol. 5, 32; and Sandys 1901, 108.

[43] Chouet 1684 (Karlsruhe). Niebler 1969, 38–39 reports the provenance of this manuscript from St Peter monastery in Freiburg, Germany. Niebler explains that the St Peter library burnt down in 1644 and 1678, and was sacked in 1713 by French troops in the War of Spanish Succession. St Peter acquired many parchment and paper manuscripts to compensate for these losses, ranging from medieval manuscripts to works of enlightenment science (Niebler 1969, XI-XII). The paper manuscripts were bought mostly in the second half of the eighteenth century, and we suppose that St Peter acquired the Chouet manuscript during this time (Niebler 1969, XV-XVI). On Franconis who went on to become a minister, see Stelling-Michaud 1959, vol. 3, 357.

[44] The first part covers physica generalis and runs a bit shorter than the second part (physica specialis).

[45] "Opera physica: authore domino Johanni Roberto Chouëto in Acadimiâ Genevinsi Philosophice professore dignissimo" Houghton Library, Ms Lat. 452. The book came to Harvard through the acquisition of George Sarton (1884–1956), professor of history of science there, who mentioned the manuscript in one of his articles. Sarton 1950, 171, fn 32: "…I have a contemporary MS. (292 pages in a good hand) of his 'Opera physica,' which as far as I can judge from a few samplings, is very scholastic in tone. According to Bayle himself Chouet made curious experiments; yet, it is clear from his own writings that he could not give his students any concrete idea of the scientific spirit. Bayle did not devote any article to his old teacher in his 'Dictionnaire,' but he published an 'astronomical' letter of him in the Nouvelles de la R. d. L., March (1685). It was Chouet who introduced Cartesianism in Geneva and was responsible for the considerable increase of Genevese interest in mathematics and physics at the beginning of the eighteenth century. He was more important, however, as a teacher and administrator than as an author." We are grateful to Alex Csiszar for this reference. On Pierre Bayle's encounter with Chouet first at Saumur then likely later in Geneva when he served as preceptor for the family of the brother-in-law of Michel de Normandie, see Labrousse 1963, 102. More generally on Bayle and Chouet see Heyd 1982, 53 and *ad indicem*.

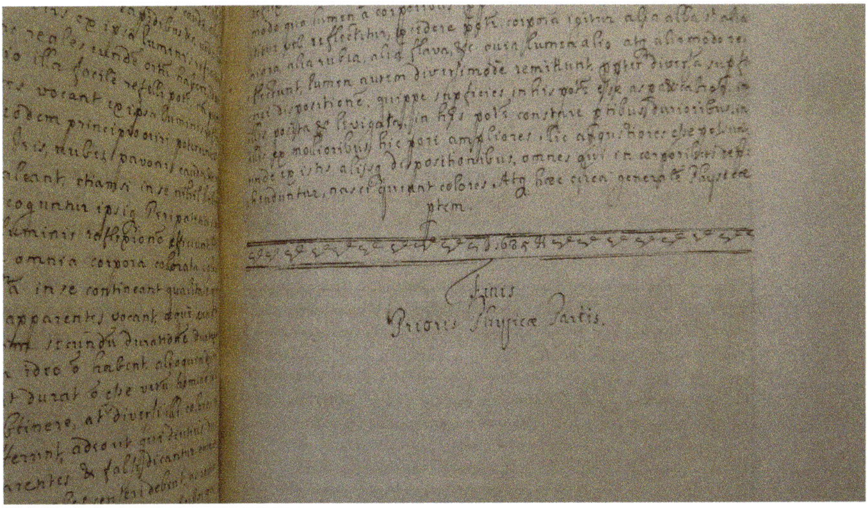

Fig. 7.3 Detail from a student transcript of Chouet's lecture on natural philosophy ("Opera physica") at the academy of Geneva. The decorative frieze occurs at the end of the first part of the manuscript; it contains "1685," presumably the year in which the manuscript was made, and the initials I.H. (straddling the date)—presumably those of the student (or possibly a hired scribe) who wrote it but who remains unidentified. Houghton Library, Harvard University, MS Lat 452, p. 123

note on the front flyleaf offers a date of 1756, a close look reveals a more plausible date inserted into the ornamental border at the end of the first part of the book—1685—which places the manuscript within Chouet's period of teaching at Geneva. There and again at the end of the manuscript one can also read "I.H." –presumably the initials of the writer.[46] Snippets of the decorative frieze recur at other break points in the manuscript, serving to mark the beginning or end of a section and here also to discreetly convey information about the circumstances of the making of the manuscript (see Fig. 7.3). No student is recorded at the Academy with those initials in the appropriate span of time; but we also know, as for example from the case of de Livron, that those matriculation records were not kept rigorously. Another possibility, given the unusually regular hand and inking, is that this copy was the work of a professional scribe who left a barely visible trace of his initials. The Karlsruhe and Houghton manuscripts are virtually identical, except for variations among a very few words. Another striking sign of their close similarity is that they both contain the same subchapter on space (*spatium*) which is present in none of the other Chouet manuscripts we have found.[47]

[46] Houghton Library Ms Lat. 452, 123, 271.

[47] Chouet 1684 (Karlsruhe), 197–223; Houghton Library, Ms Lat. 452, 36–41.

Overall the tenor of Chouet's teaching remained quite consistent, but he varied the presentation of the material, by changing the number and nature of the chapter headings and subheadings and occasionally the order of topics and some of his formulations. The differences between 1679 and 1684 which we will study in more detail also show that Chouet developed a new interest in space by adding a chapter on the topic in the Karlsruhe manuscript.

The Syntagma physicum is divided into two parts —general and special. In the general part Chouet treats of bodies in general, their principles and characteristics. The general topics include quantity, place and void (*vacuum*), motion and time; the course then covers the qualities of bodies that one senses—through touch, taste, smell, and hearing, and ends with a section on light/color instead of the sense of sight. In the second part of the Syntagma Chouet treats individual bodies, especially the heavenly bodies, elements and *meteora* (e.g. winds, clouds, rain and rainbow). Only the Wellcome manuscript of 1667 features a few explanatory diagrams here and there. The most consistently present graphical element in the manuscripts is the depiction of the three cosmologies in play at the time –Ptolemaic, Copernican and Tychonic. From 1667 to 1684 Chouet discussed objections and responses to these systems, but without dictating a conclusion. This was a common solution in teaching about a topic that was touchy not only for Catholics (due to the condemnation of heliocentrism in 1616 based on a traditional interpretation of the Bible) but also for Protestants (given their tendency to prefer a literal interpretation of the Bible).[48] Thus he left the task of assessing whether the Tychonic system accorded with the "true principles of physics" to his "readers" –a choice of term which also indicates that Chouet expected that his courses would find readers thanks to the listeners who recorded his teaching in writing.[49] The students made careful drawings of the Copernican, Ptolemaic and Tychonic systems at this juncture (see the drawing of the Copernican system in de Livron's script in Fig. 7.4). Making these required somewhat different skills from taking dictation (including the use of a compass) and were likely done at a later time, and perhaps even by someone else if a student was wealthy and insecure about his drawing skills and might hire someone to make them in his stead. In one of the Chouet manuscripts blank spaces were left for these illustrations which were never completed.[50]

[48] Jean-Cecile Frey made a similar choice to discuss the topic without reaching a conclusion, see Blair 1993, 126. On other hand the Franciscan Scotist Claude Frassen flatly rejected Copernicanism, although he noted that it was hard to refute on mathematical or philosophical grounds. See Schmutz 2008, 405–6. On the Catholic reception of Copernicanism, see Blackwell 1991; for an example of the Calvinist reception see Vermij 2002.

[49] "Sequitur nunc ut disquireremus an hoc Tychonis systema admitti possit. Verum quoniam diutius iam quam par erat in hisce quae ad caelos spectant immorati fuimus ideo observasse sufficiat illius ope omnia astrorum phaenomena rite explicari atque adeo nihil obstare quominus eo veluti hypothesi quadam utamur. Sed vero an cum veris physicae principiis stare possit, id lectoribus (examinandum) relinquimus." Chouet 2010, vol. 2, 158–59. Heyd argues that Chouet leaned toward the Copernican hypothesis though he did not adopt it explicitly; this stance also fit into Chouet's general avoidance of controversy; Heyd 1982, 82, 86.

[50] See BGE, Ms Lat. 292 (scribe unknown, undated), 263, 270, 293; see Jeger 2016, 1256–60.

Fig. 7.4 Depiction of the Copernican cosmological system in Abraham de Livron's manuscript. Except for the earth the planets are depicted like the fixed stars on the outermost sphere; notice the four moons circling around Jupiter (as per Galileo's description of them in his *Siderius nuntius*, 1610). The moons of Jupiter did not prove the motion of the earth although they alleviated the objection that the earth alone has a moon in the Copernican cosmology. BGE, Ms. Lat 323, f. 84r (see Jeger 2016, pp. 1399–1406)

Each part of the Syntagma is further subdivided in an elaborate hierarchical sequence: into three sections, then each section into 2–4 books, and each book into 3–7 chapters, each of which contains multiple subheadings.[51] Within this structure which remained consistent throughout his career, Chouet could vary the extent of treatment of a topic by expanding some books with extra chapters like an accordion, and reducing others by taking away or shortening chapters. Between 1667 and 1684 Chouet generally reduced the amount of attention devoted to classical Aristotelian topics, especially those at the beginning of the course, in order to expand the later sections that discussed new views associated with Cartesian thought. But he made these changes within the familiar structure of topics of Aristotelian physics.

For example in 1667 and in 1679 the Proemium contains three subchapters: "1. Quae fuerit Physicae origo? 2. Quodnam sit Physicae obiectum quaenamue illius definitio? and 3. Physicae Partitio." (1. What was the origin of physics? 2. What is the object of physics and what is its definition? 3. Division of physics) but these are dropped in 1684.[52] After the Proemium the first section on the principles of bodies also becomes progressively compressed between 1667 and 1684 and scholastic-style questions structuring the discussion are increasingly eliminated in favor of shorter titles in the form of "de…".

Wellcome Ms (1667)	Geneva Ms Lat. 323 (1679)	Chouet 1684 (Karlsruhe)
Liber primus De principiis corporum in genere cap. 1. quid sit principium in genere et quotuplex? (p. 13)	Cap. 1. De principiis corporum in genere (fol. 31v)	Cap. I. De principiis corporum in genere (p. 17)
Lib 1, ch 2: De intrinsecis corporum naturalium principiis (p. 15)	Cap. 2. De corporum naturalium materia (fol. 33r)	Cap II. De materia corporum... (p. 26)
Lib 1, cap. 3: de quibusdam principiorum physicorum affectionibus (p. 20). **Liber secundus:**— de corporum principiis in specie: cap.1: quid sit materia prima et forma corporum naturalium? (p. 22) cap. 2: praecipuae philosophorum sententiae de materia prima et forma (p. 26) ch: 3 quid de materia et forma statuendum sit? (p. 30)	Cap. 3. De corporum formis (fol. 35r)	Cap. III. De Formis corporum (p. 53)

(continued)

[51] See the detailed contents of Ms Lat. 292 in Jeger 2016, 1256–58.

[52] Wellcome Ms 1633, 1, 6, 11. Compare with Chouet 2010, vol 2, 269.

Wellcome Ms (1667)	Geneva Ms Lat. 323 (1679)	Chouet 1684 (Karlsruhe)
Lib 2 Cap. 4. Recentiorum Peripateticorum opinio circa naturam materiae et formae examinantur et refellitur (p. 38)	Cap. 4. Recentiorum Peripateticorum opinio circa naturam materiae et formae examinantur et refellitur (fol. 36r)	Cap. IV. Variae Philosophorum opiniones de materia et forma discutiuntur (p. 73)
Lib 2 cap. 5 quae sunt praecipua materiae primae et formae attributa? (p. 45)	Cap. 5. Praecipuae materiae et formae affectiones (fol. 38r)	
Lib 2 cap. 6 De unione et composito (p. 49)	Cap. 6. De primaria materiae diuisione simplicioribusque corporum formis (fol. 39v)	Cap. V. De primaria materia divisione, simplicioribus que corporum formis (p.101)
Lib 2 cap. 7 De natura (p. 52)	Cap. 7. De natura (fol. 41v)	

Thus this introductory topic that was initially subdivided into two books totaling ten chapters in 1667 is treated in one book with seven chapters in 1679 and in just five chapters in 1684. That compression of the chapters of the introductory part of the course is accompanied by an expansion of chapters in the later part of the course, especially concerning matter, form, quantity, and space. All of these new sections explained ideas from the Cartesian tradition, though they were included within the same overall structure of section and book topics characteristic of Aristotelian philosophy, and Descartes was never named directly in them.

For example Part I, section 2, book 3 on motion first shrinks from four chapters to three in 1679 and then grows by two new chapters in 1684, when the treatments on the nature and reasons of movement in the 1679 version are supplemented by a chapter on the determination of motion and one on the communication of motion. In the first of these Chouet teaches for example that "any body of itself strives to proceed to be moved in a straight line, and never in a curved line."[53] And in the second of these new chapters he discusses the various interactions of bodies when they are of unequal or equal size and velocity, following the approach of Descartes.[54]

Wellcome Ms (1667)	Geneva Ms Lat. 323 (1679)	Chouet 1684 (Karlsruhe)
Cap 1. Quid sit motus? (p. 111)	Cap. 1. Quid sit motus? (fol. 56r)	Cap. I. De natura motus & quietis (p. 277)
Cap. 2. De causa motus variisque quae ad eum spectant regulis (p. 117)	Cap. 2. De causis motus et variis quae ad eum spectant regulis (fol. 58r)	Cap. II. De causis motus (p. 302)
		Cap. III. De determinatione motus (p. 316)

(continued)

[53] "Unumquodque corpus ex se tendere ut semper secundum lineam rectam pergat moveri, numquam secundum curvam." Wellcome Ms 1633, 320. This statement is also written in a larger font, presumably to call attention to it.

[54] For Descartes' laws of motions see *Principia Philosophiae*, II, section XXXVI, and following in Descartes 1996, VIII, 61ff.

Wellcome Ms (1667)	Geneva Ms Lat. 323 (1679)	Chouet 1684 (Karlsruhe)
		Cap. IV De communicatione motus (p. 330)
Cap. 3 Variae diuisiones motus (p. 122)	Cap. 3. Variae diuisiones motus (fol. 60r)	Cap. V. Variae diuisiones motus (p. 362)
Cap. 4 An alii praeter localem dentur motus? (p. 126)		

In tracking the discussions within a chapter we find that Chouet reuses many of the same sentences but also sometimes changes his metaphors. For example the beginning of chapter 1 of Part I, section 2, book 2, entitled "De loco et vacuo" (on place and the void) remains unchanged from 1667 to 1684: "Quantitatem proxime sequitur locus, quippe duae illae corporis affectiones tantam inter se connectionem habent ut locus, sicuti ex sequentibus patebit, pro ipsa corporis quantitate sive magnitudine plerumque vsurpetur."[55] ("Place follows quantity closely, indeed those two affections of the body have such a great connection between them that place, as will be clear from the following, is frequently used for the quantity itself or the magnitude of a body.") Chouet goes on to explain the characteristics of place (*locus*), treating the topic through a succession of oppositions: common versus proper, total and adequate versus partial and inadequate, external versus internal, and natural versus artificial. That order remains the same in 1679 and 1684 but Chouet's later treatments are shorter, as he eliminates some repetition of examples and chooses different ones. In the Wellcome manuscript, he mentions as examples for the common place not only the air for the bird, as in the later Geneva or Karlsruhe manuscripts, but also the sea for the fish, an example he does not take up in later lecture scripts.

Metaphors illustrating oppositions:	Wellcome Ms (1667)	Geneva Ms Lat. 323 (1679)	Chouet 1684 (Karlsruhe)
Common place vs proper place	Air to bird and Sea to fish vs Vein to blood	Air to bird vs Vein to blood	Air to bird vs Bone to marrow
Total vs partial	Air that surrounds the human completely vs that air, where the human is partially in water and partially in air. (He can be in two partial places at once)	Air that surrounds the human completely vs that air, where the human is partially in water and partially in air. (He can be in two partial places at once).	Air that *together with the underlying earth* surrounds the human completely vs that air, where the human is partially in water and partially in air. (He can be in two partial places at once).

(continued)

[55] Wellcome Ms. 1633, 92; Chouet 2010, vol. 2, 59; Chouet 1684 (Karlsruhe), 183ff. Two minor changes are made in 1679 and Chouet 1684 (Karlsruhe): seu for sive; saepissime for plerumque.

[56] "Internus est spatium quod vnumquodque corpus occupat; externus… nihil est aliud praeter corpus quodpiam a quo aliud corpus ambitur." Wellcome Ms, 92–93.

Metaphors illustrating oppositions:	Wellcome Ms (1667)	Geneva Ms Lat. 323 (1679)	Chouet 1684 (Karlsruhe)
Internal vs external	No metaphors in this paragraph: "internal [place] is the space that any body occupies; external [place] is nothing but the space that surrounds a body."[56] (divided then in natural vs artificial place)	No metaphors in this paragraph: as in Wellcome ms	The discussion of external and internal place is missing in this paragraph. It is moved to the following paragraphs and the next subchapter (on space). That discussion is thus completely separated from the division into natural vs artificial place.
Natural place vs artificial place	Bone to marrow vs Barrel to wine	Bone to marrow vs Sheath to sword	Vein to blood vs sheath to sword (but this distinction is placed at the beginning of the section rather than the end)

At a broader scale in comparing Chouet's Syntagma to the Physica of his teacher Wyss, some of these examples are repeated, but Wyss's discussions and methods of presentation are noticeably more scholastic. In Wyss's treatment chapters were often divided into questions which are addressed with a succession of responses, objections and responses to the objections. Chouet's course was distinctively his own in its organization and emphases though it also contained elements drawn from his teacher's courses.

7.4 Chouet's New Emphases 1679–84

A close study of the changes in Part One of the Syntagma between 1679 and 1684 reveals few major changes, but among them Chouet's development of a new distinction between locus and spatium. Given the similarity between the two versions, it is striking how book 2, on place and void (De Loco & Vacuo) stands out in the Karlsruhe manuscript because it is extended into two additional chapters, including one new chapter on space. (See the table of contents of the Karlsruhe manuscript in Fig. 7.5.) In this way Chouet followed a clearer sequence for discussing locus and spatium separately. We take this development as a sign of Chouet's engagement with Cartesian thought.[57]

Within chapter two of the Karlsruhe script, the topic of "fuga vacui" (fear of a vacuum) is unfolded and separated into two chapters, one about motion and another on the question of why nature avoids a vacuum as a rule.

[57] On locus and spatium, see Descartes, *Principia Philosophiae*, II, section XIV and following, in Descartes 1996, VIII, 47ff.

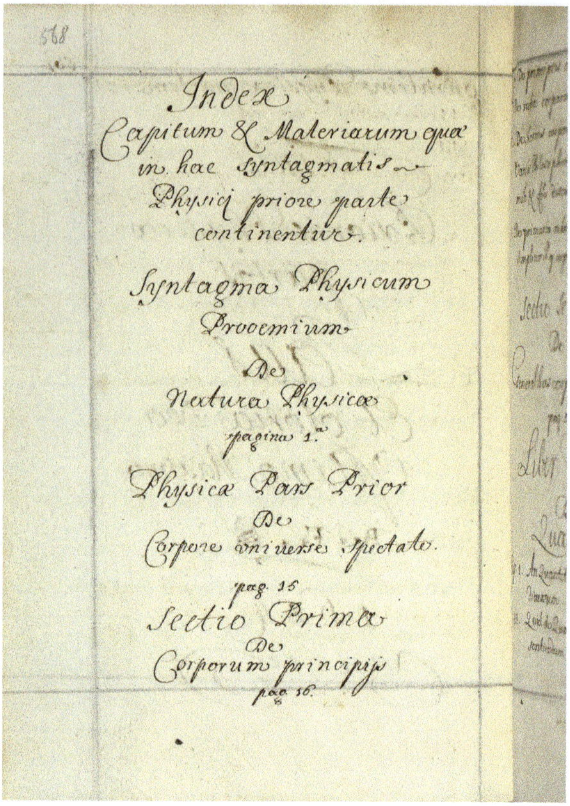

Fig. 7.5 Table of contents of Bartholomäus Franconius's transcription of Chouet's natural philosophy from Geneva, 1684. It is called "index" in a common use of that term at the time, although it follows not an alphabetical order but the order of appearance of the different sections: *Syntagma physicum. Proemium.* On the nature of physics, page 1. On the body considered universally, page 15. Section 1: on the principles of bodies, page 16. Badische Landesbibliothek Karlsruhe, St. Peter pap. 22, f. 284v

Geneva Ms Lat. 323 (1679)	Chouet 1684 (Karlsruhe)
Cap. 1. Quid sit locus? (fol. 49v)	Cap. I. De natura loci (p. 183)
	Cap. II. De spatio (p. 197)
Cap. 2. An detur vel dari possit vacuum? (fol. 52r)	Cap. III. An in rerum natura extet vel extare possit vacuum? (p. 224)
Cap. 3. De his motibus qui ob metum vacui fieri dicuntur (fol. 53v)	Cap. IV. De Motibus qui metui ac fugae vacui vulgo tribuuntur (p. 237)
	Cap. V. De aliis effectibus qui ad fugam vacui vulgo referuntur (p. 262)

The text of "De Loco et Vacuo" in the Karlsruhe manuscript makes alterations that help to separate and augment the arguments containing spatium from those about locus and vacuum. We can observe how Chouet is seeking to develop and clarify his views. In both 1679 and 1684 Chapter 2, subchapter 1 "De Natura Loci" begins with the same first sentence (also unchanged from 1667). And in the remainder of the first paragraph the two versions share about 80% of the same words, but they are organized differently. In the last sentence of the first paragraph in the 1679 manuscript, Chouet economically explains the common notion of twofold state of the place (external and internal) in introducing space as commonly set equal to the internal place. Extracting all explanations concerning space from this step of the argument in the Karlsruhe manuscript helps Chouet to be clearer with his argument. Chouet only mentions internal and external as the two states of each place, saving the discussion of space for later.

	Geneva Ms Lat. 323 (1679) vs Chouet 1684 (Karlsruhe)
First sentence	Identical in both manuscripts
Last sentence of the paragraph	Geneva distinction "natural vs artificial" is moved to an earlier part of the first paragraph in the Karlsruhe ms
Remainder of first paragraph excluding first sentence	Similar wording in both manuscripts, but different organization

Starting with the second paragraph the Karlsruhe manuscript deviates from the argument offered in 1679. Chouet explains at length the internal and external place, offering careful explanations and examples, such as, in the third paragraph, describing what lies outside a tower, bodies that are external places, "houses, trees, hills." In this third paragraph, he also introduces the term "spatium" for the first time, giving a first definition: "Atque hinc apparet, locum internum nihil aliud esse quam spatium uniuscujusque corporis cum enim affirmamus locum angustum vel amplum, …." ("It seems that the inner place is nothing else than the space of each body, because we agree the place to be either narrow or wide.") In the following chapter, on space, he discusses the philosophical term "extensio" to explain space in relation to body. In 1679 he had mentioned "extensio" in the chapter on place.[58]

In this way Chouet introduces Cartesian concepts within the framework of a traditional course of physics. The differences are subtle and gradual, and Chouet never calls attention to them. Nonetheless Chouet adjusted his teaching at least at these intervals that we are able to observe, leaving different generations of students with somewhat different versions of his course.

7.5 The "Compendium Physicae" and the Yale Manuscript (1684)

In addition to the "Syntagma physicum" a number of manuscripts (Houghton, Yale, BGE Mss Lat 322, 323) contain a separate "Compendium physicae" or summary. The Chouet manuscript at Yale's Beinecke Library contains only this shorter work, though this Compendium is different from and much longer (at 144 pages) than the Compendium in the Geneva manuscripts which comprise only 18 pages in the script

[58] On internal and external place, see Descartes *Principia Philosophiae* II, section XV, in Descartes 1996, VIII, 48ff

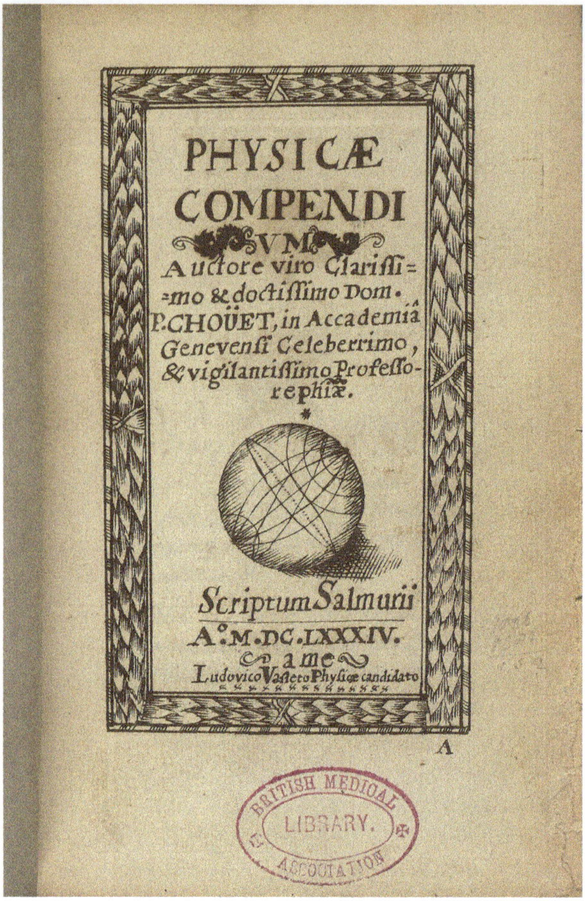
PHYSICÆ
COMPENDI
VM.
Auctore viro Clariſſi=
=mo & doctiſſimo Dom.
P.CHOUET, in Accademiâ
Genevenſi Celeberrimo,
& vigilantiſſimo Profeſſo-
re phiæ.
Scriptum Salmurii
A°. M.DC.LXXXIV.
a me
Ludovico Vaſleto Phyſicæ candidato
A

Fig. 7.6 Title page of Louis Vaslet's transcription of Chouet's "Physicae compendium" from Geneva, 1684. Note how the form imitates that of a printed title page, with varied font sizes and registers, and the claim that it was written in Saumur appears where one would find a printer's name. The decorative border framing the information resembles those produced in printed books by woodblocks. Beinecke Rare Book and Manuscript Library, Yale University, Gen Ms Vol 535

by Morel de la Pize for example. The Yale manuscript features an elaborate title page identifying it as written by Ludovicus Vasletus in 1684 in Saumur (see Fig. 7.6).[59] Louis Vaslet (1666–1731) was 28 years old in that year and a BA candidate in physics at the Academy in Geneva. Later, in 1713, he became master of the "French School" in Hampstead, Britain. He moved the school to Fulham near London in 1716.[60] But Chouet was not in Saumur in 1684 and the Academy there

[59] Chouet 1684 (Yale).

[60] Agnew 1886, 289; Anonymous 1969, 241–242.

was hardly thriving in the wake of Louis XIV's increasing persecution of Huguenots (including a policy of forced conversions and dragooning enacted in 1681); it was officially closed in 1685 with the Revocation of the Edict of Nantes. The mention of Saumur is therefore puzzling. Was Vaslet in Saumur copying from another Chouet manuscript? Was Vaslet creating a memorial of Chouet's teaching in Saumur well after the fact? With no further evidence, we are left with speculations.

The manuscript appears to have been prepared for publication although no printed book ever materialized that we know of. The manuscript was carefully edited, as someone added correct spellings in the margins with an asterisk indicating which word to change in the text. The ninety-two emendations also gave instructions for adding and deleting, using "adde" and "dele" just as a professional corrector might have done. The ink and pen-strokes closely resemble those of the text, suggesting that Vaslet made the corrections himself. In addition this manuscript presents the most elaborate finding devices of all: a table of contents and a 10-page alphabetical topical index at the back, both keyed to page numbers present throughout the text. Those references made the manuscript useful in itself, regardless of plans to publish which would have created a new pagination.

The chapters follow a similar order to the other Chouet manuscripts. However, given the process of abbreviation, the writer has eliminated whole topics, such as the two chapters and their subchapters on the quality of the bodies, and how different senses like tactility are experienced. Chouet had not altered the chapter in preceding lectures (as recorded in the student notebooks) but evidently he or Vaslet found it no longer worth including in this abridged Compendium.

Even though the Yale manuscript dates from 1684, a careful comparison shows that it is more closely related to the Geneva manuscripts of 1679 than to the Karlsruhe manuscript of the same year. Although Vaslet uses very short chapter titles as in the Karlsruhe manuscript, he does not elaborate on space or extension as in the Karlsruhe text. Instead he extracts sentences in their entirety from the 1679 edition. De loco & vacuo (pp. 27–32) starts with "Vt planè intelligamus quis sit verus ac germanus corporum locus consentaneum rectae videtur rationi nonnullas praemittere distinctiones quibus ipsius vocis ambiguitas tollatur." ("So that we clearly understand what the true and genuine place of the bodies is, it seems agreed upon by straight reckoning to present a few distinctions by which the ambiguity of the expression itself is lifted.") This sentence is not part of the Karlsruhe ms, but can be traced to the 1679 ms (BGE, Ms. lat 323, f. 49v); and some of its expressions can already be found in the Wellcome ms (Wellcome manuscript, p. 92). The sentence is in both cases the second sentence of the paragraph. The 1679 version reads (adding italics to highlight the variation in expression): "Ut *autem clare peruideamus* quis sit verus ac germanus corporum locus, *necesse est ut* nonnullas praemittamus distinctiones, quibus ipsius vocis ambiguitas tollatur." ("In order for us to clearly see what is the true or german place of bodies, it is necessary to first make some distinctions by which the ambiguity of the expression itself is removed.") This version is easier to understand than the Yale sentence, because it uses "it is necessary", instead of the very polite phrase that the Yale sentence prefers (one could speculate whether the social standing of the student being addressed might affect Chouet's choice of formulation). In a table, a comparison of sentences with the Yale ms shows the following.

Wellcome Ms (1667)	Geneva Ms Lat. (1679)	Chouet 1684 (Yale)	Chouet 1684 (Karlsruhe)	Houghton Ms (1685)
First sentence	First sentence	Dropped	First sentence	First sentence
Second sentence	Second sentence with alterations	=First sentence as in 1679	Dropped	Dropped
Third sentence	Third sentence, with alterations	=Second sentence following 1679	Sense and many of the words are unchanged, but the sentence structure is different	As in 1684 (Karlsr.) (a few words changed)
Fourth sentence	Fourth sentence, one word changed and one clause dropped	=Third sentence as in 1679	The content is similar to the fourth sentence in the Wellcome and Geneva mss, however the syntax and structure are different: the third, fourth and fifth sentences are merged into one long one.	Follows 1684 (Karlsr.)
Fifth sentence	Fifth sentence, with alterations	=Fourth sentence as in 1679.	Part of one long sentence	Follows 1684 (Karlsr.)

The word-for-word match starting with the second sentence of the 1679 manuscript and the first sentence of the Yale Compendium continues until sentence seven; there the Yale script returns to the Wellcome's example of "barrel to wine" for the artificial place, instead of using the sheath to sword example of the other manuscripts. Vaslet uses the example of barrel to wine again, when he writes about the internal vs external place, as a barrel is easily imagined filled with wine. Sadly it is hard to discern the reasons behind these variations in the examples Chouet used.

The Yale manuscript omits some sentences, such as a proposition that appears in 1679 but that Chouet explicitly rejected ("opinio ideo a nobis non admittitur" "for this reason we do not accept this opinion"), then copies one Peripatetic statement that Chouet had classified as not satisfactory enough, summarizing a paragraph of Geneva 1679 into one long sentence.[61] After borrowing a total of six paragraphs from a text that resembled the twelve introductory paragraphs to De Loco & Vacuum in 1679, the Yale manuscript briefly discusses the vacuum in three paragraphs extracted from what were originally 24 paragraphs in the 1679 version. Omitting the details of how and why the vacuum is made, and the problems of fear of the vacuum, the writer starts the next chapter on motion in eleven paragraphs (extracted from roughly 32 paragraphs given in the 1679 manuscript). The goal was to shorten the 1679 Syntagma, we surmise for publication, without eliminating any particular topic.

[61] Chouet 2010, vol. 2, 59; Chouet 1684 (Yale), 27–28.

These comparisons of Chouet's texts across the eight surviving manuscripts illustrate that Chouet modified his teaching gradually without modifying the overarching structure of his treatment of the topic. Rather than radical changes of content, Chouet introduced new formulations and emphases, choosing to offer more explanation to some terms like "spatium" and extension while reducing attention to more traditional Aristotelian topics. The production of new manuscripts every time Chouet taught gave him great flexibility to adjust his teaching, even if he did not avail himself of the option to change his positions as much as or as explicitly as we, his modern counterparts, might expect.

7.6 Student Decisions

The system of making course manuscripts also gave students great flexibility, as we can see in the manuscripts that survive from the philosophy teaching of Chouet as well as his immediate successors, Antoine Léger and Jean-Antoine Gautier.[62] These manuscripts come in a variety of formats, bindings, and presentations, depending on the expense and care that each student invested in the making of the manuscript. (For a view of the different formats and bindings of these manuscripts see Fig. 7.7.)

Fig. 7.7 Stack of student coursebooks from seventeenth-century and early eighteenth-century Geneva, illustrating the range of their formats and bindings. Courses by Chouet, Antoine Léger the younger, Jean-Antoine Gautier (see bibliography for details): BGE, Mss Lat 221 (Jeger 2016, 926–33), 294 (Jeger, 1268–72), 295 (Jeger, 1273–76), 296 (Jeger, 1277–81), 226 (Jeger, 949–51)

[62] See above note 15.

The Chouet manuscripts discussed here range in size from 16 cm (BGE, Ms. Lat 323) to 24.5 cm (Wellcome).[63] The students wrote in a range of script sizes, and the volumes contained different combinations of texts, so the manuscripts also vary widely in the numbers of pages they comprise.[64] In some cases the pages were carefully prepared for writing by outlining the margins (e.g. Karlsruhe); others were messier. (For a manuscript of the messier kind see Fig. 7.8.)

The most elaborate student manuscript from Geneva was made by Nicolas Fatio de Duillier (1664–1753) who took Chouet's course in philosophy in 1678–80. Later in life, Fatio de Duillier would become an exceptional natural philosopher, mathematician, and close friend to Isaac Newton. In his time with Chouet when he was a teenager of 14–16 years old, Fatio devoted exceptional effort to the presentation of his manuscript, with careful calligraphy, layouts, and abundant paratexts, including a frontispiece which is unique among the student manuscripts we have seen. The title page of his two-volume philosophy coursebook mimics that of a printed book not only in its contents (as in other Chouet manuscripts) but also in its layout, with variations in the size and fonts of the lettering and a box where a decorative printer's mark would appear in a printed title page. In this case Fatio seems to have forgotten to complete his plan for this space, presumably meant to hold a device of his choice. Beneath, where the printer's information would normally appear, Fatio has offered an unusually enthusiastic scribal inscription: "Nicolas Fatio wrote and heard [this] with the greatest pleasure in the years 1678, 1679 and 1680."[65] (See Fig. 7.9.) After the title page follows a list of contents, though without the page numbers that would be the norm in a printed book, thus reducing its effectiveness as a finding device (see Fig. 7.10). In addition the frontmatter includes a "Tree of Porphyry" which draws on a long tradition, dating back to Boethius, of representing the kinds of being in a logical hierarchy (see Fig. 7.11).[66]

Facing the first page of the text, an ink-drawn frontispiece that rivals the detail of a copper engraving conveys the moral virtues of education and wisdom through the study of philosophy in a pious mindset. Athena/Minerva goddess of wisdom (recognizable from her lance and helmet) guides the best of all pupils—an eager youngster of royal rank with fleurs de lys on his fur cape— in reading the stone inscription bearing the title, author, and date of the coursebook. (See Fig. 7.12). The image resembles a copper engraving because it was in fact copied from one—a frontispiece engraved by Claude Mellan (1598–1688) which appeared in La Mothe Le Vayer, *De l'instruction de Monseigneur le Dauphin* (Paris: Cramoisy, 1640)

[63] The other manuscripts have these heights: Chouet 1684 (Yale) 17 cm; BGE, Ms Lat. 322, 17.3 cm; Chouet 1684 (Karlsruhe): 18 cm; Houghton: 22 cm; BGE, Ms Lat. 292 23.6 cm.

[64] For example: Wellcome 337 pages; BGE, Ms Lat. 322, 294 pages; Chouet 1684 (Karlsruhe) 594 pages; Houghton 292 pages.

[65] "Scribebat et audiebat summa cum voluptate Nicolaus Fatio. Annis 1678, 1679, et 1680." BGE Ms Lat. 221, title page.

[66] For a recent and well illustrated article on the tree of Porphyry see Verboon 2014.

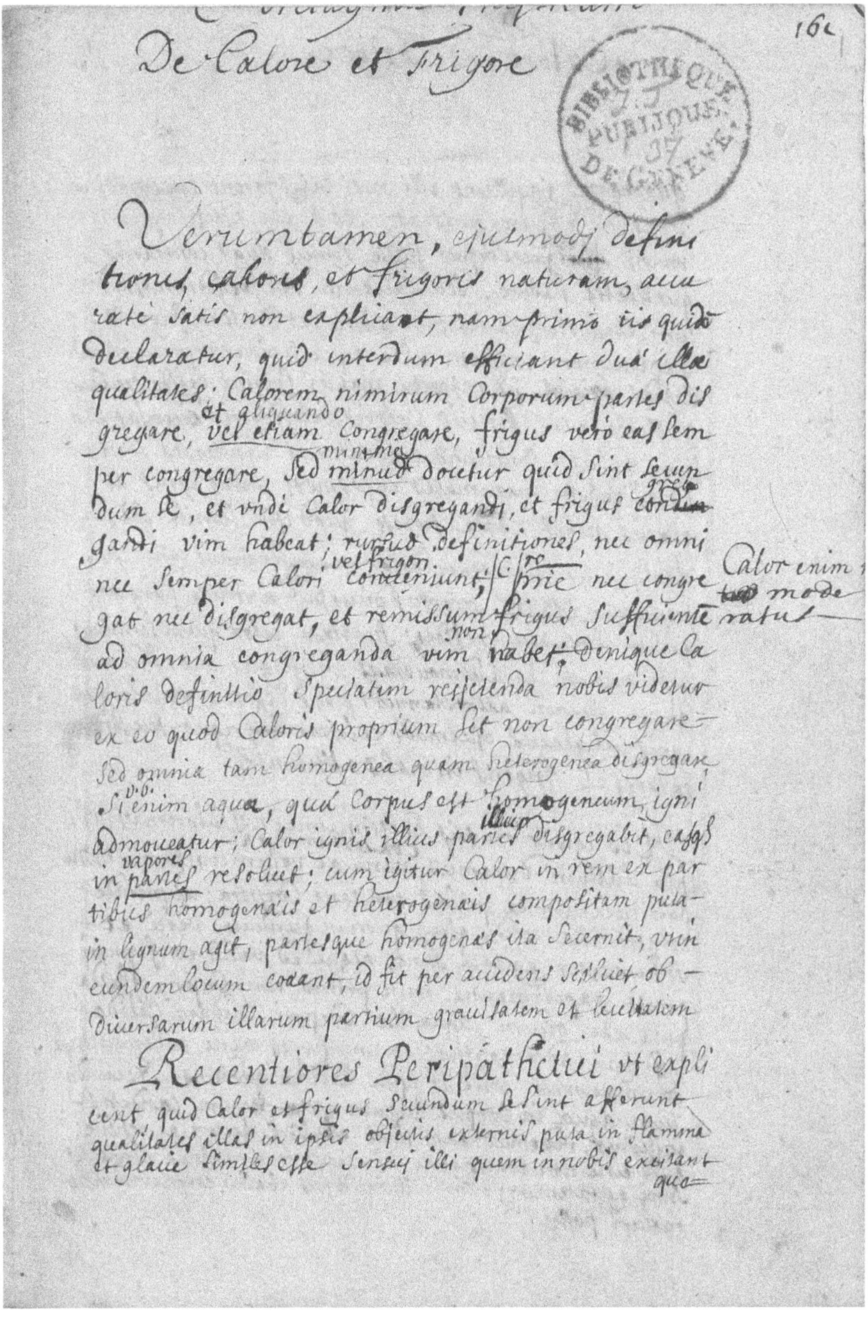

Fig. 7.8 This partial manuscript of Chouet's *Physica* starts with a page numbered 162 and the chapter on heat and cold. The writing includes later additions and corrections, and the use of larger script for the beginning of paragraphs. BGE, Ms. Jallabert 37, first page

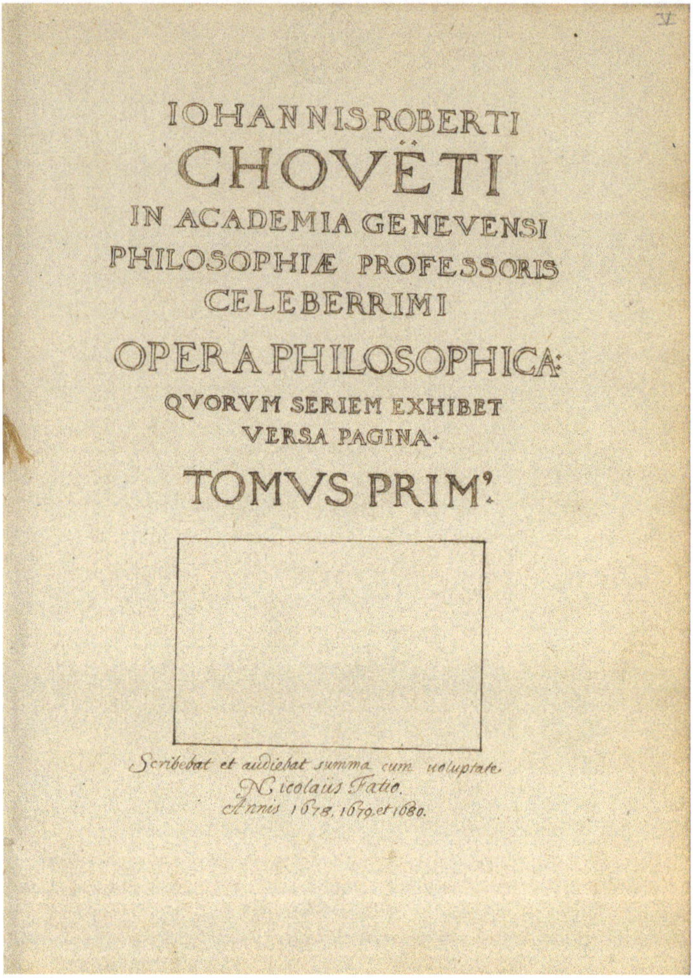

Fig. 7.9 Fatio de Duillier, *Chouet Opera philosophica*, 1678–79, title page. The title page mimics the layout typical of printed books, including a box to contain a decorative device which is left blank and incomplete. BGE, Ms. Lat 221, f. 5r (see Jeger 2016, pp. 927–33)

(see Fig. 7.13). François de La Mothe Le Vayer, author (elected to the Académie française) and pedagogue, wrote the book as a bid to be appointed tutor to the future Louis XIV who was two years old at the time (he got the job. But only later, from 1652 to 1660). Fatio copied the age and garb of the very young prince, but omitted many of the symbols present in Mellan's engraving to designate the fields that a "dauphin" should learn. Fatio omitted for example the instruments of war and of music, but retained nonetheless a few which have little to do with Chouet's course —the caduceus, the clarion (or narrow-tubed trumpet), the compass, and the painter's palette on which Fatio has inscribed the date (30 April 1678) where the

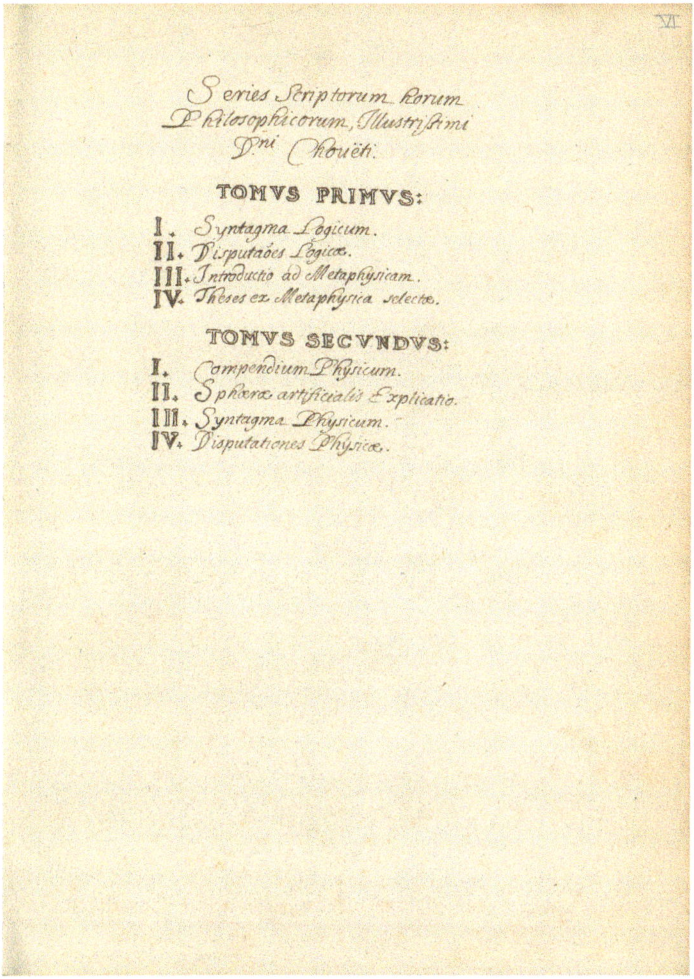

Fig. 7.10 Fatio de Duillier, *Chouet Opera philosophica*, table of contents of the two volumes, only the first of which is extant; but it is not helpful as a finding device since no page numbers are provided. BGE, Ms. Lat 221, f. 6r

original carried the note "with royal privilege." In Fatio's image the quill is more prominent (as the main tool used by pupils in Chouet's physics course), an owl has appeared on the branch above (reinforcing Athena's wisdom) and an inscription in the open book on the left. Rather than blank pages, in Fatio's frontispiece the book offers the end of a prayer, setting a tone of piety for the study of philosophy: "In nomine Domini nostri. Amen" ("In the name of our Lord. Amen"). Whereas La Mothe Le Vayer (who presumably commissioned Mellan's engraving) was suspected of irreligion in his life (though he remained a Catholic), Fatio emphasized

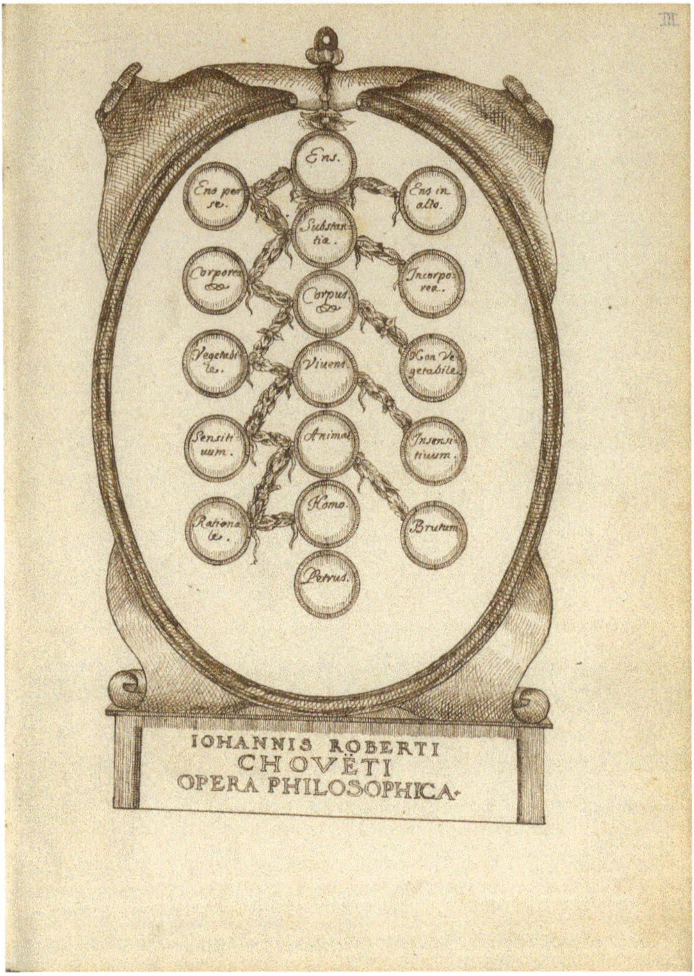

Fig. 7.11 Fatio de Duillier, "Tree of Porphyry." This elaborate rendition of a long traditional logical tree precedes the title page in Fatio's manuscript. BGE, Ms. Lat 221, f. 3r

that Chouet pursued physics with a pious spirit. This message was presumably also welcome to Chouet who tread carefully throughout his career to avoid offending those colleagues and religious leaders who held more conservative religious positions than his own.[67]

[67] For example a dispute within Geneva's Petit Conseil in 1669 over a statement of faith required of professors was resolved so that Chouet did not have to assert "sic sentio" ("this I believe") and instead only signed "sic docebo…": "This I will teach and I will not teach anything contrary to it publicly or privately." See Labrousse 1963, 105–6.

Fatio went on to become an internationally reputed natural philosopher and one of Jean-Robert Chouet's most famous students, and already as a student he devoted unusual skill and effort to making his coursebook in philosophy, presumably due to his enthusiasm for the discipline, whether or not he already had in mind a career-long commitment to it.[68] Volume one of Fatio's coursebook survives in Geneva containing logic and metaphysics, but sadly volume two devoted to physics does not; one can speculate about the reason for this loss— including the possibility that the volume in this, Fatio's field of predilection, got more use, e.g. perhaps travelled with him on his multiple stays in London, and did not survive as a result.[69]

Although none of the other surviving manuscripts are so elaborate, many were given the features typical of a printed book: title page, table of contents, and pagination and occasionally an alphabetical index. Students called attention to the divisions and subdivisions of the text by using larger and thicker scripts to highlight them, in addition to blank space, as modelled in printed books. These manuscripts were not just personal notes but well-designed books, carefully kept and bound. They were meant to be read and consulted after they were written whether by the student or a friend or relative. A few show signs of later annotation, including corrections (in the Yale manuscript) and marginal summaries, though it is probably not possible to determine whether these were added at the moment of writing the text or during a later reading. (See the marginal summaries in the Wellcome manuscript, in Fig. 7.14, made either at the moment of production or afterward on re-reading.)

7.7 In Closing

The increased ease and availability of digitizations have played a crucial role in making it possible to compare manuscripts scattered across two continents and four countries. Digitization also invites more in-depth comparisons than the initial observations we are able to present here—one could compare the Chouet manuscripts more systematically with one another as well as with those of Chouet's predecessor and successors. Chouet was no doubt peculiar in some respects—he was a Calvinist and pursued an unusual blend of Aristotelian and Cartesian physics, but we expect that his teaching was rather typical of philosophy courses elsewhere in the French cultural area and perhaps beyond. His method of teaching by dictation matched cultural expectations about handwriting as a form of retention, the transmission of authoritative knowledge from master to student, and the value of making a finished

[68] For recent work see Iliffe 2012; Mandelbrote 2005 and idem 2018; Recous 2016; Speziali 1997, 21–24. Chouet's other noteworthy students included Jean Le Clerc (see note 39) and Pierre Bayle (see note 46).

[69] Jeger 2016, 932 explains that the surviving volume entered the library in Geneva only in the nineteenth century.

Fig. 7.12 Fatio de Duillier, Chouet *Opera philosophica*, 1678–79, frontispiece. This expert ink drawing reiterates the title of the course inscribed on the stone monument. The scene is saturated with symbols of wisdom: owl, Athena instructing a princely child while the open book reminds the reader of the pious motive for the study of philosophy. BGE, Ms. Lat 221, f. 9v

course manuscript to save and consult later again. Thanks to the copies surviving from different (and sometimes the same) versions of Chouet's courses, we have especially investigated how much Chouet and his students used the flexibility of manuscript to customize respectively each iteration of a course (not very much) and

Fig. 7.13 "Minerva instructs the Dauphin," frontispiece engraved by Claude Mellan (1598–1688) for François de La Mothe Le Vayer, *De l'instruction de Monseigneur le Dauphin* (Paris: Cramoisy, 1640). Freely available from the Metropolitan Museum of Art, Harris Brisbane Dick Fund, 1953

each student notebook (somewhat more so). Presumably these tendencies too are rather typical of other contemporaries. Many questions remain unanswered about the classroom and learning experience in early modern Europe, but student manuscripts are a rich source which may still hold some answers if we continue to ply them with new lines of investigation.

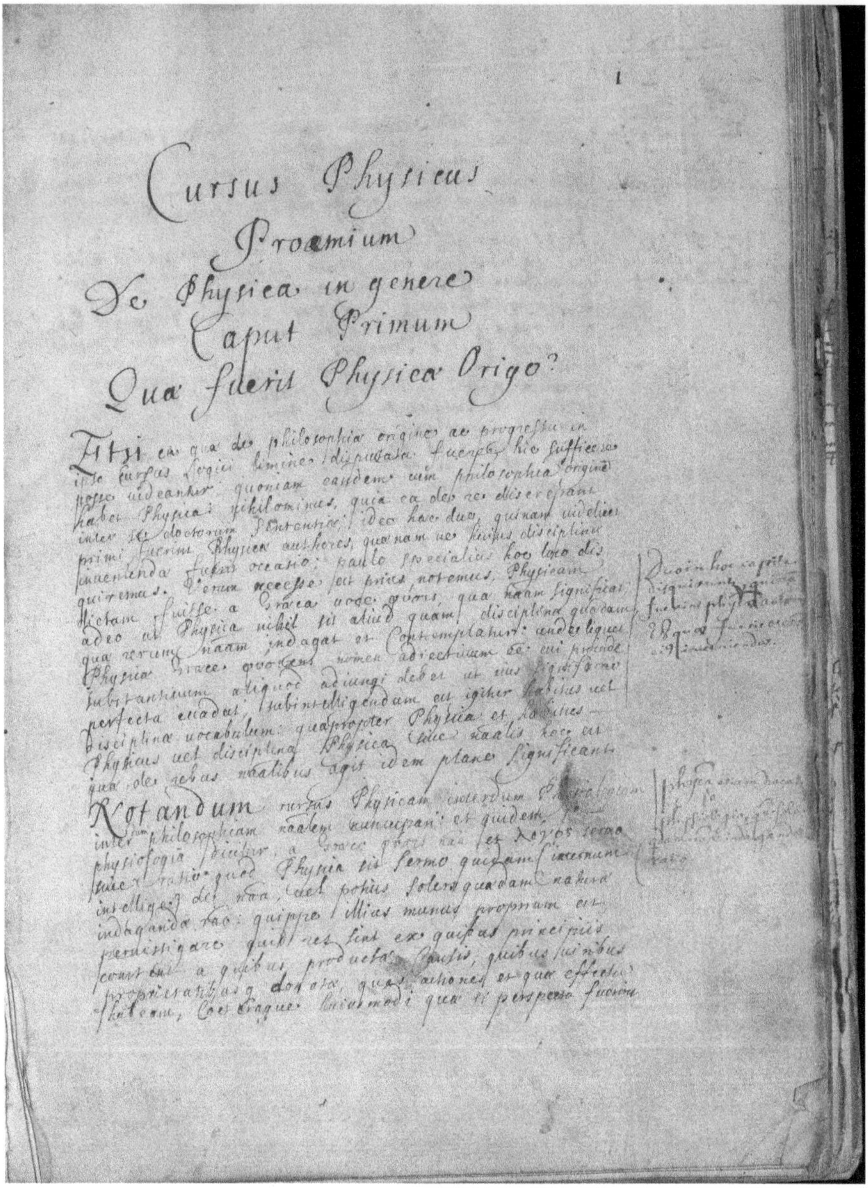
Cursus Physicus
Proœmium
De Physica in genere
Caput Primum
Quæ fuerit Physicæ Origo?

Fig. 7.14 First page of Jacob de Normandie's transcription of Chouet's *Cursus physicus* from Saumur, 1667. The marginal summaries could have been made during or close to the time of copying or afterward on re-reading. Wellcome Institute (London) Ms. 1633

Bibliography

Manuscripts

Cambridge Mass. Houghton Library, Harvard University

Chouet, Jean-Robert. "Opera physica." Script by I.H. (unknown), [Geneva], 1685. Ms Lat. 452.

Geneva, Bibliothèque de Genève (BGE)

Chouet, Jean-Robert, incomplete "Syntagma physicum," script by Abraham de Livron (c. 1660–1730), Geneva, 1679. Ms Jallabert 37.

———. "Syntagma logicum, Brevis introductio ad metaphysicam," script by Jean Le Clerc (1657–1736), Geneva, 1673–4. Ms Lat 220. See Jeger 2016, 922–25.

———, incomplete. "Syntagma logicum, Disputationes logicae, Introductio ad metaphysicam, Theses ex metaphysica selectae," script by Nicolas Fatio de Duillier (1664–1753), Geneva, 1678–79. Ms Lat 221. See Jeger 2016, 926–33.

———. "Syntagma physicum," scribe unknown, no place [Geneva], n.d. Ms Lat 292. See Jeger 2016, 1256–60.

———. "Syntagma logicum, Metaphisica compendium, Phisicae compendium, Sintagma physicum (including a third part entitled Tractatus de rebus viventibus seu animatis), Sphaerae artificialis sive armillaris brevis explicatio," script by Morel de la Pize, Geneva, 1678–79. Ms Lat 322. See Jeger 2016, 1390–98.

———. "Brevis introductio ad metaphysicam, Compendium physicae, Syntagma physicum, Brevis introductio ad doctrinam de corporibus viventibus sive animatis, Sphaerae artificialis seu armillaris brevis explicatio, Tractatus de magnete," script by Abraham de Livron (c. 1660–1730), Geneva, 1679. Ms Lat 323. See Jeger 2016, 1399–1406.

Léger, Antoine (son), "Ethicae compendium," scribe unknown, Geneva, n.d. (18th century), Ms Lat 225. See Jeger 2016, 947–48.

———. "Breve metaphysicae compendium," script by Firmin Abauzit, Geneva, 1694. Ms Lat 226 (in Fig. 7.7). See Jeger 2016, 949–51.

———. "Syntagma physicum," scribe unknown, Geneva, after 1694. Ms Lat 294 (in Fig. 7.7). See Jeger 2016, 1268–72.

Gautier, Jean-Antoine. "Syntagma physicum," script by Jacques Eynard (1700–73), Geneva, 1716. Ms Lat 295 (in Fig. 7.7). See Jeger 2016, 1273–76.

———. "Syntagma physicum," script by Jean Dentand (1689–1754), Geneva, 1707, Ms Lat. 296 (in Fig. 7.7). See Jeger 2016, 1277–81.

Karlsruhe, Badische Landesbibliothek

Chouet 1684 (Karlsruhe): Chouet, Jean-Robert. "Syntagma physicum," script by Bartholomäus Franconius, Geneva, 1684. St. Peter pap. 22.

London, Wellcome Institute Library

Chouet, Jean-Robert. "Cursus Physicus," script by Jacob de Normandie, 1667 Wellcome Library (London), Ms 1633.

New Haven, Conn., Beinecke Library, Yale University

Chouet 1684 (Yale): Chouet, Jean-Robert. "Physicae Compendium," script by Louis Vaslet. Saumur? [Geneva], 1684. Gen Ms Vol 535.

References

Anonymous. 1870–71. Liste nominative des pasteurs, proposants et autres sortis du Dauphiné, du Bas-Languedoc, des Cévennes et du Vivarais réfugiés à Genève en 1683. *Bulletin de la Société de l'Histoire du Protestantisme Français*, 19/20: 301–18.

———. 1969. Private Education from the Sixteenth Century: Developments from the 16th to the Early 19th Century. In *A History of the County of Middlesex: Volume 1, Physique, Archaeology, Domesday, Ecclesiastical Organization, the Jews, Religious Houses, Education of Working Classes To 1870, Private Education From Sixteenth Century*, ed. J.S. Cockburn, H.P.F. King and K.G.T. McDonnell, 241–255. London. *British History Online*. http://www.british-history.ac.uk/vch/middx/vol1/pp241-255. Accessed 23 Mar 2021.

Agnew, David. 1886. *Protestant Exiles from France, Chiefly in the Reign of Louis XIV. The Hugenot Refugees and Their Descendents in Great Britain and Ireland*. Third ed., vol. 2: *Refugees Naturalized in and After 1681*. For Private Circulation.

Ariew, Roger, and Alan Gabbey. 1998. The Scholastic Background. In *Cambridge History of Seventeenth-Century Philosophy,* ed. Daniel Garber and Michael Ayers with Roger Ariew and Alan Gabbey, 423–53. Cambridge: Cambridge University Press.

Arnaud, Eugène. 1979. *Histoire des Protestants du Vivarais et du Velay (1888)*. Genève: Slatkine Reprints.

Belloni, Annalisa. 1986. Professori Giuristi a Padova Nel Secolo XV: Profili Bio-Bibliografici e Cattedre. In *Studien zur Europäischen Rechtsgeschichte: Veröffentlichungen des Max-Planck-Instituts für Europäische Rechtsgeschichte Frankfurt am Main*, 63–106. Frankfurt am Main: Klostermann.

Blackwell, Richard. 1991. *Galileo, Bellarmine, and the Bible*. Notre Dame: University of Notre Dame Press.

Blair, Ann. 1991. Ovidius Methodizatus: The Metamorphoses of Ovid in a Sixteenth-Century Paris College. *History of Universities* 9: 73–118.

———. 1993. The Teaching of Natural Philosophy in Early Seventeenth-Century Paris: the Case of Jean-Cécile Frey. *History of Universities* 12: 95–158.

———. 2008. Textbooks and Methods of Note-Taking in Early Modern Europe. In *Scholarly Knowledge: Textbooks in Early Modern Europe*, ed. Emidio Campi, Simone de Angelis, Anja-Silvia Goeing, and Anthony Grafton, 39–73. Geneva: Droz.

———. 2010. *Too Much to Know: Managing Scholarly Information Before the Modern Age*. New Haven: Yale University Press.

———. 2012. Principes et pratiques de la pédagogie humaniste et réformée. In *Enseignement secondaire formation humaniste et société, XVIe-XXIe siècle*, ed. Charles Magnin and Christian Alain Muller with Blaise Extermann, 39–67. Geneva: Slatkine.

———. 2015. Reflections on Technological Continuities: Manuscripts Copied from Printed Books. *Bulletin of the John Rylands Library* 91-1: 7–33.

Borgeaud, Charles. 1900. *Histoire de l'Université de Genève, vol. 1: L'académie de Calvin 1559–1796*. Geneva: Georg.

Brockliss, Laurence. 1987. *French Higher Education in the Seventeenth and Eighteenth Centuries: A Cultural History*. Oxford: Clarendon Press.

Budé, Eugène. 1899. *Vie de Jean Robert Chouet, professur et magistrat genevois*. Geneva: Maurice Reymond.

Chouet, Jean-Robert. 1662. *Theses ex universa philosophia selectae*. Nîmes: Eduardus Rabanus.

———. 2008. In *La corrispondenza di Jean-Robert Chouet Professore di Filosofia a Saumur et a Ginevra, Con documenti inediti relativi al suo insegnamento filosofico*, ed. Mario Sina. Florence: Olschki.

———. 2010. *Corsi di filosofia*, ed. Mario Sina, Marco Ballardi, Elena Rapetti, 2 vols. Firenze: Olschki.

Chouet, Jean-Robert, and Timotheus Rovierus. 1667. *Theses ex universa philosophia selectae*. Dissertation of T. Rovierus. Saumur: Ioannis Lesnerus. Facsimile reprint in Chouet (2010), vol. I, LXXIII-C.

Compère, Marie-Madeleine, Marie-Dominique Couzinet, and Olivier Pédeflous. 2009. Eléments pour l'histoire d'un genre éditorial. La feuille classique en France aux XVIe et XVIIe siècles. *Histoire de l'éducation* 124: 27–49.

Couzinet, Marie-Dominique, and Jean-Marc Mandosio. 2004. Nouveaux éclairages sur les cours de Ramus et de ses collègues au collège de Presles, d'après les notes inédites prises par Nancel. In *Ramus et l'Université*, 11–48. Paris: Editions rue d'Ulm.

Descartes, René. 1996. *Principia Philosophiae* (1644). In *Œuvres de Descartes*, ed. Charles Adam and Paul Tannery, vol. 8. Paris: Vrin.

Dufour, Alain. 2014. Gautier, Jean-Antoine. In *Historisches Lexikon der Schweiz*. Online http://www.hls-dhs-dss.ch/textes/d/D11242.php. Accessed 13 Dec 2017.

Fatio, Oliver. 2014. Léger, Antoine. In *Historisches Lexikon der Schweiz*. Online http://www.hls-dhs-dss.ch/textes/d/D11242.php. Accessed 23 Mar 2021.

———. 2015. *Louis Tronchin: Une Transition Calvinienne*. Paris: Classiques Garnier.

Feldhay, Rivka, and Michael Heyd. 1989. The Discourse of Pious Science. *Science in Context* 3-1: 102–142.

Firode, Alain. 2008. Le cartésianisme dans le cours de philosophie au début du XVIIIe siècle. *Histoire de l'éducation* 129: 55–76.

Gabbey, Alan. 1998. New Doctrines of Motion. In *Cambridge History of Seventeenth-Century Philosophy,* ed. Daniel Garber and Michael Ayers with Roger Ariew and Alan Gabbey, 649–79. Cambridge: Cambridge University Press.

Gagliardi, Ernst, and Ludwig Forrer. 1982. *Neuere Handschriften seit 1500.* Zurich: Zentralbibliothek Zürich.

Galiffe, J.-B.-G. 1884. *Notices généalogiques sur les familles genevoises*. Geneva: Jullien.

Garber, Daniel, and Michael Ayers, ed. with Roger Ariew and Alan Gabbey. 1998a. *Cambridge History of Seventeenth-Century Philosophy.* Cambridge: Cambridge University Press.

Garber, Daniel, John Henry, Lynn Joy, and Alan Gabbey. 1998b. New Doctrines of Body and Its Powers, Place, and Space. In *Cambridge History of Seventeenth-Century Philosophy,* ed. Daniel Garber and Michael Ayers with Roger Ariew and Alan Gabbey, 553–623. Cambridge: Cambridge University Press.

Germiquet, J. 1889. Clerus Rauraciae reformatus IIe partie: Les pasteurs d'Erguël, 1530-1888. *Actes de la Société Jurassienne d'Emulation* 2: 30–48.

Goeing, Anja-Silvia. 2007. Schulausbildung im Kontext der Bibel: Heinrich Bullingers Auslegung des Propheten Daniel (1565). In *Heinrich Bullinger, Life—Thought—Influence. Zurich, Aug. 25–29, 2004, International Congress Heinrich Bullinger (1504–1575)*, ed. Emidio Campi and Peter Opitz, vol. 1, 437–458. Zurich: TVZ.

———. 2016. Paduan Extracurricular Rhetoric, 1488-1491. In *For the Sake of Learning: Essays in Honor of Anthony Grafton*, ed. Ann Blair and Anja-Silvia Goeing, 542–560. Leiden: Brill.

———. 2017. *Storing, Archiving, Organizing: The Changing Dynamics of Scholarly Information Management in Post-Reformation Zurich.* (*Library of the Written Word*). Leiden: Brill.

———. 2021. The Genevan Academy. Scrutinizing European Connections in the Time of Theodore Beza. In *A Companion to the Reformation in Geneva*, ed. Jon Balserak, 277–299. Leiden: Brill.

Grafton, Anthony. 1981. Text and Pupil in the Renaissance Class Room. A Case-Study from a Parisian College. *History of Universities* 1: 37–70.

Hajnal, István. 1959. *L'enseignement de l'écriture aux universités médiévales*. Budapest: Maison d'édition de l'Académie des Sciences de Hongrie.

Hall, David. 2008. *Ways of Writing: The Practice and Politics of Text-Making in Seventeenth-Century New England.* Philadelphia: University of Pennsylvania Press.

Heyd, David. 2020. My Brother Michael Heyd (1943–2014): A Personal Intellectual Profile. *The European Legacy, Toward New Paradigms*, 25-7-8: 722–732.

Heyd, Michael. 1973. Jean-Robert Chouet et l'introduction du cartésianisme à l'Académie de Genève. *Bulletin de la Société d'histoire et d'archéologie de Genève* 15: 125–153.

———. 1979. From a Rationalist Theology to Cartesian Voluntarism: David Derodon and Jean-Robert Chouet. *Journal of the History of Ideas* 40: 527–542.

———. 1982. *Between Orthodoxy and the Enlightenment: Jean-Robert Chouet and the Introduction of Cartesian Science in the Academy of Geneva*. The Hague: Nijhoff.

Iliffe, Rob. 2012. Servant of Two Masters: Fatio de Duillier Between Christiaan Huygens and Isaac Newton. In *Newton and the Netherlands: How Isaac Newton Was Fashioned in the Dutch Republic*, ed. Eric Jorink and Ad Maas, 67–92. Leiden: Leiden University Press.

Jeger, Isabelle. 2016. *Catalogue des manuscrits latins 1–376*. Geneva: Bibliothèque de Genève.

Keller, Vera, Anna Marie Roos, and Elizabeth Yale, eds. 2018. *Archival Afterlives: Life, Death, and Knowledge-Making in Early Modern British Scientific and Medical Archives*. Leiden: Brill.

Klauber, Martin I. 1998. Eighteenth-Century Geneva: Natural Theology from Jean-Alphonse Turretin to Jacob Vernet. In *The Identity of Geneva: The Christian Commonwealth, 1564–1864,* ed. John B. Roney and Martin I. Klauber, 129–142. Westport: Greenwood Press.

Labrousse, Elisabeth. 1963. *Pierre Bayle*, vol. 1 *Du Pays de Foix à la Cité d'Erasme* (Archives internationales d'histoire des idées: 1). The Hague: Nijhoff.

Lacki, Jan. 2007. The Physical Tourist: Geneva from the Science of the Enlightenment to CERN. *Physics in Perspective* 2: 231–252.

Leonhardt, Jürgen. 2008. Classics as Textbooks. In *Scholarly Knowledge: Textbooks in Early Modern Europe*, ed. Emidio Campi, Simone de Angelis, Anja-Silvia Goeing, and Anthony Grafton, 89–112. Geneva: Droz.

Lewis, Charlton T., and Charles Short. 1879. *A Latin Dictionary*. Oxford: Oxford University Press. Accessed online through the Perseus Digital Library.

Magnin, Charles, and Christian Alain, Muller, ed. with Blaise Extermann. 2012. *Enseignement secondaire formation humaniste et société, XVIe-XXIe siècle*. Geneva: Slatkine.

Mandelbrote, Scott. 2005. The Heterodox Career of Nicolas Fatio de Duillier. In *Heterodoxy in Early Modern Science and Religion*, ed. John Brooke and Ian Maclean, 263–296. Oxford/New York: Oxford University Press.

———. 2018. Fatio, Nicolas, of Duillier (1664–1753). In *Oxford Dictionary of National Biography* (online ed.)

Marti-Weissenbach, Karin. 2012. Kaspar Wyss. In *Historisches Lexikon der Schweiz.* Online http://www.hls-dhs-dss.ch/textes/d/D26290.php. Accessed 23 Mar 23.

Niebler, Klaus. 1969. *Die Handschriften der Badischen Landesbibliothek in Karlsruhe, vol 10.1: Die Handschriften von St. Peter im Schwarzwald.* Wiesbaden: Harrassowitz.

Ragni, Alice. 2020. L'ontologie à Genève: de David Derodon à Jean-Robert Chouet. *Les Études Philosophiques* 203-3: 59–77.

Recous, Noémie. 2016. S'intégrer dans la République des Lettres. Le cas de Nicolas Fatio de Duillier (1681-1688). *Revue historique* 677: 83–112.

Sacchini, Francesco. 1786. *Moyens de lire avec fruit.* Trans. Durey de Morsan. The Hague/Paris: Chez Guillot.

Sandys, Arthur, ed. 1901. *Annals of de Normandie as preserved in documents, notes, private papers, public records, genealogies, the writings of old authors, and the registers of the city of Geneva.* Cambridge: Riverside Press.

Santschi, Catherine. 2005. Chouet, Jean-Robert. *Dictionnaire historique de la Suisse.* Online http://www.hls-dhs-dss.ch/textes/f/F25650.php. Accessed 23 Mar 2021.

Sarton, George. 1950. Boyle and Bayle: The Sceptical Chemist and the Sceptical Historian. *Chymia* 3: 155–189.

Schmutz, Jacob. 2008. Le petit scotisme du Grand Siècle: Étude doctrinale et documentaire sur la philosophie au Grand Couvent des Cordeliers de Paris, 1517-1771. *Quaestio* 8: 365–472.

Sharratt, Peter. 1975. Nicolaus Nancelius, 'Petri Rami Vita', Edited with an English translation. *Humanistica Lovaniensia* 24: 161–277.

Sina, Mario. 2008. Introduzione. In *La corrispondenza di Jean-Robert Chouet Professore di Filosofia a Saumur et a Ginevra, Con documenti inediti relativi al suo insegnamento filosofico*, ed. Mario Sina, VII–XL. Florence: Olschki.

———. 2010. Introduzione. In Jean-Robert Chouet, *Corsi di Filosofia*, ed. Mario Sina, Marco Ballardin and Elena Rapetti, vol. 1, XIII–LXX. Florence: Olschki.

Speziali, Pierre. 1997. Physica Genevensis. In *La vie et l'œuvre de 33 physiciens genevois, 1546–1953*, ed. Charles P. Enz. Geneva: Georg.

Stelling-Michaud, Suzanne. 1959. *Le livre du recteur de l'Académie de Genève*. Geneva: Droz.

Verboon, Annemieke R. 2014. The Medieval Tree of Porphyry: an Organic Structure of Logic. In *The Tree: Symbol, Allegory, and Mnemonic Device in Medieval Art and Thought*, ed. Salonius Pippa and Andrea Worm, 95–116. Turnhout: Brepols.

Vermij, Rienk. 2002. *The Calvinist Copernicans: The Reception of the New Astronomy in the Dutch Republic, 1575–1750.* Amsterdam: Koninklijke Nederlandse Akademie van Wetenschappen.

Waquet, Françoise. 2003. *Parler comme un livre: l'oralité et le savoir, XVIe-XXe.* Paris: Albin Michel.

Wyss, Kaspar, and Jean-Robert Chouet. 1661. *Disputationum Physicarum Nona, De Motu. Cum Thesibus ex universa Philosophia depromptis.* Geneva: Miege.

Yale, Elizabeth. 2016. *Sociable Knowledge: Natural History and the Nation in Early Modern Britain.* Philadelphia: University of Pennsylvania Press.

Chapter 8
Pierre Bayle as a Teacher of Philosophy

Martine Pécharman

Abstract As a teacher of philosophy at the Protestant Academy of Sedan (1675–1681), Pierre Bayle composed a course arranged according to the usual quadripartite schema of the curriculum, logic, ethics, physics, metaphysics. Bayle's *Systema totius philosophiae*, however, was not an ordinary textbook in philosophy. Instead of being predetermined by the curriculum requirements, its inner structure was strictly commanded by an original perspective on philosophical training in the post-Cartesian period. The central objective was to teach a 'disputational' physics, leading by the use of the logical patterns of Scholastic dialectic to the destruction of the Aristotelian legacy in natural philosophy. Taking into account the context and genesis of Bayle's course of philosophy, the present paper outlines the methodological decision supporting its composition. The originality of Bayle's academic introduction to natural philosophy, it notes, does not consist only in applying the Scholastic style of argumentation to defend anti-Scholastic physical principles. For Bayle, in addition to being the most valuable ally in the promotion of the mechanical philosophy, the technical framework furnished by Scholastic logic is able to satisfy a 'conceivability' criterion of Cartesian inspiration.

Keywords Argumentation · Bayle, Pierre · Conceivability · Dialectical style · Matter and form · Natural philosophy · Physical principles · Quantity · Scholastic logic · Substantial form

---

M. Pécharman (✉)
CRAL, CNRS-EHESS, Paris, France
e-mail: martine.pecharman@ehess.fr

S. Berger, D. Garber (eds.), *Teaching Philosophy in Early Modern Europe*, Archimedes 61, https://doi.org/10.1007/978-3-030-84621-3_8

Pierre Bayle (1647–1706) started teaching philosophy at the Protestant Academy of Sedan in November 1675.[1] His teaching there lasted until the suppression of the Academy by order of Louis XIV in July 1681. It was organized as a two-year "campaign" (Bayle's word) with the same team of students: logic and ethics the first year, physics and metaphysics the second year. Bayle had to deliver daily two lessons of two hours each, plus one hour tutoring.[2] In his lessons, first he dictated a pre-written Latin course and then he explained what he had dictated.[3] This quadripartite course, logic, ethics, physics and metaphysics, arranged according to the order of the curriculum was completed by early 1681, when Bayle began writing his first published work, the *Lettre […]. Où il est prouvé par plusieurs raisons tirées de la Philosophie, & de la Theologie, que les Cometes ne sont point le presage d'aucun malheur* (1682).[4] Unpublished during Bayle's lifetime, the course composed for his Sedan professorship was included in the volume IV (1731) of his posthumous *Œuvres diverses* published at The Hague.[5]

This early writing, entitled *Systema totius philosophiae* by its editors,[6] has usually been neglected in interpretations of Bayle: Bayle as a teacher of philosophy at Sedan has been disregarded on behalf of Bayle as the so-called "philosopher of

---

[1] For the circumstances that led Bayle to apply to this position, see his letters to his friend Jacques Basnage (1653–1723), who had informed him about this vacant chair of philosophy and encouraged him to be a candidate: Letter 83, 3 April 1675; Letter 90, 5 May 1675 (CB II, 105, 160–161). Cf a letter to his elder brother Jacob in the same period: Letter 91, 18 May 1675 (CB II, 167–168). His brilliant disputations on the imposed topics of "time" and "virtues in general" on 22–23 October 1675 led Bayle to be chosen on 2 November 1675 as the successor at Sedan of Claude Pithoys (1587–1676), who was compelled to retire because of his age. For the details of the competition, see Letter 112, 29 October 1675, CB II, 291–292. The other professor of philosophy at the Sedan Academy was Etienne Brazi (1618–1681).

Please note that "OD" abbreviates Bayle, *Œuvres diverses* (Bayle 1964–68 and 1982), and "CB" abbreviates Bayle, *Correspondance* (Bayle 1999–2017), with the Roman numeral following designating the relevant volume.

[2] Bayle's audience at Sedan consisted of very few pupils: see Letters 113, 11 November 1675 (first day of Bayle's teaching), 142, 14 September 1677 (beginning of the "second campaign"), 144, 16 November 1677, CB II, 294, 436–437, 447. Even so, Bayle was regularly complaining that to be a "Professor of Philosophy" was to him an unbearable burden: in addition to Letter 113, see Letters 115, 17 December 1675, 117, 6 February 1676 ("I am obliged to work as hard as a convict"), CB II, 303, 311.

[3] Bayle's teaching position at the Rotterdam Illustre School from the fall of 1681 required a different way of teaching philosophy: to dictate a written course was no longer possible with an audience of learned people (Letter 200, 26 March 1682, CB III, 287).

[4] Republished in an augmented and different form in 1683: *Pensées diverses à l'occasion de la comète qui parut au mois de décembre 1680*. Both versions were anonymous.

[5] Volumes I-III of the *Œuvres diverses* were published in 1727. Bayle 1964–1968 (OD I-IV) is the reprographic reprint of the edition of 1727–1731.

[6] The editors' full title reads: Institutio Brevis et Accurata Totius Philosophiae, *in quatuor praecipuas partes distincta,* Logicam, Ethicam, Physicam, & Metaphysicam, *in usum Studiosae Juventutis*. The abbreviated title Systema totius philosophiae appears at the top of the Prooemium's first page. See OD IV, 199, 201a. The Latin text and its French translation occupy some 320 pages in OD IV, 199–520.

Rotterdam." His 1675–81 professorial lectures have gained scholarly attention only in retrospect, from the standpoint of his later non-academic writings. The *Systema* was generally viewed as unable to provide more than vague indications of the philosophical themes which were to become dominant in Bayle's mature thought.[7] Todd Ryan, however, has defended another approach. Ryan shows that Bayle's first writing (the most technical he would ever make) proves to be "an indispensable aid for reconstructing Bayle's understanding of numerous philosophical issues."[8] For example, Ryan claims, the *Systema* is extremely important to understanding some metaphysical arguments which are too briefly exposed in Bayle's *Dictionnaire historique et critique* (1697). Ryan's approach constitutes a decisive step forward. Yet, it still only judges Bayle's *Systema* in terms of its usefulness for our comprehension of his philosophical development. I propose that Bayle's course of philosophy should also be considered in its own right: not as merely the prefiguration of Bayle's full-fledged philosophy, but also as the specimen of philosophical training that the young Sedan professor had planned to work out.[9] From this viewpoint, the *Systema totius philosophiae* should be acknowledged as an original textbook of philosophy in the second half of the seventeenth century. My essay attempts to reconsider the *Systema* from this perspective. After detailing the context which preceded the composition of the *Systema*, namely, Bayle's personal formation in philosophy, I discuss the purpose of Bayle's academic introduction to philosophy. The *Systema*, I show, was an argument in the Scholastic style defending anti-Peripatetic physical principles.

8.1 In the Background of the *Systema*

8.1.1 Studies

Two opposing collections of philosophical theses fix the bounds of Bayle's intellectual evolution from the conclusion of his studies in philosophy with the Jesuits in Toulouse to the conclusion of his philosophy teaching in Sedan. When he took the degree of Master of Arts in August 1670 at Toulouse, Bayle defended a series of anti-Cartesian theses, published under the title *Selecta Peripateticae Philosophiae Pronunciata*.[10] But over the decade his views had changed: the *Theses philosophicae* defended at Sedan in August 1680 support Cartesianism against a recent Jesuit

[7] See for example Mori 1999, 91–92.

[8] Ryan 2009, 6.

[9] See Gros 1989, 75–99.

[10] Bayle 1670 (reproduced in Bayle 1982, 2–11).

attack.[11] This volte-face, however, does not involve a complete rejection by the professor Bayle of his Jesuit education. In particular, a comparison made by Elisabeth Labrousse of the *Physica* of the *Systema* with the course on physics delivered in 1669–1670 at Toulouse by the Jesuit Pierre Rome (?-1693) reveals some indebtedness to Father Rome.[12] Though Rome's teaching included some rudiments of Cartesian and Gassendist physics, the philosophical theses Bayle defended for graduating on 7 August 1670 were uncompromisingly directed against the new philosophy. Divided into the traditional four rubrics of the collegiate curriculum, *Ex Logica, Ex Physica, Ex Ethica, Ex Metaphysica* (in this order), their conclusive subsections enumerate erroneous doctrines of the *recentes Philosophi*, the Cartesians and the neo-Epicureans or Gassendists. For instance, the 1670 theses reject as fallacious the Epicurean canonic based on sense. They also reject the Cartesian method of universal doubt as prohibiting any access to truth and plead for Peripatetic logic as the best method for the acquisition of all sciences. Another proposition in the same subsection distinguishes the material composition of natural bodies from Epicurean atoms. It affirms the reality of substantial forms as irreducible to the properties and diverse arrangements of the parts of matter. It also affirms that some accidents (the sensible qualities) are really separable from all substances and that the contrary view in Descartes is just temerarious. In the final subsection, the penultimate proposition protests against Descartes that if extension constitutes the *ratio corporis*, then we can doubt the real presence of the body of Christ in the Eucharist, since Christ has no extension in the Eucharist; or, at least, Bayle argues, the Cartesian thesis on extension makes it difficult to deny that the same body (namely, bread) remains in the Eucharist, since the same extension remains. The last proposition returns to the topic of substantial forms to denounce the rejection by the *recentiores Philosophi* or

---

[11] They were printed in association with another writing, the *Dissertatio in qua vindicantur a Peripateticorum exceptionibus rationes quibus aliqui Cartesiani probarunt essentiam corporis esse in extensione*. The *Dissertatio* was a response to anti-Cartesian attacks by the Jesuit Louis Le Valois (1639–1700), as discussed below. It developed the arguments of the *Theses philosophicae* supporting the position that extension is the essence of body. In 1684, Bayle had his 1680 Latin theses and dissertation both incorporated into a *Recueil de quelques pieces curieuses* which he published in defense of Cartesian philosophy: Bayle 1684, 138–266 (also OD IV, 109–145). For Bayle's presentation of the *Dissertatio* and the *Theses philosophicae* as complementary writings, see the *Recueil*'s "Avis au Lecteur" (OD IV, 186–188); cf. Letter 186, 28 October 1680, CB III, 229, for the characterization of the Sedan philosophical theses as directed against "Mr de La Ville" (the pseudonym used by Le Valois). Bayle published a review of the *Recueil* in the first issue of the *Nouvelles de la République des Lettres*, March 1684, art. III (OD I, 11a).

[12] On the influence of Rome's physical teaching on the content of the *Physica* of Bayle's *Systema*, see Labrousse 1964, 142–143. Cf eadem, 1963, 95–96. A manuscript notebook of this physics course by Rome is conserved in the Toulouse municipal library, MS 906: "Phisica (philosophiae secunda pars), a R. P. Rome, Soc. Jesu, anno 1670, Tholosae tradita, Petro Malefette audiente" (332 folios in 8°). Bayle's other Jesuit teachers at Toulouse were Antoine Bonnet (1634–1700) for logic and Joseph-Louis Germain (1633–1722) for metaphysics.

"new philosophers"[13] of the "eduction" of forms from matter and their argument that nothing in nature is produced anew.

According to the Huguenot theologian Pierre Jurieu (1637–1713), Bayle's ferocious adversary, especially after the anonymous publication of the *Avis important aux Réfugiez sur leur prochain retour en France* (1690),[14] his Jesuit education at Toulouse had changed Bayle into an enemy of the Protestant Church. In his reply, *La Cabale chimerique, ou Refutation de l'Histoire fabuleuse, & des calomnies que Mr. J. vient de publier malicieusement* (1691), Bayle corrects Jurieu's allegation:

> It is not the case that he ["Mr. Bayle"] was passionate against Reformed religion [...] If he was then passionate against anything, it was against the new philosophy; on occasion, indeed, he disputed vehemently in favor of the Scholastic philosophy recorded in his notebooks.
>
> It is not the case that he learned from Jesuits anything other than Peripatetic philosophy, which he abandoned shortly afterwards.[15]

As a matter of fact, just after defending Peripatetic theses with Jesuits, Bayle abjured Catholicism on 21 August 1670 and returned to Protestantism. He had formerly abjured Protestantism in March 1669 because of his disappointment with the weakness of Calvinist arguments in the Eucharist controversy. The allegiance of his Toulouse theses to the Catholic doctrine of transubstantiation, though, was a façade. The Jesuit attacks on Descartes's doctrine of body and extension had drawn Bayle to view Catholic objections to Calvinist arguments as mere sophisms.[16] Hence his second conversion, or return to Protestantism. Being a "*relaps*," he was punishable by imprisonment. Bayle had to flee to Geneva, where he arrived in early September 1670. He enrolled there as a student of theology. Now, the reformed Academy at Geneva was then becoming a Cartesian stronghold. In his refuge, accordingly, Bayle was separated from the massive anti-Cartesian campaign launched in France by midsummer 1671.[17] Cartesianism was quite successful at the Academy of Geneva in the 1670s, insofar as both theology teaching by Louis Tronchin (1629–1705) and philosophy teaching by Jean-Robert Chouet (1643–1731) were under Cartesian patronage. In a letter to his father on 21 September 1671, Bayle portrays Tronchin

[13] Even though the editors of the *Systema* translate Bayle's *recentiores Philosophi* by *les philosophes modernes*, I will understand it as meaning more strictly *the new philosophers*. The opposition is not between Ancients and Moderns, but between *Peripatetici* (neo-Aristotelians) and *recentiores Philosophi* (Cartesians and Gassendists).

[14] See the critical edition of the *Avis* by Gianluca Mori: Bayle 2007. The Revocation of the Edict of Nantes had been decided by Louis XIV on 22 October 1685. Bayle and Jurieu had been previously very good friends at the Academy of Sedan.

[15] OD II, 759b.

[16] *Cabale chimerique*, OD II, 759a. Against Jurieu's accusation, Bayle claims that through his studies in philosophy at Toulouse he acquired a clearer view of "the impossibility of transubstantiation."

[17] The target was the propagation of doctrines refusing the separability of accidents from substance. On this decree defending the Aristotelian underpinnings of the Eucharist, see Schmaltz 2017, ch. 1. Ariew 2011, ch. 9, 267–294, has nicely documented its consequences on the academic destiny in France of the new Cartesian philosophy.

as "the most judicious theologian in Europe" for the way he applies the principles of Cartesian physics to advance the cause of Calvinism:

> You just cannot imagine how many advantages he draws from the philosophy of Mr Des Cartes (to which he pays a rather explicit tribute) to war against Roman Catholics.[18]

The same letter describes Chouet, "an extremely refined mind," also very "solid," as a strictly pro-Cartesian professor of philosophy.[19]

Bayle mainly details Chouet's experiments of physics (including a series of experiments on the vacuum and the air pressure). As is obvious from his precise observations, most of Chouet's acclaimed experimental protocols at the Academy of Geneva in 1671 were no other than the "new experiments" fully described in the *Traité de physique* published by Jacques Rohault (1617/1618–1672) at the beginning of the same year.[20] However, the most remarkable fact about Chouet was his ability to inject Cartesian concepts into the standard structure of teaching, and to replace the Aristotelian categories with them.[21] No doubt Bayle found in the teaching of natural philosophy by Chouet at Geneva the model of a smooth transition from Scholasticism to the philosophy of the *recentiores*—the philosophy he himself had condemned in the 1670 Toulouse theses.

8.1.2 Readings

Bayle abandoned his theology studies, presumably for financial reasons, in June 1672 to become a full-time preceptor of young children. The twenty months spent at the Academy of Geneva were followed with three years of tutoring.[22] Over this five-year period before he competed for the Sedan professorship, Bayle himself already deserved the label of *helluo librorum*, "devourer of books," which he applied

---

[18] Letter 11, 21/11 September 1671, CB I, 47.

[19] Ibid, 47–48. Bayle specifies that Chouet had previously gained a wide reputation as professor of philosophy at the Academy of Saumur. See Armogathe 2007, 240. (Chouet had become famous in 1664 as a champion of Cartesian physics by defeating Pierre de Villemandy (1636–1703), his Aristotelian competitor for the vacant chair of philosophy at Saumur, with the question: "Why is red always the highest color in the rainbow and green always the lowest?").

[20] When Bayle reports the experiments performed by Chouet, he merely adds that it would be too long to describe with all the particulars their "order and manner" (CB I, 48). His letter does not say whether Chouet's teaching emphasized the main characteristic of Rohault's physics: not so much the imagination of a wide variety of experiments, but the dependency of those experiments on a geometric method. Rohault's "new experiments" confirm conclusions derived from principles (definitions, axioms, hypotheses). See *Journal des savants*, 22 June 1671, 26: Rohault, the reviewer writes, "shows that upon some hypothesis, things must necessarily happen in the way we see they actually happen," so that his book "consists only of a series of reasoned experiments methodically arranged."

[21] On this, see Heyd 1982.

[22] From June 1672 to May 1674, Bayle was preceptor for the Count of Dohna's two young sons at Coppet. He became tutor to other children from mid-June 1674 at Rouen and from early April 1675 at Paris.

at that time to his friend Jacques Basnage.[23] Yet, during the interval between his arrival at the Academy of Geneva as student of theology and his arrival at the Academy of Sedan as professor of philosophy, Bayle's correspondence mentions few books of philosophy among his many and eclectic readings.[24] A letter on 2 November 1670 makes reference to "l'*Idee des sciences* en 3 petits tomes," that is, *Le portrait de la sagesse universelle, avec l'idée générale des sciences* (1655) by Jean Macé, the Carmelite Léon de Saint-Jean (1600–1671). To Bayle's eyes, the book was not "universally good" even though it contained some good things. Instead, at the beginning of his theology studies at Geneva, Bayle's "major study" in philosophy was according to this letter "something from the *Philosophie* of Mr. De Rodon."[25] David Derodon (1600–1664) had been Chouet's professor of philosophy at the Academy of Nîmes. Bayle may be alluding to "something" either from Derodon's *Philosophia contracta* (1664), composed of a *Logica* and the first part of a *Metaphysica*, which was used as a textbook at the Geneva Academy, or from the *Opera philosophica hactenus in lucem edicta* (1668) collected after Derodon's death.[26] As the article "RODON (David de)" of the *Dictionnaire historique et critique* corroborates in 1697, Derodon was famous from the mid-seventeenth century both as a Scholastic logician and as a Gassendist natural philosopher.[27] When in the fall of 1670 at Geneva he read "something from the *Philosophie* of Mr. De Rodon," Bayle inevitably had already heard of a virulent controversy generated among Protestants in the late 1650s by Derodon's denial of the notion of continuous creation.[28] Interestingly, when a few years later he composes the lectures on metaphysics for his course at the Academy of Sedan, the axiom to which Bayle devotes the greatest attention in Descartes's metaphysics is the axiom stating that the distinction between conservation and creation is only a distinction of reason.[29] We may conjecture that Bayle's "major study" of Derodon's *Philosophie* at Geneva played a

[23] Letter 18, 2 July 1672, CB I, 112; Letter 79, 9 March 1675, CB II, 83 (*helluo librorum* is from Cicero, *De finibus bonorum et malorum*, III, ch. 2, 7—applied there to Cato). Bayle's own insatiable curiosity for new learned publications would result later in his editorial activity of the *Nouvelles de la République des Lettres* (printed at Amsterdam), from the first issue of the journal in March 1684 until February 1687, when illness would force him to resign.

[24] For Bayle's self-portrait as an eclectic reader, see Letter 13, 21 September 1671, CB I, 67.

[25] Letter 10, 2 November 1670, CB I, 33.

[26] Derodon was a prolific author. Among his other publications: *Disputatio de supposito* (1645), *Logica restituta* (1659), *Metaphysica* (1659), *Traité démontrant par raisons naturelles qu'il y a un Dieu* (1659), *Disputatio de atomis* (1661), *Disputatio de ente reali* (1662), *Disputatio de libertate* (1662), *Compendium logicae* (1663), *Discours contre l'astrologie judiciaire* (1663).

[27] The article of the *Dictionnaire* depicts Derodon as a highly subtle dialectician, on par with the most sophisticated Scholastic logicians on universals, beings of reason, categories, and syllogistics. On physical topics, it places Derodon largely above Scholastics, insofar as he was a Gassendist atomist, able to explain many natural effects merely by mechanical principles.

[28] See Heyd 1979, 533.

[29] In the *Synopsis metaphysicae* forming the fourth and final part of the *Systema*, a subsection of ch. ii *De principiis entis* of the *Metaphysica generalis* examines the most basic axioms of Cartesian metaphysics. For this specific axiom and Bayle's comments on the conservation of substances, see OD IV, 477a–479a.

substantial role in securing an important place for the continuous creation thesis in the *Systema* composed at Sedan.[30]

When in the autumn 1671 he tells his elder brother Jacob about his manner of reading, Bayle confesses: "no flighty lover (*amant volage*) has as often changed mistresses as I have changed my readings."[31] However, in spite of the fact that books of philosophy were proportionally a tiny part of Bayle's readings over the years 1670–1675, his philosophical choices were not at all random. In philosophy, Bayle gave preference to two kinds of readings: reports of physical experiments and categorizations of the history of philosophy. His knowledge of the advancement of experimental physics was in particular drawn from a publication that challenged the *Journal des Savants* (of which he was a constant reader), the *Memoires concernant les arts et les sciences*.[32] Through the *Memoires*, Bayle was acquainted with lectures about experimental natural philosophy given in Paris by a Cartesian, Jean-Baptiste Denis (1640–1704), whom he describes as a resolute partisan of the new physics (*fort enteté de la physique*).[33] In addition to some issues of the *Memoires* that Basnage lent him, Bayle knew the collection Denis had published of his lectures, the *Recueil des memoires et conferences qui ont esté présentées à Monseigneur le Dauphin pendant l'année M.DC.LXXII*.[34] Concerning the classification of philosophers, Bayle found himself disappointed with the *De philosophia et philosophorum sectis libri II* (1657–1658) by Johannes Gerardus Vossius (1577–1649), published posthumously by his son Isaac.[35] In a substantial letter in January 1673 to Vincent Minutoli (1639–1709) on the topic of typologies in the history of philosophy, Bayle expressed the judgment that in his presentation of the ancient sects of philosophers, Vossius does not completely satisfy our curiosity. Our curiosity (namely, Bayle's curiosity) focuses on three primary questions: *when* did the different sects begin to be constituted? *who* were their supporters and followers? *how* did they propagate themselves in many places? Bayle cannot understand why Vossius did not try to elucidate these questions: "Vossius has neglected to clarify this for us, I don't know why."[36] The kind of classification that, on the contrary, Bayle particularly praised was "the usual most general division of all the sects of philosophers" into Dogmatics,

[30] The exposition of categories in the *Logica* of the *Systema* already gives Bayle an opportunity to contend that the conservation of a substance is continuous creation: not only to begin to exist, but also to continue to exist, requires divine omnipotence. See OD IV, 228b ("the common opinion of philosophers that conservation is a continuous production of the thing.")

[31] Letter 13, 21 September 1671, CB I, 67.

[32] Letter 23, 9 October 1672, CB I, 134.

[33] Denis had published in 1665 at Paris a *Discours sur les cometes, suivant les principes de M. Descartes*.

[34] Bayle mentions appreciatively Denis's *Recueil* in Letter 98, 26 June 1675, CB II, 202; cf. Letter 103, 13 July 1675, CB II, 231. On the importance of experiments by Denis in the history of medicine, see Hamraoui 2008, 330b–333a.

[35] On the numerous learned publications by Vossius known to Bayle, see Letter 13, 21 September 1671, CB I, 67.

[36] Letter 31, 31 January 1673, CB I, 185.

Academics, and Pyrrhonians or Skeptics.[37] Among the ancient philosophers, he insists in his letter to Minutoli, there are those (Aristotelians, Stoics, and Epicureans) who believed they had discovered truth, those who asserted that no truth can be discovered, and those who carried on investigating and searching for truth. Obviously, to Bayle's eyes, the tripartite categorization of philosophers in Montaigne's *Apologie de Raimond Sebond* and its model in Sextus Empiricus's *Outlines of Pyrrhonism* were essential to determine the most general trend in the history of philosophy from antiquity to the seventeenth century. In Bayle's opinion, even though none of the ancient philosophical sects can be deemed to have overcome the others, the greatest ancient dogmatic and academic philosophers fundamentally tended to Pyrrhonism, as do the Gassendists or neo-Epicureans in recent times.[38]

Bayle highly valued the comparative history of philosophical doctrines. In contrast with his dissatisfaction with Vossius, he considered that *La Comparaison de Platon et d'Aristote*, published in 1671 by the Jesuit René Rapin (1621–1687), was an "excellent" book for improving our knowledge of those "two great sect leaders."[39] Later on, his correspondence shows a permanent interest in publications attempting a conciliatory comparison between the new and ancient philosophers. The multiplication of such books was no doubt of great worth in his eyes. In a letter in March 1675 mentioning that a second edition of the *De Consensu veteris et novae Philosophiae* by Jean-Baptiste Du Hamel (1623–1706), the secretary of the Academy of Sciences, was at that moment in print at Rouen, Bayle was keen to emphasize that already a "*Parallele de la philosofie de Des Cartes et d'Aristote*" had been recently published in French—namely, the *Parallele des principes de la physique d'Aristote, & de celle de René Des Cartes* published in 1674 by René Le Bossu (1631–1680).[40] However, Bayle did not neglect reading some non-ecumenical books of philosophy, especially those written by Gassendists. In the months preceding his election to the chair of philosophy at Sedan, his correspondence of June 1675 mentions the recent publication of the second tome of the *Abregé de la philosophie de Gassendi* by François Bernier (1620–1688)—which, by a funny *lapsus calami*, Bayle miswrites *Abregé de la philosophie d'Epicure*—and of Gilles de

[37] Ibid., 185–187.

[38] Ibid., 187.

[39] Ibid., 192 ("ces 2 grands chefs de parti"). The full title of Rapin's book reads: *La Comparaison de Platon et d'Aristote, avec les sentimens des Pères sur leur Doctrine, et quelques reflexions chrestiennes*. At Sedan, while he works on his course of physics, Bayle, who continues to read the *Journal des Savants*, has his attention drawn again to a recent publication by Rapin, his 1676 *Reflexions sur la philosophie ancienne et moderne* (Bayle quotes *Reflexions sur l'ancienne et nouvelle philosophie* as the title). See Letter 126, 25 August 1676, CB II, 353 (cf a first mention of this book in Letter 123, 19 July 1676, CB II, 340). The same letter mentions a three-volume book by the Capuchin friar Casimir de Toulouse (c. 1633–1674), his *Atomi peripateticæ* published in 1674 (full title of this book: *Atomi peripateticæ, sive tum veterum, tum recentiorum Atomistarum placita, ad Neotericae Peripateticae scholae methodum redacta*).

[40] Letter 79, 9 March 1675, CB II, 83. This *Parallele* by Le Bossu based the explication of similarities between the principles of physics in Descartes and in Aristotle on Rohault.

Launay's *Introduction à la philosophie*,[41] both Gassendist tracts. In particular, his correspondence during the second year (on physics and metaphysics) of his Sedan teaching contains an appreciative description of Bernier's compendium of Gassendi's physical doctrines.[42]

Yet, Bayle was sometimes behind in acquiring firsthand knowledge of the most famous philosophical writings of the time. Stunningly, in 1674 he did not attempt to obtain Malebranche's *De la Recherche de la vérité*.[43] His correspondence mentioned only in an oblique way this publication: Bayle, who had just enjoyed reading the anonymous treatise *De l'egalité des deux sexes* (1673), reported to his elder brother that "an Oratorian Father, great Cartesian," the author of "a rather beautiful book" of the year "entitled *La Recherche de la verité*," was thought to be also the author of this treatise.[44] More than a year after its publication, Bayle had not yet read the first volume of Malebranche's *Recherche*. The judgment in a letter of July 1675 that "*La Recherche de la verité* by Father Malabranche, a priest from the Oratory," is "a most excellent book" is mere hearsay.[45] The same letter indicates that a *Critique de la Recherche de la vérité* attributed to "the Abbé Fauché, a native of Dijon" has been published some three or four months earlier.[46] Bayle, who had read neither the *Recherche*, nor its *Critique* by Simon Foucher (1644–1696), simply conjectures in

[41] Letter 101, 29 June 1675, CB II, 213. The first volume of Bernier's *Abregé de la philosophie de Gassendi* had been published in 1674. Bayle judges this *Abregé* most useful to "save the trouble of browsing Gassendi's big volumes." De Launay's *Introduction à la philosophie, contenant une ample Dissertation sur ses Questions préliminaires* is a corrected and augmented version of his *Dissertation de la philosophie en general* published in 1668. Bayle's letter mentions laudatorily a series of public lectures through which de Launay (? -1695) has been teaching "for a long time" Gassendi's doctrine. On the importance of de Launay as an advocate of Gassendist philosophy, see Lennon 1993, 66–73.

[42] See Letter 135, 28 March 1677, CB II, 402: "The philosophy of Mr. Bernier is very good. It is an *Abbregé*, which will be composed of several tomes, of the philosophy of Mr. Gassendy. So far he has procured two of them only. The first tome deals with the principles and general qualities of natural things. The second tome deals with heavens. Mr. de Gassendy devoted himself, properly speaking, only to physics, so that physics is his best achievement. His system of logic must be acknowledged as very comprehensive, but it is a little too concise in comparison with ordinary logics. Also his ethics is a most beautiful and learned work. As to his physics, we may assert that this work is infinitely profitable. It is necessary, though, to enjoy a life of leisure to read the whole of it. This is the reason why Mr. Bernier, one of his beloved friends and disciples, provides us with some excerpts. To my mind, what is best explained in them is qualities such as savors, odors, hardness, fluidity, etc."

[43] Nicolas Malebranche (1638–1715), *De la Recherche de la Verité. Où l'on traitte de la nature de l'esprit de l'homme, & de l'usage qu'il en doit faire pour éviter l'erreur dans les Sciences*. This volume, published in May 1674, contained Books I-III.

[44] Letter 66, 12 November 1674, CB II, 4. The true author was François Poullain de la Barre (1647–1723): *De l'Egalité des deux sexes. Discours physique et moral, où l'on voit l'importance de se défaire des Préjugez*.

[45] Letter 105, 21 July 1675, CB II, 245.

[46] Ibid. Simon Foucher had published anonymously in early 1675 his *Critique de la Recherche de la Verité. Où l'on examine en même-temps une partie des Principes de Mr Descartes. Lettre par un Academicien* [that is, by an Academic philosopher]. See Letter 83, 3 April 1675, CB II, 106: "A Critique de la Recherche de la verité (this book by Father Malabranche, of the Oratory, which was so highly esteemed) has been made. So far its author remains unknown."

this letter that Malebranche, who, he tells, is "presently occupied with the second part of his book," will give in it a reply to his opponent.[47] Until mid-1678, Bayle's correspondence no longer contains any additional information about Malebranche's *Recherche*. Bayle gives no information on the Preface against Simon Foucher included in September 1675 in the first edition of the second volume of the *Recherche*. He does not mention mid-1676 the second edition of this second volume, in which Malebranche adds to the Preface against Foucher an Advertisement against Robert Desgabets (1610–1678).[48] The evolution of the polemics on the *Recherche* initiated in early 1675 by Foucher remains unmentioned. Bayle's astonishing silence on Malebranche only ends after the publication in June 1678 of the third volume of the *Recherche*, *Contenant plusieurs Eclaircissemens sur les principales difficultez des précédens Volumes*. Bayle writes at that moment to his elder brother Jacob:

> As for the last news on books, I will tell you that Father Malabranche has provided a third volume of the *Recherche de la verité* in which he examines many difficulties proposed against his prior volumes. He is a great man, of profound judgment and extraordinary discernment, so this third part will be an Oedipus.[49]

The analogy between Malebranche's "elucidations" of the objections to the *Recherche* and Oedipus solving the Sphinx's enigma of mankind does not imply, however, that Bayle's favorable opinion on Malebranche's capacity to unriddle philosophical conundrums results from a direct knowledge of the *Recherche*. Bayle might be simply desirous to express a bit emphatically Malebranche's excellent reputation. It remains thus somewhat uncertain whether the young professor of philosophy at Sedan had read the first two volumes of Malebranche's *Recherche* before the publication of the third volume in early summer of 1678. Bayle's correspondence of the previous period only refers to Malebranche on the basis of external information.

The first indirect mention of the *Recherche* on 12 November 1674 (as a book whose author is supposed to have also penned *De l'egalité des deux sexes*) contrasts in particular with the mention in the very same letter of "a treatise of Mr. Rohaut … in which he explains transubstantiation by the principles of Descartes, and shows that beasts are automata" as a recent personal reading.[50] Bayle had direct knowledge of Rohault's *Entretiens sur la philosophie* (1671) reprinted in 1674 by Michel Le Petit. Later on, a letter of August 1675 underlines that when Bayle was attending the

---

[47] Ibid. Actually, Malebranche provides in 1675 a second revised and augmented edition of volume I of the *Recherche*, with a number of corrections derived from Foucher's critique. He published the volume II of the *Recherche* in September 1675, appending to it a Preface against Foucher.

[48] The Benedictine Dom Robert Desgabets had published anonymously at the end of 1675 the book *Critique de la Critique de la Recherche de la Verité, où l'on découvre le chemin qui conduit aux connoissances solides. Pour servir de Réponse à la Lettre d'un Academicien.*

[49] See Letter 155, 28 August 1678, CB III, 58. In May 1676, Foucher has published his *Réponse pour la Critique à la Preface du second volume de la Recherche de la verité, où l'on examine le sentiment de M. Descartes touchant les Idées, avec plusieurs remarques pour les sciences.*

[50] Letter 66, 12 November 1674, CB II, 4 ("I have just read a treatise of Mr. Rohaut…, etc").

Academy of Geneva, Chouet had discussed the manner in which Rohault, in his *Entretiens*, "applies the principles of his [*Cartesian*] philosophy to the hypothesis of transubstantiation."[51] Chouet, this 1675 letter insists, had proposed then to Bayle a demonstration which was quite effective against Rohault's explication of transubstantiation.[52]

Admittedly, over the decade 1670–1680, Bayle pursued a careful investigation of the use of the principles of Cartesian physics as arguments either *contra* or *pro* the Catholic account of transubstantiation.[53] However, this systematic inquiry led him to examine more generally the value of the Cartesian physical principles in comparison with Aristotelian principles in natural philosophy. Strikingly, it is from this viewpoint, with the design to defend the new principles of natural philosophy against neo-Aristotelian positions, that Bayle makes explicit in 1680 for the first time his own reading of Malebranche's *Recherche*. When the Jesuit Louis Le Valois (using the pseudonym Louis de La Ville) publishes his *Sentimens de M. Des Cartes touchant l'essence & les proprietez du corps, opposez à la doctrine de l'Eglise, & conformes aux erreurs de Calvin sur le sujet de l'Eucharistie* in the spring of 1680, Bayle declares in his correspondence that with this treatise, "the author proves indisputably what he intends to prove, because in fact it was not very difficult to do so."[54] Yet, as is obvious from a later letter in October 1680, after the publication of the *Sentimens*, Bayle held that this book had to be discussed at a level distinct from that of the Eucharistic controversy.[55] In the *Physica generalis* of his *Systema*,

[51] Letter 107, 3 August 1675, CB II, 268.

[52] Bayle specifies (ibid.) that Chouet's demonstration was on the contrary ineffective against "the method of Father Maignan." The article "MAIGNAN (Emanuel)" in the *Dictionnaire* will highlight in 1697 that Maignan, "one of the greatest philosophers of the seventeenth century," explained the conservation of accidents without a subject in the Eucharist better than Descartes had done (Maignan's "manner" for this explication was "more happy than that of Mr. Descartes".) For Bayle, the Minim Emmanuel Maignan (1601–1676) had seen the difficulties stuck to Descartes's explication of the Eucharist, whereas Rohault had not. (The article was much expanded in the *Dictionnaire*'s second edition in 1702.)

[53] For example, Bayle's constant interest in the two-sided argumentation on the relation between Cartesian physical principles and transubstantiation (arguments for their incompatibility and arguments for their compatibility) led him to welcome with equal consideration two manuscripts by the same author, Jean Terson (c. 1653–17..), who sustained at first the incompatibility thesis and then the opposite thesis: see Letter 157, October 1678?, Letter 194, 27 September 1681, CB III, 65, 267. Terson had defended in 1674 his *Theses theologicae* at the Academy of Puylaurens. He abjured Protestantism in 1681. See *Objections proposées à Monsieur Clerselier au mois de mai 1681 par Monsieur Terson, savant calviniste, que l'on destinait à être ministre de Charenton à qui on avait fait voir ce que le sieur Clerselier avait écrit sur le Saint Sacrement lequel ensuite se convertit*, reproduced from MS Chartres 366 in Agostini 2009, vol. 2, 171–181.

[54] Letter 183, 24 March 1680, CB III, 222. His theology studies with Tronchin have taught Bayle that on the subject of the Eucharist, the Cartesian philosophy does not support the transubstantiation hypothesis, it supports Calvinism.

[55] See Letter 186, 28 October 1680, CB III, 229. After mentioning that "a friend of Father Malebranche" has composed a writing replying to the Jesuit with an explication of the Eucharist "according to the new philosophers," Bayle adds: "Since I had no interest in reconciling my principles with the decisions of the Council of Trent, I have taken a different approach to repel Mr. de La Ville." In this passage, Bayle's "principles" are manifestly those of "the new philosophers," namely, the principles of the new physics.

written in the summer of 1676, the professor Bayle had incorporated the controversial schema opposing the *Reformati* and the *Pontificii* into the discussion of quantity ("is *quantitas* distinct from the *res quanta*?").[56] But that to which Bayle aims to react when he replies to the *Sentimens* with his 1680 *Dissertatio in qua vindicantur a Peripateticorum exceptionibus rationes quibus aliqui Cartesiani probarunt essentiam corporis esse in extensione* is the Jesuit's philosophical charge against the most fundamental principle of Cartesian physics.[57] Le Valois, indeed, does not attack Descartes with regard to his specific explications on the Eucharist in the *Quartae responsiones*. He attacks in a more radical way the first principle of Descartes's doctrine of natural bodies in the *Principia philosophiae*, the principle that extension constitutes the essence or nature of the corporeal substance. The *Sentimens* quotes and discusses in detail a series of articles from the *Principia*, as well as large excerpts from authors (including "L'autheur de la Recherche de la Verité")[58] embracing Descartes's thesis of the identity of body with extension. Accordingly, when Bayle undertakes to confute Le Valois, the target of his pro-Cartesian *Dissertatio* is strictly the Scholastic denial of the principle that the essence of body is extension. Neo-Aristotelians such as Le Valois, Bayle claims in the first article of the *Dissertatio*, view this principle as "an offense against the lights of natural philosophy."[59] Bayle intends to show instead that the proofs of the truth of this principle in the Cartesian writings assailed by Le Valois are rationally coherent. The *Dissertatio* responds to the objections and difficulties raised in the *Sentimens* against Clerselier (1614–1684), Rohault, and Malebranche.[60] In each case, the *Dissertatio* defends the Cartesian under assault, but it makes a special and more complex case with the defense of Malebranche, in particular because Le Valois alters and mutilates the text of the *Recherche*. Thus, in the summer of 1680, against the Jesuit's tricky arguments, Bayle has become a militant advocate of the argument concerning the essence of matter in the text of Malebranche's *Recherche*.[61]

[56] See OD IV, 287–289.

[57] Letter 185, 27 October 1680, CB III, 226.

[58] Le Valois 1680, 74.

[59] OD IV, 109a.

[60] See OD IV, 110–114, § II-VI for Clerselier, Preface to his 1664 edition of *L'Homme de René Descartes*; 114–124, § VII-XIV for Rohault, *Traité de physique*, Part I, ch. VII-VIII; 124–129, § XV-XIX for Malebranche, *Recherche de la vérité*, Book III, Part II, ch. VIII. Bayle singles out only these three Cartesian authors. Le Valois attacks a number of other Cartesians and also includes some Gassendists among them: see his *Sentimens*, Part I, ch. IV.

[61] The beginning of the *Dissertatio*'s subsection concerning Malebranche applies emphatically to "the highly celebrated author of the book *de disquisitione veritatis*" the first lines of Horace's famous "Archytas Ode" (*Odes*, I, XXVIII). Bayle glorifies Malebranche's doctrine of divine exemplars for the creation of the visible world ("the intelligible nature and the Archetypal world", OD IV, 124b).

8.2 Inside the *Systema*

8.2.1 A "Means-to-End" Structure

Bayle's main concern after deciding to apply to the vacant chair of the Sedan Academy was to gather good models of course-building in philosophy. His intent was not to plagiarize the content of some printed course of philosophy. But he repeats in his letters of May 1675 that not only during his studies of theology he had already "begun to neglect philosophy" and "left Monsieur Descartes for Homer and Virgil," but moreover, his three years of tutoring with young children have "reduced [him] to such a state" that he has forgotten even "the first elements of logic"; more generally, he has "forgotten philosophy."[62]

Even so, Bayle envisioned philosophical training along a structured plan. His recommendations during the summer of 1676 to his young brother Joseph, who was about to attend the philosophy course at the Academy of Puylaurens, reflect the precise schema the professor Bayle applied for teaching philosophy:

> The best choice regarding philosophy is first of all to focus on scholastic philosophy, to get one's mind accustomed to the chimeras of universals and to the metaphysical abstractions, and then to consider the systems of physics, ancient and modern. You have sufficient knowledge when you know the history of philosophy, namely what is said in every sect, because actually all are almost equally uncertain.[63]

The imperative to comply with a traditional mode of exposition in his professorship at Sedan actually met Bayle's personal view of teaching (or symmetrically learning) philosophy.

For Bayle, a sharp distinction had to be drawn between the contents taught in a course of philosophy and the method for teaching them. For this reason, a book such as the *Introduction à la philosophie* by the Gassendist Gilles de Launay, which he had applauded in June 1675 when it was first published, was to his eyes of no benefit in academic teaching. Bayle judges de Launay's *Introduction* inappropriate for a course of philosophy because this book openly claims to be specifically neo-Epicurean, whereas its defense of neo-Epicureanism should conform to the Scholastic pattern, in which one sets out several opposite theses. Thus in September 1676 he dissuades Joseph from reading this *Introduction* to learn philosophy:

> Reading Mr de Launay's *Introduction de la philosophie* would make little sense for your purpose. People like you must follow through with Scholastic philosophy, and after that they become philosophers in what seems the best way in their eyes, either Gassendists, or Cartesians, or Pyrrhonians, or Peripatetics. The methods of Moderns are useful only to those who have already lost their virginity, that is to say, who have already understood common philosophy.[64]

---

[62] See Letter 90, 5 May 1675, Letter 91, 18 May 1675, CB II, 160, 168.

[63] Letter 126, 25 August 1676, CB II, 353.

[64] Letter 128, 12 September 1676, CB II, 362.

A year earlier, in November 1675, shortly after Bayle had begun simultaneously teaching and writing day by day the course he had to teach, he had declared to his elder brother Jacob:

> I am a great friend of the new philosophers, and I am longing for the year of physics, when I will throw myself into Cartesianism and Epicurus's atoms which the great Mr Gassendi has so well reestablished. For the present year, I will have to stick to the old spectrum and to be Peripatetic.[65]

To assume from this declaration that the curriculum constraints at the Academy of Sedan doomed Bayle's course to an artificial juxtaposition between a Scholastic logic and an anti-Aristotelian physics would be erroneous, though. Bayle did not deplore the obligation to start teaching philosophy with lessons of traditional logic. The composition of a standard Peripatetic logic was not at all for him a boring routine. Instead he viewed the relationship between his teaching of Scholastic logic and the promotion of non-Scholastic physics as a "means-to-end" relationship. The best method to teach or to defend the physics of new philosophers against Peripatetics was according to him to place the physics of Cartesians and Gassendists within its polemical context and to use to that end the argumentative tools provided by Scholastic logic. Consequently, the links between the *Logica* and the *Physica* of Bayle's *Systema* cannot be reduced to a contingent relationship between a Scholastic logic and a non-Scholastic physics, demanded by the curriculum arrangement alone. For Bayle, the Aristotelian logic supplies the necessary means for a dialectical defense of anti-Aristotelian physics. This is still the credo expressed in 1680 as the guiding principle of the *Theses philosophicae*. The *Thesis I* proclaims:

> We believe that Logic is much more useful, and worthy of being studied by those who apply themselves to philosophy, than the new philosophers inconsiderately think it is.[66]

The dialectical value of the common Aristotelian logic should not be rejected when the new philosophers reject the common Aristotelian doctrines. Needless to say, the excessive use of the subtleties of dialectic by Scholastics to entangle any philosophical question with captious distinctions, their "rage for disputation,"[67] must be avoided. But the legitimate use of the dialectical method of reasoned arguments must be preserved. The disputational framework furnished by Aristotelian logic is not useless and unprofitable, especially when the aim is to demonstrate the superiority of the mechanical philosophy over neo-Aristotelian doctrines in explaining natural phenomena. As *Thesis I* claims in 1680, there are in the Scholastic books of logic "many things of abundant yield" which ought to be made use of in the philosophy of the *recentiores*: "the use of common Dialectic can bear excellent fruit."[68] Thus, teaching philosophy requires teaching the use of the Scholastic dialectical "style" in

[65] Letter 114, 25 November 1675, CB II, 299. This constituted a message to be transmitted by Jacob to Thomas de Martel (1618/1619–1685), who had been a friend of Hobbes (1588–1679) and Gassendi (1592–1655).

[66] OD IV, 132b.

[67] Ibid.

[68] Ibid.

the beginning. This is a leitmotif in Bayle's description of philosophical training. On 6 May 1677, he writes to Joseph:

> Take the early habit to stir your imagination, and to exercise it about concepts and formalities. This will be useful for the future, and will make you accustomed to method and accuracy. A clever man who is a good logician is the most fearsome adversary. He is able to overthrow the most solid books, and unless you are yourself a good logician, it is impossible to cope with him.[69]

Again in a letter dated 16 November 1677, Bayle insists that without a complete mastery of Aristotelian logic, it is impossible to defend the doctrines of the new philosophers:

> Go ahead from good to better, become a great philosopher, have a complete knowledge of scholastic philosophy, and be confident that by this way you will use the doctrine of Moderns more forcefully.[70]

For Bayle, thus, in order to become the best supporter of anti-Peripatetic physics, you should be at first the best Peripatetic logician. Bayle holds that a course of philosophy must supply the capacity to examine arguments and counter-arguments, and to frame objections and replies to objections. The use of the Scholastic method of argumentation is for him quite the opposite of a constraint. This Scholastic pattern does not impose a formal straightjacket on the mind, on the contrary it allows everyone not only to scrutinize arguments and check their consistency, but also to find out by oneself the reasons sustaining a certain opinion against possible objections. The professor Bayle conceives of logic as necessary to acquire the mental ability for a critical and analytical evaluation of any philosophical question. To his eyes, a course of philosophy must not provide prefabricated arguments, but the method for discovering arguments. Such is the competence to be acquired by Joseph at Puylaurens:

> On every matter, you must aim at a fine knowledge of the state of the question, of reasons *pro* and *contra*, and of possible replies on both sides. […] I commit you to your own invention; […] always reduce your reasoning to the syllogistic form, because to accustom the mind to a determined and precise formal order which never fails to be accompanied with accuracy when it is maintained in its true bounds constitutes almost the only way common philosophy is fruitful. As for me, I am Peripatetic almost everywhere with the exception of Physics, in which I am totally against Aristotle and in favor of Mr Des Cartes.[71]

This solemn endorsement of Cartesianism in a letter to Joseph comes in March 1677, a few months before the completion of Bayle's first "campaign" of philosophy at Sedan. It casts light on the way Bayle envisages possible corrections and additions to his textbook for the following two campaigns. His academic course, he asserts, belongs to the kind of work that "always have some poor parts, like some cities whose fortifications require perpetual repairs."[72] Now, when he looks for

---

[69] Letter 138, CB II, 418.

[70] Letter 144, CB II, 446.

[71] Letter 135, 28 March 1677, CB II, 400–401.

[72] Letter 180, 4 February 1680, CB III, 212.

books that might suggest to him some improvements, it is specifically for the physical part of his course that Bayle is most eager to get some new material. He considers this part of his teaching as a section necessarily in progress, in which he must attempt to perfect again and again his arguments in favor of anti-Aristotelian positions. To widen his knowledge of new trends in physics appears as a recurrent concern in his correspondence. In July 1677, after reading excellent reviews of the book *Problemata physica et medica*, just published by the physician François Bayle (1622–1709), in three journals (*Journal des savants*, *Acta eruditorum*, *Philosophical Transactions*), he looks for additional information about his namesake, who, he remembers, had delivered some Cartesian philosophical lectures around ten years earlier in Toulouse with "un nommé Mr Regis"—that is, with Pierre-Sylvain Régis (1632–1707.)[73] Later on, in March 1678, he expresses his intention to read a book which, he thinks, should be "a pretty good treatise of anatomy and of Cartesian physics," the *Explication mechanique et physique des fonctions de l'âme sensitive ou des sens, des passions et du mouvement volontaire*, published in 1677 by the physician Guillaume Lamy (1644–1682).[74] In November of the same year, Bayle also reads Lamy's *Discours anatomiques* (1675).[75] In parallel, only textbooks on philosophy from which either the "means-to-end" arrangement of his lessons of logic and physics or the comparison between Aristotelian and anti-Aristotelian physical principles might draw support are courses that appeal to him. In this respect, the *Cursus philosophicus, recognitus et auctior* published by Emmanuel Maignan in 1673 occupies a prominent place in Bayle's collecting of information about the best methods of teaching philosophy. Over the whole year of his first physical lessons, Bayle makes a great deal of effort to obtain Maignan's course. He eventually gets "the *Philosophie* of Father Maignan" in September 1677.[76] We may regard the description of the Minim in the article "MAIGNAN (Emanuel)" of the *Dictionnaire historique et critique* of 1697 as a representation of that which Bayle twenty years earlier had held in high esteem in Maignan when he was looking for models of philosophy teaching:

> He abandoned the opinions of the Schools, and combated them very solidly. He was neither Cartesian, nor Gassendist, but agreed with the two heads of these two sects in rejecting accidents, qualities, substantial forms, and in cultivating experimental Physics. […] He had to sustain many disputes against the Peripatetics; and he was all the more able to prevail against them as he preserved a great part of the method of the Scholastics in his writings.[77]

The other textbooks on philosophy that Bayle favored were the *Novum totius philosophiae syntagma in IV. partes distributum ad usum scholae* (1669) by Etienne de

[73] See Letter 140, 24 July 1677, CB II, 425.

[74] Letter 150, 24 March 1678, CB III, 28.

[75] Letter 159, 3 November 1678, CB III, 75–76.

[76] See Letters 128, 12 September 1676, 140, 24 July 1677, 143, 19 September 1677, CB II, 365, 425, 439. For Bayle's judgment after reading the *Cursus*, see Letter 146, 19 November 1677, CB II, 457: "I find the *Philosophie* of Father Maignan most beautiful."

[77] Bayle 1697, I, 500–501.

Melles (1620–1683)[78] and the "very good" course in six volumes by Pierre Barbay (?-1664) published in 1675.[79] He was even more interested in the course *Philosophia vetus et nova ad usum scholæ accommodata* by Du Hamel, published in 1678: the book, he estimates, is "full of new inventions, and yet it also deals with all that which is the most well-known in the old philosophy."[80] In 1678 too, Bayle received from Pierre de Villemandy an *Introduction* of some one hundred pages to a course aiming at "a constant parallel between the three sects, of Aristotle, Epicurus, and [*Descartes*]."[81] In contrast to this set of textbooks envisioning philosophy teaching as a way to reconcile the new philosophers and Scholastic philosophy, Bayle shows complete disdain for a unilaterally scholastic course by the doctrinaire Jean Vincent (1606–1677?), his *Cursus philosophicus in quo totius scholae quaestiones, fere omnes, […], in utramque partem propugnantur* (1658–1671). Because he already knows of this *Cursus philosophicus*, the publication by Vincent in 1677 of his *Discussio peripatetica in qua philosophiae cartesianae principia per singula fere capita, seu articulos dilucide examinantur* contains nothing disturbing for Bayle:

> The little book in 12° of Father Vincent against Des Cartes does not alarm me. I have seen by this author a thick scholastic *Cours de philosophie* in four volumes in 4°, which is a real hodgepodge.[82]

The pattern of philosophy teaching represented by Vincent is for Bayle the exact antithesis of that of Emmanuel Maignan.[83] Would Bayle hold, on the contrary, that

---

[78] See Letter 153, 16 July 1678, CB III, 45. Even though Bayle mentions in 1676 that he was told that "the course of Mr. de Mellez … is very good," he has not read it immediately (Letter 128, 12 September 1676, CB II, 362). It seems that a new publication of this course was planned in 1678.

[79] Barbay had studied philosophy with Antoine Arnauld in 1639–1641 at the College du Mans in Paris. The six-volume series of his teaching at the University of Paris, *In universam Aristotelis Philosophiam Introductio*; *Commentarius in Aristotelis Logicam*; *Commentarius in Aristotelis Metaphysicam*; *Commentarius in Aristotelis Moralem*; *Commentarius in Aristotelis Physicam* (2 vols) was published in 1675 by de Servigny; a second augmented edition was made in 1676.

[80] Letter 153, 16 July 1678, CB III, 45 (see Letter 152, 5 June 1678, CB III, 39). This book is dedicated by the printer Etienne Michallet to Jacques-Nicolas Colbert (1655–1707), from whose notes when the course was dictated at the College of Bourgogne in Paris it was composed. Bayle had read the review of the *Logica*, the *Ethica* and the *Metaphysica* in the *Journal des Savants* on 2 May 1678, the review of the *Physica generalis* on 23 May 1678 and that of the *Physica particularis* on 4 July 1678.

[81] Bayle writes "Gassendy" instead of "Descartes." Villemandy never completed this course. He only published in 1678 the introduction sent to Bayle, under the title *Philosophiae Aristoteleae, Epicureae & Cartesianae parallelismus* (reprint of *Manuductio ad philosophiam vetero-novam, in qua triplicis philosophiae […] Aristoteleae nempe, Epicureae et Cartesianae placita […] adducuntur et […] conferuntur*, 1674). The *Manuductio* was augmented in 1679 solely with the logical part, even though the new title announced also a general metaphysics and a pneumatology. After defending Peripatetic philosophy, Villemandy had become open to the new philosophy and was a conciliationist. See *Nouvelles de la République des Lettres*, October 1685, art. IX, for Bayle's positive review of a new edition of the *Manuductio* (OD I, 399).

[82] Letter 147, 12 January 1678, CB III, 6.

[83] See on these opposite sides of philosophy teaching in the late seventeenth century, Roux 2013, 55–95.

a unilaterally Cartesian textbook is acceptable? Interestingly enough, the response is no. The Latin course of philosophy published in England by the Cartesian Antoine Le Grand (1629–1699), his *Institutio Philosophiae, secundum Principia Domini Renati Descartes: Nova Methodo Adornata & Explicata. In Usum Juventutis Academicae* (1672) reprinted in 1678, does not constitute a model for Bayle. To his eyes, Le Grand totally fails to assess the necessary relationship in philosophy teaching between dialectical argumentation with Aristotelian tools and the defense of the new physics. Instead, Le Grand merely reproduces a Cartesian logic and a Cartesian physics which remain disconnected from each other within his whole course of philosophy: "his physics is almost just a translation from Rohault and his logic only a translation from *L'Art de penser*."[84]

At first sight, however, the arrangement of Bayle's *Systema* presents nothing original. When looked at superficially, it is organized like any ordinary textbook as a reflection of the philosophical curriculum.[85] Furthermore, one might argue that this course is uninventive from the very point of view of its author, since Bayle, when he composes the ethical part of the *Systema*, makes no attempt to reform the poor status of ethics in the curriculum, which he had earlier denounced in a letter to Minutoli of September 1674:

> [W]hat man does least is that which man should do mostly. Also the philosophers, not only the common people, are guilty of this defect. They dedicate the best of their time about logic and metaphysics, neglecting ethics, whereas ethics should be the main subject of their studies. Assuredly, there are countless ethical books. But what I have just said concerns the usual practice in colleges and universities. Professors hardly spend one month dictating their treatise on ethics, whereas seven or eight months are needed for dictating logic. This is not the worst, however. What is especially blameworthy is that they do not deal with the excellency of virtue and the necessity of its acquisition. Instead, they merely dispute over the nature of virtue and its definition. How could then the lessons on ethics dictated in the philosophy class benefit any schoolboy's moral conduct?[86]

The *Ethica* of the *Systema* does not discuss the different definitions of virtue in general; it concerns our knowledge of those eternal truths or axioms that are moral principles, and treats of the efficient causes of our actions. However, insofar as this part of his Sedan course is condensed into about ten pages, Bayle might be viewed as himself guilty of the usual defect he condemned in other professors of philosophy. Yet, the fact that the *Systema* leaves moral teaching to its ordinary academic fate must be placed within a larger context. Despite the resemblance between the structure of Bayle's *Systema* and the normal organization of the academic curriculum, the former is not a mere reflection of the latter. There is another kind of

---

[84] Letter 168, April or May 1679, CB III, 168.

[85] According to the quadripartite arrangement of studies in Protestant Academies, the *Systema* includes first a *Logica* followed by an *Ethica* (*Brevis et Accurata Ethicae Delineatio*) and then a *Physica* (*Syntagma Physicum, in generalem & particularem divisum*) followed with a *Metaphysica* (*Synopsis Metaphysicae*). The *Syntagma Physicum* is disproportionately large by comparison with the other parts of Bayle's *Systema*, since this oversized physics covers twice the number of pages than the three other parts put together.

[86] Letter 65, 27 September 1674, CB I, 322–323.

structure internal to Bayle's course of philosophy. It is governed by the very specific relationship that Bayle sees between Scholastic logic and anti-Scholastic physics, as I said earlier. The inner structure of Bayle's *Systema* is that of a course which aims to use Scholastic-style argumentation to teach the superiority of the natural philosophy of the "new philosophers," the *recentiores*, in comparison with that of the Peripatetics.

Using a number of definitions and examples, Bayle places some signposts in his *Logica* that prepare for the demonstration in the *Physica* of the truth of the principle of the new philosophers, the identity of matter and extension. In chapter II, *De Universalibus*, the exposition of that which medieval logicians called the *Quinque Voces*, Porphyry's predicables—the five different kinds of universals: genera, species, differences, properties, accidents— provides Bayle with an interesting opportunity to anticipate what he will argue in the *Physica*. This occurs in the definition of the *proprium quarto modo*.[87] In his *Introduction* to Aristotle's *Categories*, Porphyry divides the universals called "properties" into four. He defines the fourth sort, which, he says, is that of properties in the strict sense, by the coincidence among "of all," "alone," and "always."[88] The *proprium quarto modo* holds of *all* the individuals of the species of which it is the property, of this species *alone*, and *always*. Accordingly, a species and its fourth property convert. The example in Porphyry is that of the property "to laugh" for humankind. If humankind, then laughter, and vice versa, if laughter, then humankind: the fourth property is counter-predicated of its species. However, in this characterization of the fourth property (or property strictly speaking) by Porphyry, a property holding of a thing does not indicate what it is to be the thing of which this property holds. For Porphyry, a property is never the definition of its subject. But for Bayle, the characterization of the *proprium quarto modo* implies that this necessary attribute is also an essential attribute really identical with the *differentia*. This property is the same, he claims, as the inseparable essential difference constitutive of a species. He gives the proof for their real identity with the example of the *proprium quarto modo* of the corporeal substance, *divisibile*. This example, not the traditional one of *risibilis* for humankind, conforms according to Bayle's *Logica* to the signification of the *proprium quarto modo*. It is impossible to conceive of *risibilis* and *rationalis* as identical: the capacity to laugh cannot be conceived of as the essence of mankind, as rationality is. On the contrary, there is no real distinction between *to be divisible* and *to be extended*, that which constitutes the *differentia* of body. The property of divisibility is therefore really the essence of its subject. In this reasoning, the proof of the antecedent (the identity of *divisible* and *extended*) is a proof *per impossibile*. If *divisible* and

[87] Interestingly enough, the way Bayle's *Logica* understands the *proprium quarto modo* differs from Descartes's use of this logical notion when he rejects the definition of body by impenetrability proposed by Henry More (1614–1687) against its definition by extension in the *Principia philosophiae*. See Descartes, *To More*, 5 February 1649, Descartes 1996, V, 269; trans. in Descartes 1984–91, III, 361.

[88] The first three modes are successively defined by the combinations "alone-not of all"; "of all-not alone"; "of all-alone-not always." See Bayle's summary: OD IV, 224b.

extended were really distinct, then that which is extended would not be by itself divisible. It would be divisible on the condition of something joined to it. This constitutes an absurdity, insofar as we conceive clearly and distinctly of a thing extended as a thing divisible without supposing anything joined to it: when we conceive of some thing as extended, we have the idea of this thing as consisting of parts outside parts (*partes extra partes*). The only difference between the *proprium quarto modo* and the *differentia* lies in our manner of thinking. We conceive of the fourth property as following from the difference, that is to say, as being in the subject subsequently to the difference, whereas they are really identified. Extension is not in body prior to divisibility—a body is simultaneously extended and divisible—, but our conception of body as extended is prior to our conception of body as divisible.[89]

Other instances in which the *Logica* prepares the section on *Physica* are found in the exposition of categories. The only example proposed by Bayle for the category of substance, *ens per se subsistens*, is body. In order to distinguish substance, whose existence does not require the inherence relation, from accident, which cannot exist without being present in a subject, Bayle develops the example of that which differentiates body from motion. Body is a substance for which no other created thing is needed for it to persevere in being, but motion is just an accident, it cannot be without a subject and needs to be sustained forever by a body. The *Logica* examines from this viewpoint an objection drawn from the thesis commonly agreed (with the exception of Gassendists such as Derodon, as we saw) that perseverance in being is continuous creation. According to this objection, the thesis that not only to begin to exist, but also to persist in being requires divine omnipotence as its cause, would imply the falsity of the proposition *body is a thing subsisting in itself*. Bayle refutes this conclusion. The independence involved in the persistence of a thing by itself must not be viewed as this objection does. The notion of substance signifies exclusively independence from any subject: a self-subsistent thing is a thing conceived of as a thing that does not stand in an inherence relation. To assert that body could not persist beyond one single moment without a continual divine production of its being does not make God a subject in which body inheres. The ontological dependence of body on God is a dependence on the efficient cause of its existence, not on a subject of inherence.[90]

The *Logica* goes further in the way of a defense of the *recentiores* with the following definition of the category of quantity: "Quantity in its formal concept entails

[89] This finds moreover an application in the subsection on demonstrative syllogism in the *Logica*'s chapter IX *De Argumentatione* (OD IV, 250b). Bayle's example for a demonstrative syllogism reads: *Omne extensum est divisibile. Omne corpus est extensum. Ergo Omne corpus est divisibile.* This reasoning proves that divisibility holds of body because extension is the necessary cause of divisibility. What makes this conclusion a scientific conclusion and the knowledge of a thing by its necessary cause, is not however that extension is "a real and physical cause" of the divisibility of body. A physical cause is really distinct from its effect and it exists prior to its effect, whereas extension and divisibility are really identical. What is meant, thus, with this syllogism is that when we conceive of extension, we conceive of an attribute of body from which we conceive that divisibility necessarily follows.

[90] On this, see OD IV, 228b.

multiplicity of parts."[91] After saying that according to this definition quantity is nothing other than number, since we understand by number a multitude of units,[92] Bayle introduces an observation that makes his *Logica* anticipate the affirmation of the identity of matter and quantity later in the *Physica*. The essential point concerning the category of quantity for the *Logica* is not that the equivalence of its definition with the Euclidean definition of number allows it to include any multitude of non-corporeal substances. Bayle immediately adds that the most common notion of quantity is rather that of quantity insofar as quantity pertains to extended things, that is to say, to matter. A number of philosophers, he then states, hold that quantity is an accident distinct from matter that results in matter having parts outside parts (*partes extra partes*), namely, in matter being extended. This doctrine, he claims, is false. It is impossible to hold that quantity is an accident separable from matter and that extension does not pertain intrinsically to matter. Matter "by itself," *per se*, consists of parts outside parts, *partes extra partes*. Thus, Bayle contends, we must conclude from matter's composition as many parts outside parts that quantity is nothing other than matter itself. This derives necessarily from the definition of the category of quantity. The quantity of matter is, by definition, the multiplicity of parts of matter; now, undeniably the parts of matter are themselves matter; therefore, the quantity of matter must be identified with matter itself.

The exposition of the category of quantity in the *Logica* paves the way for the discussion raised later in the *Physica* with the examination of the question *An quantitas distinguatur a re quanta* (*Is quantity distinct from the thing quantified?*).[93] Admittedly, as I said, in the *Physica* of the *Systema* Bayle does not constrain his discussion to physics proper. The issue in the *Physica generalis* is more importantly theological, so that the two antithetical positions, *quantity is distinct from matter* and *quantity is not distinct from matter*, are presented as those of *Pontificii* for the former, of *Reformati* for the latter, rather than as those of Peripatetics on the one side and of the new philosophers on the other side. Yet what is of notable significance in this section of the *Physica* is that Bayle presents a dense argument intended to invalidate the claim some philosophers have made that the categorial or general notion of quantity (quantity is a multitude of parts) by itself cannot define quantity. In the position that Bayle attacks, the "formal concept" of quantity requires, he says, more than a multiplicity of parts: these many parts must possess actual extension.[94] The aim in Bayle's *Physica* is accordingly to demonstrate that the actual extension of parts cannot be differentiated from matter, so that even in such a reconceptualization of quantity, it is necessary to conclude that quantity is matter itself. Bayle's complex argumentation in this section of the *Physica*[95] attests that for him the

---

[91] OD IV, 229a.

[92] See Euclid, *Elements*, VII, Def. 2: "a number is a multitude composed of units."

[93] See OD IV, 287a–289a.

[94] OD IV, 287a. Bayle attacks the neo-Scholastic distinction between the internal quantity according to which a material substance possesses aptitudinal extension (the capacity to extend itself in space) and the external quantity according to which a body is actually extended in tridimensional space. This distinction supports the Catholic doctrine of transubstantiation.

[95] Bayle piles up five arguments to prove that actual quantity, or actual extension, is the essence of matter. Compare with the passage of the *Metaphysica specialis*, chapter II *De Substantia &*

formalized technique of Scholastic disputation is necessary to make unobjectionable the anti-Scholastic principle of the *recentiores* concerning quantity.

8.2.2 A Disputational Physics

Thus, the *Physica* is central to Bayle's *Systema*, and the *Logica* is structured to support it. For Bayle, the main objective in his course of philosophy was to give his students the ability to apply the Scholastic procedure of argumentation to the physical debates opposing the new philosophers to the Peripatetics. The disproportion or asymmetry between the different parts of the *Systema* does not mean that their hierarchy is merely determined according to the conventional organization of the curriculum. If the third part, the *Physica* is much larger than the other three parts summed together, it is because Bayle considers physical questions through formal argumentation, organized in accordance with the patterns of traditional logic. In Bayle's course of philosophy, the *Physica* consists in a series of *disputationes*. There are in the *Systema* no *disputationes logicae* and no *disputationes ethicae*, and if there are some disputations in the *Synopsis metaphysicae*, they are not designated as such in the titles of the subdivisions of the two parts of the metaphysical lessons, *Metaphysica generalis* and *Metaphysica specialis*. On the contrary, the two parts of the physical lectures, both the *Physica generalis* and the *Physica particularis*, explicitly deal with disputational questions.[96] The most original feature of this *Physica* is its abundant use of an iterative schema of objections and replies, that is to say, its dialectical style.[97]

Accidente, examining the question *An accidens possit per miraculum subsistere sine suo subjecto (Can an accident miraculously subsist without its subject?)*, in which Bayle attacks "Papists" for their distinction between an accident's "aptitudinal inherence" (the essence of accident, inseparable from it) and its "actual inherence" (separable from it): see OD IV, 503a–504b.

[96] The *Physica generalis* (OD IV, 271–329) is composed of two disputations, *Disputatio prima. De principiis corporis naturalis*; *Disputatio secunda. De generalibus corporis naturalis affectionibus*. Of the two treatises which compose the *Physica particularis*, the *Tractatus primus. De corpore inanimato* (OD IV, 329–426) and the *Tractatus secundus. De corpore animato* (OD IV, 426–462), only the first treatise on inanimate body is divided into disputations. It contains three disputations: *Disputatio prima. De elementis*; *Disputatio secunda. De corpore mixto inanimato*; *Disputatio tertia. De caelis* (the discussion of the principal cosmological systems, Ptolemaic, Tychonian, and Copernican). In both parts of the *Physica*, general and particular, each disputation is divided into chapters subdivided into articles—yet, as we shall see, the *Disputatio prima* of the *Physica generalis* consists of a unique chapter composed of four articles.

[97] To defend the new natural philosophy, the *Physica generalis* in the *Systema* proves to be more disputational, or more logical, than Cartesian. If on some physical issue Bayle cannot find arguments allowing him to conclude in favor of Cartesianism, he is prepared to abandon Cartesianism. This occurs, for instance, in the definitions of place and motion, the questions central for mechanical philosophy. On this, see Pécharman 2017, 505–520 (514–517). Cf in the 1680 *Theses philosophicae*, *Thesis V* on the lack of explication of motion among Cartesians (OD IV, 135b–136a).

In *Disputatio prima* of his *Physica generalis*, which treats "the principles of natural body," Bayle follows a specific strategy for defending in the *Disputatio secunda* (which treats "the general affections of natural body") the anti-Peripatetic thesis of the new philosophers on the lack of a distinction between quantity and the quantified thing.[98] In this essay, I will focus only on the manner in which this *Disputatio prima* subverts the Aristotelian legacy of the principles of natural substances. The traditional framework for this first disputation is what we usually call hylomorphism, the doctrine that physical objects are compounds of matter and form. Now, the procedure followed by Bayle in the *Articulus primus* and *Articulus secundus* of the *Disputatio prima*, before arriving at the examination of "what pertains to matter" in the *Articulus tertius* and finally "what pertains to form" in the *Articulus quartus*, already orients the discussion toward a concluding anti-Aristotelian position. Bayle's course in this disputation is carefully ordered into a series of steps which lead to aporetic arguments requiring the elimination of the Peripatetic notion of substantial form. What the Sedan students learned from professor Bayle's first class in physics was the unavoidable cleavage between the new philosophy and the Peripatetic legacy concerning the understanding of form.

The Subversion of Prime Matter

First of all, the *Articulus primus* shows why necessarily the principles of composition of natural body can be neither more nor less than two, matter and form. When a body undergoes transmutation and is changed into another body (e.g. wood changed into fire), in this process there is "something common" to the body undergoing transmutation and to the body into which it is transmuted. The name *matter* designates this "something common."[99] But in the same process of natural change there is also "something distinctive" by which matter is *this* (fire) rather than *that* (wood), something which, as long as it remains, matter perseveres in its being such a body rather than such another. The name *form* designates this "something distinctive" by which bodies are not just body, but specific bodies.[100] These prefatory remarks on the number of physical principles and the signification of their names might *prima facie* seem to constitute an orthodox Aristotelian view based on the consideration of substantial change in the perceptible bodies as opposed to their mere alteration. But these remarks aim instead at conveying a pro-Cartesian sentiment. Bayle asserts that a body's being not this or that body, but absolutely a body, means a body's being composed of "parts that are situated outside one another and that fill up some space."[101] This claim about the identity of being a body and being *partes extra partes* (in other words this claim about extension as the *ratio corporis*)

[98] I am quoting the titles of these two disputations: see OD IV, 271a and 286b.

[99] OD IV, 273a.

[100] Ibid.

[101] OD IV, 272b.

implicitly grounds in the new natural philosophy the lexicological statement: "that by which all bodies are constituted in the notion of body simply (*in ratione corporis simpliciter*), and which is common to all of them, is called *matter*."[102]

The *Articulus secundus* then sketches a short history of pre-Socratic philosophy demonstrating that the ancient materialist doctrines (either monistic or pluralistic) are not the best source to get the knowledge of matter as the first principle of natural body. All of them failed to differentiate prime matter from qualified matter.[103] Bayle's final diagnosis in the *Articulus secundus* of the *Disputatio prima* is that prior to the modifications producing the different species of bodies, matter is nothing but matter as identified with extension:

> [T]he true and genuine principle of all bodies is matter insofar as it is simply extended (*materia prout simpliciter extensa*). When some modification befalls it, it becomes a species of body different from all the rest of matter.[104]

This scenario allows Bayle to state at the beginning of the *Articulus tertius* that only undifferentiated matter, not matter actually endowed with a certain form or "determined by antecedent dispositions toward a certain form rather than toward another one" can be the absolute principle sought for in the analysis of the composition of natural bodies.[105] However, this opposition is not itself the final aim. Once matter is "contradistinguished" from form, indeed, the question is: *what is matter regarded without any form whatsoever?* In other words: *what is prime matter?* Now, in Bayle's course of philosophy, the necessary response to this question lies in the definition of the essence of matter as consisting in extension. The Aristotelian doctrine of prime matter proves to be the most valuable device for introducing Descartes's doctrine of matter. To this end, it is not even necessary in the *Articulus tertius* of the *Disputatio prima* to mention Descartes or Cartesians: Bayle makes no reference to either. The mention of Aristotle is all that is needed.[106] The Aristotelian account of prime matter as deprived of all categorial determinations and the primary substratum of all natural things serves as a ground to contend that the Aristotelian notion of a wholly indeterminate prime matter coincides with the Cartesian notion of body. The latter can be substituted for the former. The Aristotelian undifferentiated prime matter is then viewed as defined by the same homogeneous structure as Cartesian matter:

> The matter to which it is an accident to be wood, or to be fire, etc., [...] is, considered in itself (*secundum se sumpta*) and in its essence, the extended substance.

---

[102] Ibid.

[103] See OD IV, 273a–276b.

[104] Ibid., 276b. The last part of this declaration already introduces an anti-Scholastic status for the principle of the differentiation of bodies: form is only a modification of matter.

[105] OD IV, 276b–277a.

[106] See OD IV, 277a. Bayle first paraphrases the apophatic description of prime matter in Aristotle's *Metaphysics* (Z, 3, 1029a 20–21). Then he summarizes the standard Aristotelian doctrine on the distinction between generation and alteration.

> Prime matter, that is to say body in general, must be therefore defined as the extended substance.[107]

For Bayle, a syllogism perfectly demonstrates the appropriateness of his redefinition of Aristotle's prime matter in terms that might be those used by Descartes when Descartes defines in his *Principia* "the nature of matter, or body considered in general."[108] The major premise of this syllogism in the *Articulus tertius* consists of a definition of essence:

> The essence of a thing is that which we conceive to be first in it, and which, when it is removed while all else remains, the thing is taken away, whereas, when it remains while all else is taken away, the thing remains.[109]

The minor premise applies this definition to extension: "Such is extension with respect to matter." The conclusion states therefore that "the essence of matter consists of extension." This Scholastic-style demonstration, however, gets its validity from our manner of conceiving of the properties of matter.[110] Any property conceived of in matter presupposes extension, it is necessarily conceived of as depending on extension. For example, we conceive of divisibility as the property of something having a multiplicity of parts outside one another. It is not vice versa from its divisibility that we conceive of a thing as extended. Therefore, we must conclude from the conceptual priority of extension over any other property in matter that extension constitutes the first attribute of matter.

Bayle confirms this conclusion with a thought-experiment in two stages.[111] When extension is taken away, no idea of matter is left, whereas when extension remains, the idea of matter survives the process of stripping off all sensible qualities. The latter are separable from matter, they are solely its accidents. We may easily conceive that matter will keep its whole essence and still exist if all of these qualities are stripped off. But in addition, we conceive of some priority between the

---

[107] OD IV, 278b. I omit in the quotation a parenthesis announcing that the next article will show that "these various states," namely to be wood, or to be fire, etc., "arise from the diversity of motion, position and figure of the parts of matter." By way of an incidental clause, Bayle prepares the audience to a rigorously anti-Peripatetic notion of form.

[108] *Principia philosophiae*, II, art. 4, (Descartes 1996, VIII-1, 42; trans. in Descartes 1984–91, I, 224): "The nature of body consists … simply in extension" (title); "the nature of matter, or body considered in general, consists … simply in its being something which is extended in length, breadth and depth."

[109] OD IV, 278b. Cf the remarks on the *proprium quarto modo* in the *Systema*'s *Logica*.

[110] The proof for the validity of a syllogism does not come for Bayle from the syllogistic form by itself, but from the conceptual support of its form. In the *Systema*'s argumentative use of Scholastic logic, a mental guarantee for syllogistic inference is always required. Conversely, as we shall see, the refutation of a thesis is very often based on what might be called the "Test of the Unintelligible" or "Test of the Unconceivable": many objections to Peripatetics are expressed in the *Physica* through the suggestion "*Try to conceive…*", doomed to meet an impossibility for our mind. This constitutes in Bayle's disputational *Physica* a striking form of combination between the Scholastic syllogistic logic and the Port-Royal logic of ideas.

[111] Compare Bayle's version of this thought-experiment with other versions in Aristotle, *Metaphysics*, Z, 3, 1029a 12–16; Descartes, *Principles of philosophy*, II, art. 4.

properties that are inseparable and essential properties of matter. Although there is no real distinction between a thing's being extended and a thing's being measurable, figured, divisible, and impenetrable, but all of these properties are "really the same entity," nevertheless, the property of extension is conceived of as the most basic. Therefore, we must conclude that the essence of matter is extension. For Bayle, this is just another way of suggesting that the Aristotelian prime matter is the same as the Cartesian extended substance. "Considered in itself," *secundum se sumpta*, prime matter indistinctly is the potentiality of all natural bodies and the extended substance.

The Case of Substantial Forms

Thus, with respect to the traditional hylomorphic framework of the *Disputatio prima* regarding the principles of natural bodies, Bayle shows that matter, the indeterminate ultimate substratum of natural bodies contrasted in Aristotelian physics to form, is to be identified with extension. The Aristotelian doctrine of *materia prima* surrenders without any resistance to a definition of matter that is recognizably Cartesian. But in *Articulus quartus* of the *Disputatio prima* there is no concord with respect to form, the second principle in the hylomorphic scheme. The redefinition of Aristotle's prime matter as extended substance inevitably results in pressing the definition of form into an overt antagonism between Peripatetics on the one side and all the new philosophers on the other side. This was prepared from the conclusion reached by the *Articulus secundus*. The assertion "the true and genuine principle of all bodies is matter insofar as it is simply extended" from the first implied that the principle of the differentiation of bodies, namely, form, consists of "some modification" of the extension common to all bodies.[112]

At the outset of the examination of the nature of form, both sides agree that form is that principle of natural body according to which individuals are of one kind or another. Beyond this, the opinions of Peripatetics and the new philosophers are "diametrically opposed."[113] For Peripatetics, Bayle relates, form is a substance distinct from matter: Peripatetics assert a real distinction between matter and substantial forms. At the opposite extreme, the new philosophers view forms as modes of matter, not as entities really distinct from it. Forms are for them nothing other than matter modified in such way or such other: any form results from the determinate figure, position, magnitude, and motion or rest of some material parts. Thus, the new philosophers explain the diversity among natural kinds in terms of a strictly mechanical diversity. For them, forms only consist of the multifarious states of matter arising from a variety in the properties of its parts. This doctrine may be

[112] See OD IV, 276b (cf note 104).

[113] OD IV, 279a.

succinctly described as "the doctrine that sets the nature of forms in a modification of the parts of matter."[114]

The two views cannot be reconciled. Whereas Bayle's teaching made the Peripatetic notion of prime matter compatible with the Cartesian definition of body in general, one cannot do the same with form. That notion engenders an insurmountable contradiction between the Peripatetic notion of substantial forms, which appeals to entities really distinct from matter, and the mechanical explanation of forms, which appeals to nothing other than the essential properties inseparable from the nature of matter (as we would say now, its "determinables" admitting of a variety of specifications into "determinates"). Thus, in the *Disputatio prima* over the principles of natural body, the mode of argumentation changes from the principle-matter to the principle-form. The arguments concerning matter constitute the degree zero of the application of Scholastic-style logic to disputes. At that stage, Bayle strategically reads Aristotle's undifferentiated *materia prima* as exactly equivalent to Descartes's extended substance. The use of the disputational model of argumentation begins with the definition of form, which allows no compatibility between Peripatetics and the new philosophers. The *Physica* of the *Systema* enters into disputation, properly speaking, with respect to the Peripatetic thesis:

> The form of a natural body is an incomplete substance educed from the potency of matter and united to matter to compose a definite species of body.[115]

In the *Systema*, Bayle presents this as the standard Peripatetic definition. But in reality, it is only found among the late Scholastics. The most common Scholastic view about substantial form, that associated with Aquinas, was that this form is a dynamic principle causing the existence *simpliciter* of a thing—not its being in some respect only, *secundum quid*, as an accidental form does. In this conception, insofar as prime matter is a pure potentiality, prime matter cannot have any being or existence without substantial form actualizing it by its being united to it in a composite. But the account of substantial form in Suárez (1548–1617) diverges from this tradition. In the section V of his *Disputatio XV*, when Suárez sums up the main points he has just stated concerning the existence of substantial forms and their mode of production, the "brief description" he provides of "form in general" constitutes in fact a redefinition of substantial form as found in Aquinas. Suárez rejects its conception as the formal cause of the being of matter:

> Form is a certain simple and incomplete substance which, as the act of matter, constitutes with matter the essence of a composite substance.[116]

---

[114] OD IV, 279a–280b.

[115] OD IV, 279a.

[116] Suárez 1965, I, 517b (*Disputatio XV*, *Sectio V*, § 1). Cf Suárez 1965, I, 499a–b (*Disputatio XV*, *Sectio I*, § 6): "By the name *substantial form*, nothing else is signified than a certain partial substance, which can be united to matter in such a way that it composes with it a complete substance which is one by itself". Helen Hattab has laid stress on the importance of Suárez's *Disputatio XV* as setting the stage for the replacement of substantial forms by mechanistic explications of the natural phenomena. Hattab 2012, 101–119, begins with the declaration: "I propose that Suárez can

Bayle portrays all Peripatetics as upholding the paradigmatic view that substantial form is so to speak a "semisubstance" which constitutes by its union to prime matter a singular substance of a certain kind.[117] After that, the *Systema* develops Suárez's contention about the existence of the substantial form as the typical argument of all Peripatetics. Peripatetics, Bayle claims, consider that without the assumption of substantial forms, it would be impossible to explain why, "when the proper constitution (*temperamentum*) pertaining to a composite is destroyed, it is restored anew" (for instance, the return of water to its natural coldness when water is no longer heated by fire).[118] This notion of a body's recovery of its original state after external circumstances affecting it have been removed proves to be central for the neo-Aristotelian statement that the function of substantial forms in their union with prime matter is to differentiate the natural compounds into particular kinds. A form, on this doctrine, is the intrinsic principle underlying the essential qualities of any single corporeal substance of a distinct sort. The existence of that which accounts for actions observed in natural bodies is inferred from the empirically known inseparability between some accident (such as coldness) and some subject (such as water). But the intrinsic principle of this inseparability is not itself known by experience. For Peripatetics, Bayle stresses,

> Each form is determined by its own nature to certain qualities rather than to other ones, and is distinct from all that which sense and imagination comprehend in body.[119]

The argument that follows in the first physical disputation in the *Systema* is constructed on the basis of this description highlighting the "mysterious" status of the internal principle held to constitute a substantial form.[120] In a first stage, the *Articulus*

---

be seen as the tragic hero in the unfolding drama we call the scientific revolution: in working tirelessly and brilliantly to save Scholastic Aristotelianism, he so transformed it that the path for its dethronement emerged." See also Hattab 2009, chapter 3, 40–64.

[117] The expression *semisubstance* is that of Eustachius a Sancto Paulo (1573–1640), quoted in Des Chene 1996, 76–77. See in Eustachius's *Physica, Tractatus primus De principiis, Disputatio secunda De principiis in specie*, § 2, qu. V, the sub-clause "*Forma est quaedam semisubstantia*" announcing the final definition of form: "*Est igitur forma, Actus quidam substantialis sed tamen incompletus, i.e. substantia incompleta seu (ut ita loquar) semisubstantia, quae conjuncta cum materia unam integram substantiam constituit*": Eustachius 1647, 124.

[118] OD IV, 279a. The example of water, which returns to its natural coldness when it is withdrawn from fire, is used by Suárez to argue that it is possible from the observation of the accidents and operations of natural things to assert the existence of substantial forms. When a body is brought back to its pristine state (*reductio in pristinum statum*), this is for Suárez a sign of its substantial form (*subtantialis formae indicium*). See *Disputatio XV De causa formali substantiali*, Sectio I *An dentur in rebus materialibus substantiales formae*, § 8 (Suárez 1965, I, 500).

[119] OD IV, 279a.

[120] I borrow this expression from Hattab 2012, 115. Hattab nicely states that Suárez's transformation of the Scholastic substantial form into a mysterious internal principle made easier its banishment by the new philosophy, which only had to denounce the recourse to this unknown entity in the causal explications of natural phenomena: "The next step, which was taken by Descartes, was to show that mechanistic causes could account for the observed changes of composite bodies in far simpler and less mysterious ways."

quartus of the *Disputatio prima* details the important objections ("many most weighty objections," Bayle insists) proposed by the new philosophers against the existence and the necessity of the substantial forms, and provides proofs for their accuracy.[121] Next, in a second stage, the disputation on form displays the manner in which the new philosophy explains forms. Bayle presents this explication as constituting by itself the most convincing proof that substantial forms are unnecessary to explain natural phenomena.[122] Finally, the last stage answers scholastic objections to the elimination of substantial forms.[123]

Interestingly enough, the last words of the *Disputatio prima* in Bayle's *Physica* are no other than the two Latin verses from the Port-Royal *Logic* (1662) that present an alternative list of predicaments. For Arnauld and Nicole, indeed, a distich can replace the Aristotelian list of categories, which they view as a merely arbitrary arrangement of "the objects of our thoughts." As they make plain:

> In fact, others have included everything in the world, viewed from the standpoint of the new philosophy, in the following couplet:
>
> *Mens, mensura, quies, motus, positura, figura:*
> *Sunt cum materia cunctarum exordia rerum.*
>
> That is, these people are persuaded that all of nature can be explained by considering only these seven things or modes:
>
> *Mens*, the mind or substance that thinks.
> *Materia*, body or extended substance.
> *Mensura*, the largeness or smallness of each part of matter.
> *Positura*, their position with respect to each other.
> *Figura*, their shape.
> *Motus*, their motion.
> *Quies*, their rest or least motion.[124]

Thus, a couplet from the Port-Royal *Logic*, which in a few words summarizes how the new philosophy eradicates the system of categories that supports Aristotelian natural philosophy, also expresses the conclusion of the disputation on the principles of natural body in Bayle's course on physics. Bayle's students are provided with a mnemonic device to help them keep in mind that matter and form cannot be regarded as equivalent to the two principles of the hylomorphic scheme, prime matter and substantial form. The notional syncretism allowed concerning matter by itself prohibits the Peripatetic understanding of form. Entities besides matter, that is to say entities besides the essential properties of the Cartesian extended substance,

---

[121] See OD IV, 279a–281a.

[122] See OD IV, 279a: that it is superfluous to admit of substantial forms will be "proved by … the explication of the doctrine for which the nature of forms lies in the modification of the parts of matter." From this doctrine, "it will become clear that the diversity of figures, the position, motion, rest, and bulk of the particles of matter are sufficient to explain natural phenomena."

[123] See OD IV, 284b–286b. Bayle transforms into objections to the new philosophers some of the principal reasons given by Suárez to prove the existence of substantial forms. Compare 284b with *Disputatio XV*, Sectio I, § 18, § 6 and § 8 (Suárez 1965, I, 504, 499 and 500)

[124] Arnauld and Nicole 1996, 34. These verses summed up the demonstration against substantial forms in the *Physiologia, sive Cognitio sanitatis* (1641) published by Henricus Regius (1598–1679) in a period when Regius was a fervent advocate of Descartes.

are useless for explaining the actions observed in the perceptible parts of natural bodies. The material structure of bodies is all that is required: "matter is that which acts; form that which modifies and determines the action."[125] A modification of matter, not a substantial form, is what explains the active power of a natural thing.

As opposed to the mysterious internal principles of natural operations featured in the Scholastic doctrine of substantial forms he rejects, Bayle describes the "hypothesis" of the new philosophers as most economic:

> They assert that the natural body is composed of a unique extended substance; according as this extended substance is or is not moved by diverse motions, and divided into parts of more or less size which obtain diverse figures and positions, the natural body is similarly affected by diverse forms, and becomes fire, water, tree, stone, etc. Consequently, the form of bodies is not a substance distinct from matter, but a certain modification of matter.[126]

Non-substantial forms must be acknowledged in physics as the necessary correlate of the notion of matter. In Bayle's teaching, Aristotle's *materia prima* itself makes the Scholastic notion of substantial form superfluous, since the production of any kind of natural body from the absolutely undetermined prime matter, *that is to say from extended substance*, is easily explained mechanically by way of the diverse properties of the particles of matter. Bayle insists that the mechanistic account of the natural kinds and of their specific properties and operations perfectly satisfies the requirement of conceivability which is the mark of a correct hypothesis:

> Nothing is easier than to conceive that matter has parts; that these parts are diversely placed, that they rest near to each other, or are moved hither and thither, and because of this motion acquire a different position, are either divided into smaller particles or joined to other particles and therefore made larger.[127]

For the new philosophers, the conceivability of this hypothesis is not limited to what we can perceive. That which is easily conceived concerning perceptible parts of bodies, is also easily conceived concerning their imperceptibly small particles, "the insensible parts of matter."[128] It is impossible to conceive of a division of matter that would result in the annihilation of its parts and of their properties of motion and figure. It may be therefore ascertained that an indefinite variety of operations arises from matter alone: a "new substantial form" is never produced, that which occurs in matter is always a "new figure and position" of its parts, nothing more than diverse modifications of its particles.[129]

In the *Disputatio prima* of Bayle's *Physica*, conceivability constitutes, as it were, the constant methodological rule for the arguments of the new philosophers on the principle-form. Not only their mechanistic hypothesis on the production of forms, but also (and most importantly) their objections to Scholastic substantial forms are structured in accordance with this criterion. Bayle's course of philosophy provides

---

[125] OD IV, 286b.

[126] OD IV, 282b.

[127] Ibid.

[128] Ibid.

[129] OD IV, 283a.

an interesting use of Scholastic-style argumentation from this perspective as well. The *Systema* utilizes a sequence of syllogisms to produce a *crescendo* of arguments from the new philosophy in objection to substantial forms. Against the Scholastic thesis that form is a substance, the new philosophers are said to present a first argument that Bayle reduces to a standard syllogistic form:

> A being that does not subsist by itself is not a substance.
> Now, form is such a being [i.e., form does not subsist by itself].
> Therefore form is not a substance.[130]

The demonstration of the validity of this syllogism is founded in the consideration of ideas. The idea of substance represents to our mind an *ens per se subsistens*, that is to say, a thing that needs no other subject to exist. But the Scholastic doctrine of the drawing of forms out of the potency of matter (the *eductio formarum e potentia materiae*) involves, Bayle insists, a threefold dependence of form on matter:

> According to the Peripatetics, the form depends upon matter both in being, in coming-to-be, and in operating (*& in esse, & in fieri, & in operari*). It is educed from the bosom of matter. This, indeed, is the signification of the word *educed*, which they understand as meaning "produced through the concurrence of matter as subject."[131]

Thus, the dependence of form on matter is not consistent with the *per se* subsistence or independence from any subject that defines a substance as an object of our conception. In the *Systema*, the representational nature of ideas therefore justifies the denial of substantial form. The legacy of the Port-Royal *Logic*, reworking in the chapter *Ideas considered according to their objects* Descartes's objective reality of ideas into a representation of things by ideas, is in the background.[132] Similarly, the Cartesian theory of clear and distinct ideas is most helpful to demonstrate the truth of a second anti-Peripatetic argument. The argument is again constructed as a syllogism:

> Entities are not to be multiplied without necessity.
> Now there is no necessity to grant substantial forms.
> Therefore substantial forms are not entities.[133]

The proof of the minor proposition (in addition, thus, to its *ipso facto* proof by the exposition of the mechanistic hypothesis) is furnished by a logical implication between two propositions, which constitutes itself the major proposition in an additional syllogism:

---

[130] OD IV, 279a.

[131] Ibid. For Suárez, "the dependence on matter in being and in coming-to-be (*in esse et fieri*), is appropriate for all substantial forms except the rational one": "the rational soul is not educed from the potency of matter" (Suárez 1965, I, 511).

[132] Arnauld and Nicole 1996, 30: "Everything we conceive is represented to the mind either as a thing, a manner of a thing, or a modified thing. I call whatever is conceived as subsisting by itself and as the subject of everything conceived about it, a thing. It is otherwise called a substance."

[133] OD IV, 279a.

> If we have no clear idea of substantial form, the substantial form is useless to account for the natural effects and properties of every composite.
>
> Now, we have no clear idea of substantial form.
>
> Therefore the substantial form is useless to account for the natural effects and properties of every composite.[134]

To demonstrate the truth of the minor "we have no clear idea of substantial form" in this additional syllogism, Bayle argues from the example of a certain natural body, wood. He introduces then a new syllogistic argument that shows the radical opposition between the Peripatetic notion of substantial form and our ideas:

> According to Peripatetics, the substantial form of wood is neither the substance that fills up a certain space, nor the color, hardness, weight, dryness, cold, etc., that can be observed in it.
>
> Now we do not have a clear and distinct idea of any other thing existing in wood.
>
> Therefore we do not have a clear and distinct idea of the substantial form of wood.[135]

If we follow the Peripatetics, substantial form should be conceived as something in a certain natural body other than the only things that are conceivable in it. This is an absurdity. The substantial form of Peripatetics is therefore unconceivable. Peripatetics explain substantial form only by "the vague and confused concept of being," not by "a clear idea or notion."[136] We must therefore conclude indisputably that substantial form is not required for the explication of the actions of natural bodies. In a fascinating symbiosis, Bayle's physical teaching on form, planned as the necessary preamble to his whole course on natural philosophy, organizes the arguments of the new philosophers against substantial forms in a cumulative syllogistic demonstration. But this Scholastic-style argument through the gradual and successive sequence of syllogisms would not succeed if its logical validity relied on the technicalities of Aristotelian logic alone. In the *Disputatio prima* of Bayle's *Physica*, the ultimate justification for the use of Scholastic logic in support of the denial of substantial forms comes from the Cartesian theory of clear and distinct ideas, that is, from a Cartesian logic.

8.3 Conclusion

In the 1670 Peripatetic theses defended at Toulouse for his graduating, Bayle's ultimate charge against the *recentiores Philosophi* was that the new natural philosophy mistakenly regards the "eduction of material forms" as inexplicable.[137] A decade later at Sedan, those of the *Theses philosophicae* that summarize Bayle's physical teaching now read:

---

[134] OD IV, 280b.

[135] Ibid.

[136] Ibid.

[137] OD V-1, 11.

> The principles of natural body are matter and form. There are no substantial forms, no material qualities distinct from the figure, motion, position, and rest of the particles of matter.[138]

The first physical lesson in the *Systema* provides consequently an uncompromising refutation of Bayle the student of philosophy at Toulouse by Bayle the professor of philosophy at Sedan. Yet, by virtue of their syllogistic construction, the arguments in the *Systema* that disprove the Peripatetic doctrine of substantial forms which was behind Bayle's attack on *recentiores* in 1670, remain arguments whose "style" pertains to the Scholastic mode of disputation Bayle learned at the Jesuit College of Toulouse. Strikingly, the history of philosophy, that is to say for Bayle the historical knowledge of the *when*, *who*, and *how* about the formation and perpetuation of philosophical sects (as is clear, I said, from his remarks on Vossius in a letter to Minutoli of 1673), was not viewed by the Sedan professor as providing an alternative method for his teaching of philosophy. To represent the antithetic positions of Peripatetic and the new philosophers on the issue of forms, the *Disputatio prima* of the *Physica generalis* anonymizes the adversaries. Suárez, the real and constant antagonist of the new philosophers in this discussion, is not for Bayle a figure with his own distinct doctrine within Scholasticism, but rather the representative of a commonly held doctrine. Similarly, the new philosophers are not identified individually, but are represented by way of a certain package of objections and arguments. However, the opposition of arguments in this *disputatio* is not all of equal value, nor is the strategy Pyrrhonian, as it were, opposing arguments that are supposed to lead the student to suspend judgment. The whole procedure is unambiguously planned in favor of the new philosophy's rejection of substantial forms. It is accordingly somewhat difficult, when reading Bayle's correspondence of the last months of his philosophy teaching at the Academy of Sedan, to take at face value Bayle's profession of Academic skepticism concerning any genre of philosophical doctrine:

> I regard Cartesianism only as a smart hypothesis that can be useful to explain some natural effects, but actually I am so little committed to it, that I would never engage anything to argue that these are the principles which rule and govern nature. The more I study philosophy, the more I find uncertainty in it: the difference between sects is limited to some higher or lower probability. None has yet hit the goal and presumably none will ever hit it [...]. I am a philosopher without any obstinateness, who regards Aristotle, Epicurus, Descartes, as the inventors of conjectures that we adopt or reject according as we look for this or that play of mind.[139]

This solemn declaration of May 1681 was dictated by Bayle's plan to go into exile and teach philosophy in the Low Countries, given the imminent suppression of the Academy of Sedan. Bayle had to reassure those who feared that his teaching might involve him in a new controversy on Cartesian philosophy in the United Provinces. But in truth, when in early December of 1681, he entered into his years of teaching philosophy and history at the Rotterdam Illustre School, Bayle had not at all ceased to regard the foundational principles of Cartesianism and, more generally, the principles of the philosophy of all the *recentiores* as deserving of defense on the basis of a close refutation of some Scholastic arguments that opposed them. From this

[138] OD IV, 144b.

[139] Letter 190, 29 May 1681, CB III, 244. On Bayle as a historian of philosophy, see McKenna 2017.

viewpoint, Bayle's *Systema* still constitutes a most stimulating testament to Bayle's willingness to engage Scholastic natural philosophy with the argumentative tools of Scholastic dialectic.

Bibliography

Agostini, Siegrid. 2009. *Claude Clerselier editore e traduttore di René Descartes*. Lecce: Conte Editore.

Ariew, Roger. 2011. *Descartes Among the Scholastics*. Leiden: Brill.

Aristotle. 1994. *Metaphysics. Books* Z *and* H. Translated with a Commentary by David Bostock. Oxford: Clarendon Press.

Armogathe, Jean-Robert. 2007. *La nature du monde. Science nouvelle et exégèse au XVIIe siècle*. Paris: Presses Universitaires de France.

Arnauld, Antoine, and Pierre Nicole. 1996. *Logic or the Art of Thinking*. Trans. Jill Vance Buroker. Cambridge: Cambridge University Press.

Bayle, Pierre. 1670. *Selecta Peripateticae Philosophiae Pronunciata*. Toulouse: Johannes Pech.

———. 1682. *Lettre à M.L.A.D.C. docteur de Sorbonne. Où il est prouvé par plusieurs raisons tirées de la philosophie, & de la theologie, que les cometes ne sont point le presage d'aucun malheur. Avec plusieurs reflexions morales & politiques, & plusieurs observations historiques; & la refutation de quelques erreurs populaires*. Cologne: Pierre Marteau.

———. 1683. *Pensées diverses, écrites à un Docteur de Sorbonne, à l'occasion de la Comete qui parut au mois de Decembre 1680*. Rotterdam: Reinier Leers.

———. 1684. *Recueil de quelques pieces curieuses concernant la Philosophie de Monsieur Descartes*. Amsterdam: Henry Desbordes.

———. 1697. *Dictionnaire historique et critique*, 4 vols. Rotterdam: Reinier Leers.

———. 1964–1968. *Œuvres diverses I-IV* [OD I-IV], ed. Elisabeth Labrousse. Hildesheim: Georg Olms Verlag.

———. 1982. *Œuvres diverses V* [OD V], ed. Elisabeth Labrousse. Hildesheim: Georg Olms Verlag.

———. 1999–2017. *Correspondance I-XV* [CB I-XV], ed. Elisabeth Labrousse and Antony McKenna. Oxford: The Voltaire Foundation.

———. 2007. *Avis important aux Réfugiez sur leur prochain retour en France*, ed. Gianluca Mori. Paris: Honoré Champion.

Des Chene, Dennis. 1996. *Physiologia. Natural Philosophy in Late Aristotelian and Cartesian Thought*. Ithaca/London: Cornell University Press.

Descartes, René. 1984–91. *The Philosophical Writings of Descartes*, ed. and trans. John Cottingham, Robert Stoothoff, Dugald Murdoch, and Anthony Kenny, 3 vols. Cambridge: Cambridge University Press.

———. 1996. *Œuvres de Descartes*, ed. Charles Adam and Paul Tannery, revised ed., 11 vols. Paris: Vrin.

Eustachius a Sancto Paulo. 1647. *Summa philosophiae quadripartita*. Leiden: Franciscus Moyardus.

Gros, Jean-Michel. 1989. Un cours de philosophie vers 1680. Le *Système abrégé de Philosophie en quatre parties: la Logique, la Morale, la Physique et la Métaphysique, à l'usage des étudiants* par Pierre Bayle. *Cahiers philosophiques* 41: 75–99.

Hamraoui, Eric. 2008. DENIS, Jean-Baptiste. In *The Dictionary of Seventeenth-Century French Philosophers*, ed. Luc Foisneau. London: Thoemmes Continuum.

Hattab, Helen. 2009. *Descartes on Forms and Mechanisms*. Cambridge: Cambridge University Press.

———. 2012. Suárez's Last Stand for the Substantial Form. In *The Philosophy of Francisco Suárez*, ed. Benjamin Hill and Henrik Lagerlund, 101–118. Oxford: Oxford University Press.

Heyd, Michael. 1979. From a Rationalist Theology to a Cartesian Voluntarism: David Derodon and Jean-Robert Chouet. *Journal of the History of Ideas* 40: 527–542.

———. 1982. *Between Orthodoxy and the Enlightenment. Jean Robert Chouet and the Introduction of Cartesian Science in the Academy of Geneva*. The Hague: Nijhoff.

Labrousse, Elisabeth. 1963. *Pierre Bayle. Tome I. Du pays de Foix à la cité d'Erasme*. The Hague: Nijhoff.

———. 1964. *Pierre Bayle. Tome II. Hétérodoxie et rigorisme*. The Hague: Nijhoff.

Le Valois, Louis [Pseud. Louis de La Ville]. 1680. *Sentimens de M. Des Cartes touchant l'essence & les proprietez du corps, opposez à la doctrine de l'Eglise, & conformes aux erreurs de Calvin sur le sujet de l'Eucharistie*. Paris: Etienne Michallet.

Lennon, Thomas M. 1993. *The Battle of the Gods and Giants. The Legacies of Descartes and Gassendi, 1655–1715*. Princeton: Princeton University Press.

McKenna, Antony. 2017. Pierre Bayle historien de la philosophie: un sondage. *Lexicon Philosophicum. International Journal for the History of Texts and Ideas* 5: 21–59.

Mori, Gianluca. 1999. *Bayle philosophe*. Paris: Honoré Champion.

Pécharman, Martine. 2017. Le *Système de Philosophie* de Bayle: quelle place pour Descartes ? In *Liberté de conscience et arts de penser (XVIe-XVIIIe siècle). Mélanges en l'honneur d'Antony McKenna*, ed. Christelle Bahier-Porte, Pierre-François Moreau, and Delphine Reguig, 505–519. Paris: Honoré Champion.

Roux, Sophie. 2013. An Empire Divided: French Natural Philosophy (1670–1690). In *The Mechanization of Natural Philosophy*, ed. Daniel Garber and Sophie Roux, 55–95. Dordrecht: Springer.

Ryan, Todd. 2009. *Pierre Bayle's Cartesian Metaphysics. Rediscovering Early Modern Philosophy*. New York/London: Routledge.

Schmaltz, Tad M. 2017. *Early Modern Cartesianisms: Dutch and French Constructions*. New York: Oxford University Press.

Suárez, Francisco. 1965. *Disputationes metaphysicae*, 2 vols. Hildesheim: Georg Olms.

Chapter 9
Literary Technology and Its Replication: Teaching the Torricellian Void and Air-Pump at the Collegio Romano

Renée Raphael

Abstract This chapter considers how seventeenth-century Jesuit professors of philosophy at the Collegio Romano depicted experiments involving the air-pump and Torricellian void. In contrast to previous scholarship, which has focused on Jesuit responses to the values of the new experimental philosophy and on issues of censorship, this chapter addresses the process of textual replication, in particular, how professors chose to recount experiments they found in printed books in the printed and manuscript texts they composed for classroom use. It argues that professors included textual details and visual images of experimental procedures to validate their competence as readers and their qualifications as participants in the wider natural philosophical community. This conclusion offers a point of contrast to previous discussions of the literary technology associated with seventeenth-century experimental texts, which, according to Steven Shapin and Simon Schaffer, served to bolster an author's credibility and to transform readers into virtual witnesses. The different role these details served in the context of teaching at the Collegio Romano likely derived from their inclusion in pedagogical texts, a genre in which the authoritativeness and reliability of the included narratives were assumed.

Keywords Air pump · Torricellian void · Collegio Romano · Experimental philosophy · Literary technology · Jean-Robert Chouet · Antonio Baldigiani · Kaspar Schott · Paolo Casati

In 1658, a graduate of the Jesuit Collegio Germanico named Andreas Portner shepherded into print the lecture notes of his professor Silvestro Mauro (1619–87), then a professor of philosophy at the Collegio Romano. After studying philosophy and theology at the Collegio Romano from 1639 to 1648, Mauro had taught first at the Jesuit school at Macerata and the Collegio Germanico in Rome before being

R. Raphael (✉)
University of California, Irvine, CA, USA
e-mail: renee.raphael@uci.edu

S. Berger, D. Garber (eds.), *Teaching Philosophy in Early Modern Europe*, Archimedes 61, https://doi.org/10.1007/978-3-030-84621-3_9

appointed a professor at the Collegio Romano.[1] There he taught theology, ethics, and sacred scripture as well as the three-year philosophy sequence, comprising logic, natural philosophy, and metaphysics.[2] In addition to his published lecture notes, which Portner had titled *Quaestiones Philosophicae* (*Philosophical Inquiries*), Mauro published a six-volume paraphrase of Aristotle's writings and a three-volume theology course.[3]

Like many of his contemporaries, Mauro did not limit his discussion of philosophy to topics treated only by Aristotle. In a discussion of the possibility of a natural void, a question considered by Aristotle, Mauro described a relatively new phenomenon:

> If a glass tube of a length beyond two cubits, in which there is a sole open hole, from which end it is filled inside with quicksilver, which is called likewise Hydrargirum or Mercurius, with air being removed, and with a finger placed on top of the hole in such a way that the quicksilver is not able to come out, the tube is turned upside down and submerged in other quicksilver prepared in a dish or bowl placed underneath, then with the finger having been removed the hole is opened so that there is an unimpeded way for the quicksilver to come out and to descend, the quicksilver descends until a certain height of the tube: then it ascends upwards to some extent, and afterwards descends, and thus it fluctuates somewhat by ascending and by descending, until at last it rests at a certain height in the tube.[4]

With this passage, Mauro recounted in general terms recent experiments with the mercury tube that were first performed by Evangelista Torricelli (1608–47), who had found that when a glass tube was filled with mercury and then inverted in an open dish, the mercury level in the tube always fell to the same height. Torricelli's claims were contested by the Capuchin friar Valeriano Magni (1586–1661), whose allegations for independent invention of the tube and his publication of a book on the device instigated a bitter priority dispute.[5] Reports of Torricelli's experiments inspired other natural philosophers, including Étienne Pascal (1588–1651) and Blaise Pascal (1623–62), to carry out their own experiments. European natural philosophers were quickly amazed to find that the height of the mercury changed with altitude but that it remained unaffected by variations in the width and inclination of

---

[1] Zanfredini 2001.

[2] Information on positions held at the Collegio Romano for Mauro and other professors is taken from García Villoslada 1954, 321–36.

[3] The two works are *Aristotelis opera quae extant omnia*, Rome, 1668 and *Opus theologicum*, Rome, 1687.

[4] Mauro 1658, vol. 2, 415–16: "si fistula vitrea ultra duos cubitos longa, in qua sit unicum foramen apertum, ex una extremitate impleatur argento vivo, quod vocatur etiam Hydrargirum, seu Mercurius penitus aere excluso, et digito supraposito foramini, ita ut argentum vivum non possit egredi fistula invertatur et demergatur in alio argento vivo parato in vase, seu catino subiecto, tum digito subducto aperiatur foramen ita ut libera sit argento vivo facultas egrediendi, ac descendendi, argentum vivum descendet usque ad certam mensuram fistulae: deinde sursum aliquantum ascendit, et post descendit, et sic aliquantulum fluctuat ascendendo, ac descendendo, donec tandem quiescat in certa altitudine fistulae."

[5] The narrative that follows is developed from Dear 1985, 180–209; Hellyer 2005, 142–58; Shea 2003, 17–185. Also see Middleton 1971; Montacutelli 2009, 59–72; Shapin and Schaffer 1985.

the tube. Most puzzling was the space above the mercury in the glass tube, which many, following Torricelli and Magni, interpreted to be a void space or vacuum.

This space above the mercury behaved as did another seventeenth-century novelty, the air-pump, a device that pumped air out of a glass receiver. Invented in the 1640s by Otto van Guericke (1602–86), mayor of the small Protestant town of Magdeburg, the device and its dissemination were intimately linked to the Jesuit order. After witnessing Guericke's demonstrations of the device, Archbishop Johann Philipp von Schönborn (1605–73) brought the pump to Würzburg, where he entrusted it to Jesuit professors at the university. There, one Jesuit mathematician, Kaspar Schott (1608–66), was immediately intrigued. He performed experiments himself, opened a correspondence with Guericke, and solicited other Jesuit scholars, including Athanasius Kircher (1602–80) and Niccolò Zucchi (1586–1670), both at the Collegio Romano, to respond to claims that the air pump created a void. Schott's description and evolving assessments of the device were printed in a series of publications, culminating in his 1664 *Technica curiosa*. In England Robert Boyle (1627–91), prompted by Schott's description of the device, built his own air pump with the help of Robert Hooke (1635–1703) and published his findings in his 1660 *New Experiments Physico-Mechanicall, Touching the Spring of the Air and its Effects*.[6]

Schott was not the only Jesuit to react and contribute to these new experimental findings. In Paris, Estienne Noël (1581–1659) responded to Pascal's 1647 claim that the height of the mercury in the tube depended on air pressure. Noël argued, against Pascal, that the upper portion of the glass tube was filled with a purified air that entered through the pores of the glass. In Rome, Gasparo Berti (1600–43) discussed his own experiments with inverted tubes of water with Zucchi and Kircher.[7] Paolo Casati (1617–1707), who taught in the Collegio Romano in the mid-seventeenth century, recounted his own experiments with the mercury tubes in the first-person in his *Vacuum Proscriptum* (*The Proscribed Void*), where he discussed the recent mercury experiments at length and refuted their supposed demonstration that a void space could exist in nature.[8] Other Jesuits, including Honoré Fabri (1608–88), also described and discussed these experiments, a collective response which may have been directed principally against Magni's strongly anti-Aristotelian interpretation of the experiment, its theological ramifications, and the strained relations between the Jesuit and Franciscan Orders.[9]

There are two dominant narratives in which the Jesuit reception of the mercury tubes and air-pump are usually slotted. The first focuses on how Jesuits responded to the values and style of the new experimental philosophy. According to Steven Shapin and Simon Schaffer, the Jesuit Francis Lines (1595–1675) was thought by

---

[6] See Hellyer 2005, 153–58; Gorman et al. 2000.

[7] Shea 2003, 28.

[8] On the reaction of Casati and other Jesuits to the experiments on the void, see Gavagna 2002, 325–38.

[9] Gorman 1994, 7–32.

Boyle to be a competent member of the experimental community, because he agreed with many of Boyle's reported matters-of-fact, recognized the value of experimentally-produced matters-of-fact, and carried out his own trials to produce such facts.[10] Schott has been described similarly. Citing Schott's enthusiastic endorsement of Boyle's writings, including his use of witnesses and his narration of both successful and unsuccessful attempts, Marcus Hellyer claims that Schott recognized and was convinced by Boyle's program of "virtual witnessing," a term developed by Shapin and Schaffer to describe Boyle's use of prolix narrative and detailed images for the purposes of convincing readers of the validity of his reported experiments.[11] Peter Dear has argued that philosophers, both Jesuit and non-Jesuit, drew on Jesuit mathematicians' tendencies to portray specific experimental claims as universal, generalized Aristotelian experiences when describing both their own and others' reported experiments, including those of the mercury tubes. In other cases, the Jesuit response has been characterized as more subdued. According to Michael John Gorman, while the larger philosophical community moved from treating the Torricellian void and air-pump as natural philosophical phenomena to discussing them as instruments, the Jesuits never made this transition because of the Order's restrictions regarding the maintenance of key Aristotelian tenets.[12]

The identification of Jesuit censorship echoes a second narrative regarding Jesuit discussion of the mercury tubes and air-pumps, namely the reception of the devices in classroom teaching. Historians have found that professors accommodated experiments and other novelties to their curriculum by slotting the new material into established questions and concerns.[13] Such a model corresponds to Charles Schmitt's description of early modern Aristotelianism as being an eclectic philosophy, one that readily drew on outside philosophical developments to strengthen and broaden its scope.[14] Less is known about how these specific experiments were taught in regular Jesuit philosophy courses in the seventeenth century. Hellyer has described a dearth of discussion in the classrooms of German Jesuit universities of the seventeenth century, arguing that the lack of discussion illustrates these professors' reluctance to engage with novel findings.[15] Hellyer's main observations include: that many professors chose to ignore the new experiments; that when they were addressed, they were discussed in the context of Book 4 of Aristotle's *Physics*; that the mercury tube was treated more often than was the air-pump; and that the tube was usually presented with few details as an experience demonstrating the impossibility of a void.[16]

---

[10] Shapin and Schaffer 1985, 156–69.

[11] Hellyer 2005, 155–156.

[12] Gorman 1994, 13–14.

[13] See, for example, Baldini 1999, 252, 2011; Brockliss 1987, 337, 1996.

[14] Schmitt 1983.

[15] Hellyer 2005, 159.

[16] Hellyer 2005, 158–61.

This article puts these two narratives of reception—one focusing on the response to the new experimental philosophy in general, the other on its inclusion (or not) in classroom teaching—into conversation. It examines the teaching of the mercury tubes and air-pump at the Collegio Romano in the second half of the seventeenth century by taking up a question posed by Nick Wilding in his study of Galileo's friend and patron Gianfrancesco Sagredo (1571–1620). In *Galileo's Idol*, Wilding commented on the curious invisibility to historians of science of literary technology, noting that the former group has tended to overlook the complexities inherent in writing, specifically the work that readers and reception play in constructing authorship. Despite the ubiquitousness of the term "literary technology" in science studies, Wilding argues, the complex social and cultural process of textual reception that make the author have largely been invisible to historians of science.[17] Indeed, though Shapin and Schaffer introduced the term "literary technology" in their *Leviathan and the Air Pump*, it and the many studies of early modern experimentation that the book inspired have emphasized experimentation and its difficulties, particularly replication.

This contribution will turn this focus on its head by considering textual replication, in particular, the recounting in teaching texts of experiments found in printed books. Using extant teaching notes and published philosophy courses, it examines how the experiments were presented to students in the second half of the seventeenth century. It then considers the sources of these Jesuit teaching texts and compares them to teaching of experimental philosophy in northern Europe. In contrast to scholarship emphasizing the role of textual details as transforming readers into virtual witnesses, I argue that authoritativeness and reliability were intrinsic to the genre of these pedagogical texts. Professors reproduced textual details and visual images of experimental procedures and apparatus not to convince readers that the experiments had been carried out but to testify to their own competence as readers and participants in a philosophical enterprise in which texts and textual practices were central.

9.1 Teaching the Torricellian Void and Air-Pump at the Collegio Romano

Professors of philosophy at the Collegio Romano described these experiments with specific visual and textual details, but their choice of particulars was inconsistent. There is little indication that these professors or their students were involved in an active program of experimentation. The close correspondence between textual details, experimental replication, and witnessing often associated with the experimental philosophy is thus absent in these teaching texts.

[17] Wilding 2014, 136.

A closer examination of Mauro's description of the mercury tubes is a case in point. While Mauro provided specific details for his students, he gave no indication that he himself had carried out the experiment or that he expected students to do so themselves. In his account, Mauro described the basic set-up: if one took a glass tube of sufficient length (more than two cubits) with one hole, filled it with mercury, and—without allowing air to enter the tube—turned it upside down in a container filled with mercury, the mercury in the tube would fall to a certain level after some moments of oscillating up and down. Arguing that nature abhors a void, Mauro explained that further experiments with the mercury tube provided convincing evidence that the space in the upper part of the tube was not a void space, for if it were heated, the level of mercury went down, and when it was cooled, the mercury level rose. Mauro interpreted this behavior as evidence that something in the upper part of the tube was rarefying with heat and condensing with cold, and since rarefaction and condensation are the properties of bodies (not voids), the upper part of the tube must contain a body, which Mauro described as "some subtle parts separated from mercury and rarefied."[18]

Mauro made no distinction with his word-choice or sentence construction, either in the initial description of the experiment or his response to it, that he was reporting a novel experiment by his contemporaries. Instead, he narrated the phenomenon of the Torricellian void the same way as he did other examples commonly employed in medieval and early modern texts. Elsewhere Mauro, for example, described the behavior of containers of water and wine to explain the position of opponents who argue for the existence of a natural void. He wrote, "A large earthenware container holds as much wine together with wine skins as without them: a dish holds as much water with ashes as without ashes, which cannot happen without a void."[19] In attributing the behavior of flint to a similar cause, Mauro included roughly the same amount of detail but elaborated on the cause of the phenomenon. He described how "While flint is broken with the blow of a hammer simultaneously its middle and end pieces are obtained; therefore a void is given in the middle for some very brief time, before air reaches the location of those [spaces] through successive motion."[20] The principle difference between his account of the mercury tubes and these other examples is in the greater detail contained in the former. Rather than being able to refer simply to the "behavior of the inverted mercury tubes," he had to offer specific information on this behavior, as if students were not entirely familiar with it. Despite the lengthier description, however, Mauro's students encountered few specifics that would have allowed them to repeat the experiment themselves or assure them of its accuracy. No information was given about the width of the tube, the amount of

[18] Mauro 1658, part 2, 415–6.

[19] Mauro 1658, part 2, 415: "Dolium tantundem vini capit una cum utribus, quantum sine utribus: vas tantundem aquae capit cum cinere quantum sine cinere, quae non possent contingere sine vacuo."

[20] Mauro 1658, part 2, 415: "Dum silex rumpitur ictu mallei simul se parantur partes mediae, et partes extremae; ergo in meiis datur vacuum pro aliquo brevissimo tempore, antequam aer per motum successivum perveniat ad locum ipsorum."

mercury, or the size of the dish holding the mercury, nor was an image provided for students to visualize the tubes. This lack of illustration in a compendium commemorating the teaching of natural philosophy at the Collegio Romano was typical. Many surviving textbooks contain no images; when they were included, often in books printed in the last decades of the seventeenth century, visual representations illustrated the different world systems, not experimental apparatus.[21] Indeed, given the lack of illustration in Mauro's text, it is likely students would not have been able to picture the inverted tubes of mercury for themselves. In a manner similar to what Hellyer argues occurred in Jesuit teaching in Germany, the experiment with the Torricellian void became in Mauro's text one of many other *experientiae* against the void.

Other professors who taught after Mauro described the experiments similarly. In his printed compendium, first published in 1674, André Semery (1630–1717) explained the experiment as follows:

> It is objected, first, the experiment of Father Magni. Take a glass tube closed in such a way from one part so that no body, even the finest one, is able to enter. Fill the tube with mercury or quicksilver and close up that part standing open with a thick finger. Then immerse it into a shell-shaped vessel filled with more mercury; and do not draw back the finger until the tube is immersed at its other end. Then indeed draw back the finger, and you will see that the quicksilver is supported from the part underneath, and in the empty space, which is of three or four finger-lengths, according to the greatness or smallness of the tube, [that it] puts forth some oscillations and undulates until a little later it is completely at rest.[22]

Although Semery's text includes images of the world systems, printed on a separate sheet at the end of the volume, it contains no images of the mercury tubes. This lack of illustration indicates that the mercury tubes were considered one of numerous examples of phenomena with which students had to be familiar when considering the question of a natural void but that they were not expected to be familiar with the details of such experiments, even when details were included in the text itself. It seems no professor envisioned that students would carry out such experiments themselves on the basis of their classroom lectures.

While following the same general narrative, Semery's account of the experimental procedure differs from that of Mauro in key ways. Semery, in contrast to Mauro, explicitly identified the novelty of the experimental claim. His description of the experiment falls within a section of his course entitled "whether a void is possible."[23] Early on in the exposition, Semery explained to readers that "Aristotle discusses the

---

[21] An example of one such text that contained images of the various world systems is Semery's *Triennium Philosophicum*, discussed below.

[22] Semery 1688, part 2, 552: "Objicitur primo, experimentum P. Magni. Accipe fistulam vitream ita clausam ex una parte, ut nullum corpus vel tenuissimum introgredi possit. Imple fistulam mercurio, sive argento vivo, et qua parte patet, crasso digito occlude. Deinde in concham alio mercurio plenam immerge; nec retrahe digitum donec fistula secundum illam extremitatem immersa sit. Tunc vero retrahe digitum, et videbis argentum vivum ex parte deorsum tuere, et in spatio relicto, quod erit trium, vel quatuor digitorum, pro magnitudine, vel parvitate tubi, edere vibrationes quasdam, et fluctuare, donec postmodum plane conquiescat."

[23] "De Vacuo ... An sit possibile," Semery 1688, part 2, 549.

void broadly in [book] 4 of the *Physics*, showing that it is simply opposed. Many out of the more recent philosophers allow the void." Semery named among these Magni and his 1647 *Demonstratio ocularis* (*Proof of the Eye*).[24] When he turned to describing Magni's experiment, Semery, however, included fewer details than Mauro, not mentioning the required length of the tube or the different names for mercury.

Finally, unlike Mauro, Semery wrote in the imperative, a verb mood which was often employed when recounting medical, alchemical, and culinary knowledge in the period and which the Jesuits generally did not use in natural philosophical writing.[25] In Semery's teaching, the description of the experiment became a recipe, rather than another commonplace example of nature prohibiting the void. Semery did not use the imperative when recounting other experiences that he claimed showed nature's repugnance to a void. Consider, for instance, his account of medicinal cupping glasses, which was written in the third person and described as being "familiar":

> Among others is that familiar [example] of cupping glasses (they call them commonly *ventossas*) which doctors everywhere make use of. For in them, when air reduces itself to the appropriate density, with the heat evidently vanishing by which it had rarefied, it is evident from experience that the flesh is raised in order to fill the space left by the air.[26]

A later example of the behavior of water and wine is similar, though composed using a combination of the second and third person:

> Thereafter if you turn upside down a flask filled with water or wine with an orifice open, often it happens that nothing flows out from it, unless you put it at an incline for some time: clearly [this is] so that the air making its way in through the small channel proceeds to the place that must be filled, which must be removed by a flowing liquid: and because if it is turned straight air does not have a way to make its way in, therefore, nothing flows. Acolytes often experienced this same thing, when from an ampulla, well-stopped up in its upper orifice, nothing flows into the chalice from the extended water-channel, even though it is open.[27]

[24] Semery 1688, part 2, 550: "De Vacuo fuse disputat Aristoteles 4. Physic. ostendens illud simpliciter repugnare. Ex recentioribus multi vacuum admittunt. Inter cæteros est P. Magnus, qui librum scripsit cui titulus est: *Ocularis demonstratio Vacui*. Et vere in illo experimento quod adducit oculi nullum corpus intra fistulam depræhendunt … Experimentum illud adducitur inferius."

[25] Gorman 1994, 28–29.

[26] Semery 1688, part 2, 550: "Inter alia est illud familiare de cucurbitulis (vulgo ventossas vocant) quibus medici passim utuntur. Nam in iis cum aer reducit se ad debitam densitatem, evanescente videlicet calido a quo raruerat, experientia constat carnem attolli, ut relictum ab aere spatium impleat. Immo si Chirurgus eas ab humeris avulsutus recta sursum attollat, diffingit potius quam avellat. Unde eas aliquantisper elevans ab una parte retrahit, aer subingrediens finit illas avelli."

[27] Semery 1688, part 2, 551: "Praeterea si lagenam aqua, vel vino plenam aperto orificio inverteris, saepe continget nihil ex ipsa effluere, nisi illam tantisper inclines: videlicet ut per canaliculum sese insinuans aer pergat ad occupandum locum, qui deferendus est a liquore effluente: et quia si recta invertatur non habet aer qua se insinuet, ideo nihil effluit. Hoc idem saepe experiuntur Acholiti, dum ex ampulla bene obturata in superiori orificio, nihil effluit in calicem ex protenso canaliculo, quamvis aperto."

Both of these examples are described in general terms and assume that readers will be acquainted with them. Semery apparently believed that few details were necessary for the reader to understand what situation is being referenced.

Professors who taught later in the century addressed both the mercury tubes and the air-pump. One such individual was Giovanni Iacopo Panici (1657–1716), who taught natural philosophy at the Collegio Romano from 1698 to 1701.[28] In his manuscript notes, dated 1696, Panici recounted several experiences drawn from the Aristotelian textual tradition regarding the void before turning to more recent experiments that purported to demonstrate one. According to Panici:

> I pick out a single and very celebrated one, which Evangelista Torricelli, Galileo's student and a mathematician of great account, proposed in the forty-third year of this century ... this author took a glass tube with a length of around three feet, sealed hermetically at one end: he filled it with *hydrargirus*, or quicksilver, or mercury (these names are used interchangeably), and with the other part of the tube stopped up with a finger placed on it, he raised the tube perpendicularly to the horizon, and immersed the finger simultaneously with that end part which it was closing, in mercury lying in a large bowl placed below as appears in the figure. Then with the finger having been removed, the mercury enclosed in the tube descended, in such a way, nevertheless, that after many undulations and vibrations, hanging, it rests in the tube at a height of nearly 27 and a half inches [digits] above the flat surface of the mercury lying in the bowl; leaving behind the space of the upper tube thoroughly empty of mercury.[29]

Panici's account differs from those of his predecessors by his inclusion of specific, key factual details. He mentioned Torricelli, identified him as Galileo's student, and recorded the year in which Torricelli first carried out his experiment. Panici also gave a specific length of the tube with a unit of measurement different from that provided by his predecessors. In addition, Panici's students learned that the tube was hermetically sealed, the multiple names of mercury, and the specific height at which the mercury rests. Finally, Panici, unlike his predecessors, provided for students an image of the experiment (Fig. 9.1). His image, which was hand-drawn, takes up an entire page. It is located, along with four other images, in the opening pages of the manuscript, though the text that accompanies it is found roughly four hundred pages later. Each of these figures was labeled (note the label "Figure B" in Fig. 9.1) so that the reader could locate the image as it was referenced later in the text. The image depicts the tube once the mercury has come to rest. It shows the mercury tube in isolation, without the experimenter's hands and fingers that are so

---

[28] Sommervogel 1960, vol. 6, 166.

[29] Panici n.d., 408^{v}: "Unum, et celeberrimum hic seligo, quod anno huius saeculi 43. proposuit Evangelista Torricéllius Galilæi discipulus, et magni nominis Mathematicus ... accepit hic author fistulam vitream trium circiter pedum longitudine, ex una parte Hermetice sigillatam: eam hydrargiro, sive argento vivo, sive Mercurio (quae nomina idem sonant) implevit; obturataque subiecto digito altera fistulae parte, fistulam perpendiculariter ad orizontem erexit, et digitum simul cum extrema ea. parte, quam claudebat, immersit Mercurio stagnanti in subiecto catino, ut apparet in figura. Amoto deinde digito, Mercurius inclusus fistula descendit, ita tamen, ut post multas undulationes, et vibrationes consisteret pendulus in fistula ad altitudinem digitorum fere viginti septem cum dimidio supra superficiem planam Mercurii stagnantis in catino; relinquens reliquum spatium superius fistulæ Mercurio penitus vacuum."

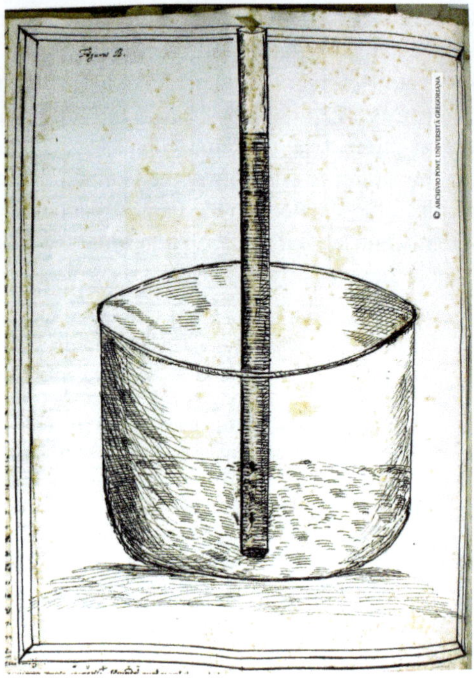

Fig. 9.1 Panici's illustration of the mercury tube (Ink on paper, 296 × 217 mm, APUG FC 1093, 4v, ca. 1698)

crucial for its set-up. Panici noted in his text that the tube was placed "in mercury lying in a large bowl placed below as appears in the figure." This image thus serves to communicate information about the relationship between the tube, the mercury, and the bowl that is not easily explained in writing. The image itself depicts the bowl, tube, and mercury in isolation. Shading gives the reader a sense of three-dimensionality and the notion that the bowl is placed on a level surface.

Panici did not use this level of detail when describing other examples drawn from the Aristotelian tradition. As one example, consider his discussion of medicinal cupping glasses:

> First the medicinal cupping glasses … indeed after the flame has been cut off inside them, the flesh immediately within the cupping glasses is vigorously raised; but this is not seen to proceed from a different source than the ceasing of that great rarity of the air that had been created from the preceding flame. And indeed with the air having been led back to itself to a smaller space, the flesh is raised so that the part of the space left by the air is filled, lest a void be given there.[30]

[30] Panici n.d., 408^{r}: "Primo cucurbitulae Medicae … postquam enim intra illas rescissa fuerit flamma, caro illico intra cucurbitulas attollitur vehementer; hoc autem non videtur aliunde procedere, nisi quod cessante illa magna raritate aeris orta et praecedente flamma: adeoque reducante se aere ad minus spatium, caro attollitur ad occupandam partem spatii ab aere relictam, ne ibi detur Vacuum."

In contrast to the detailed explanation of how the inverted tube of mercury is set up and behaves, this example was listed very briefly. He spent more time accounting for the behavior of the cupping glasses than he did recounting the observed phenomena. This abbreviated form might suggest that he assumed this example was a commonplace; additional explanation was unnecessary because the behavior of the cupping glasses should have been familiar to readers with only this brief description.

Another example, however, suggests that the distinction between Panici's narrations of these experiences has less to do with questions of novelty versus textual tradition than it does with the sources and authors of the experiences. Farther on in his enumeration of experiences drawn from the Aristotelian tradition, Panici described the following:

> [Consider] the twisted siphon having, however, unequal legs, as in the figure below. If the orifice of this siphon is submerged briefly in a dish filled with water, and then with the [experimenter's] mouth applied to the opening of the longer leg hanging outside the vessel, all the air enclosed in the siphon is driven out, at once the water descends with the air into the [experimenter's] mouth. When [the experimenter's] mouth is withdrawn, the water will advance from the siphon until either a drop flows in the vessel or the surface of the standing water within the vessel is at least lower than the leg of the siphon; for then the air will penetrate, and the flow of the water will cease.[31]

With its specific information about what the reader needed to do to elicit the reported behavior, this example of the siphon more resembles that of the mercury tubes than that of the cupping glasses. The similarities between the example of the siphon and the mercury tubes suggest that what led Panici to include greater amounts of detail was not the relative novelty of an example but who was responsible for carrying out the demonstration. Experiences carried out by trusted authorities, in the case of the cupping glasses, doctors or surgeons, required less specificity than those ostensibly performed by the reader or another interested savant. As Peter Dear and others have argued, the experiences of artisans—a group that would include the surgeons using the cupping glasses and the mechanics or miners operating water pumps—were commonly cited as believable for philosophical demonstration.[32] Such experiences could be trusted without the actions of an active reader.

Panici's text also differed from those of his predecessors in its treatment of the new air pump, whose development Panici attributed to Boyle. According to Panici, "Robert Boyle ... added other phenomena through his Pneumatic machine, which is called a Cenoscopium."[33] In the margin, he indicated "as appears in figure E below"

---

[31] Panici n.d., 408r: "Siphunculus retortus habens tamen crura inaequalia, ut in subiecta figura. Si naris brevius huius siphunculi immergatur in vas aqua plenu, et deinde applicato ore ad orificium cruris longioris extra vas pendentis, exugatur totus aer in siphunculo inclusus, statim aqua simul cum aere descendét in os exugentis. Rémoto deinde ore, tamdiu aqua a siphone prodibit, quamdiu vel gutta in vase superit, vel certe superficies aquæ stagnantis intra vas sit inferior crure siphunculi; tunc enim subintrabit aer, et deficiet fluxus aquae."

[32] Dear 1985, 144–50; Blair 1997, 104–5.

[33] Panici n.d., 410r: "alia tamen addidit phænomena Robertus Boyle ... per suam machinam Pnéumaticam, quae Cenoscopium appellatur."

(Fig. 9.2).[34] Again, this image was hand drawn, comprises a full page of Panici's manuscript, and is located in the opening pages of the treatise. Panici also referred readers directly to Boyle's own image, noting that "An image of the machine described can be seen in the front of the work which Boyle wrote, *Nova experimenta Physico-mechanica de vi aeris elastica*, or likewise in the figure below.[35] While Panici directly referenced Boyle's image (Fig. 9.3), there are differences in the two diagrams provided by these authors. Boyle's image is an *intaglio* print, included as a fold-out in his 1661 *New experiments physico-mechanicall, touching the spring of the air, and its effects*.[36] In contrast, Panici's—in a manuscript teaching treatise—was hand-drawn, perhaps by Panici himself; there is no indication in the manuscript treatise as to the author of its images. It appears that Panici (or his artist) copied from Boyle's printed image, imitating its cross-hatching and shading to provide viewers with a more naturalistic image. While the main features of the pumps are the same, Boyle used the space around the figure to depict in isolation different parts

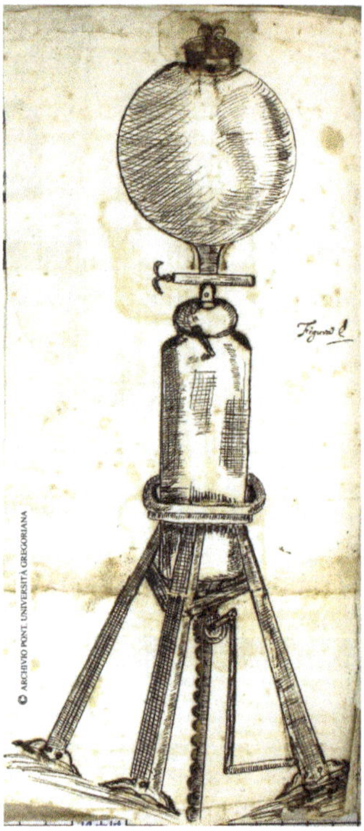

Fig. 9.2 Panici's rendering of the air-pump (Ink on paper, 296 × 217 mm, APUG FC 1093, 6v, ca. 1698)

[34] Panici n.d., 410r: "ut apparet in subiecta figura. e."

[35] Panici n.d., 410r: "Imago descriptae machinae videri potest in fronte operis quod ipse Boyle inscripsit. Nova experimenta Physico-mechanica de vi aeris elastica; vel etiam in subiecta figura."

[36] Though we know that Boyle collaborated with William Faithorne to produce other images of his air-pump for inclusion in portraits Faithorne made of Boyle with the instrument, the engraver who produced the prints for this publication is not known. On Faithorne's portraits, see Maddison 1959.

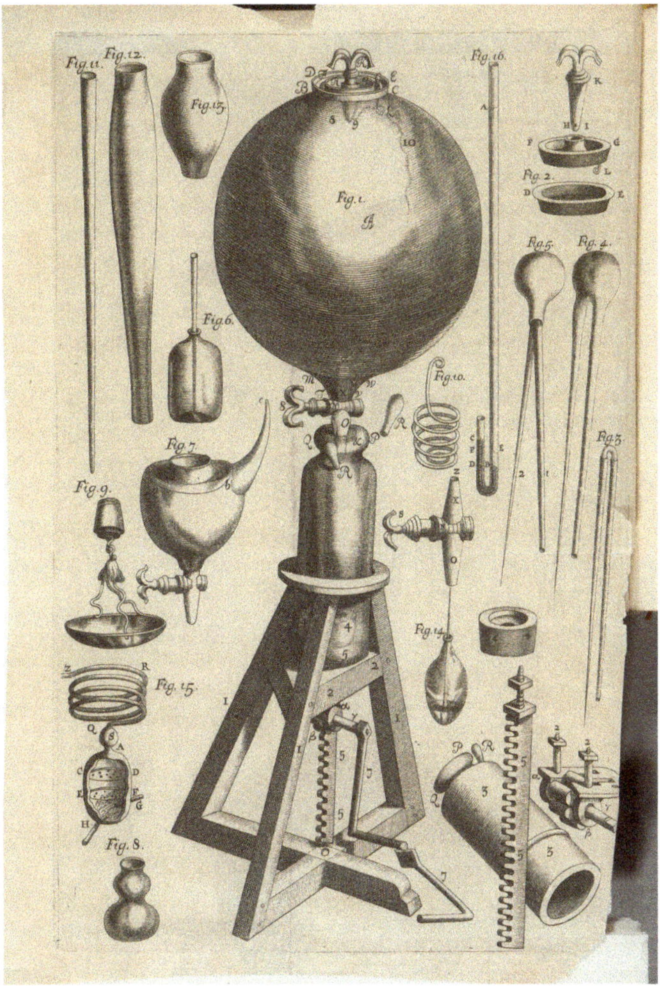

Fig. 9.3 Boyle's image of his air-pump (Engraving, 8 cm × 30 cm., Boyle, *New experiments physico-mechanicall, touching the spring of the air, and its effects* (Oxford: printed by Miles Flesher for Richard Davis, bookseller in Oxford, 1682), first fold-out before p. 1

of the apparatus and other implements employed in the experiments. Panici's drawing, in contrast, is filled with white space. In Boyle's rendition, the air pump appears as just one of many instruments; its designation as the principal instrument is made through its central position and larger size. Panici's version emphasizes the air-pump through its isolation. His image makes it easier to imagine the air-pump being used, yet the absence of the additional illustrations, many of which depict parts of the air-pump or instruments placed inside it, means that it communicates less information about how the air-pump was employed and the parts required for its operation.

Panici's text similarly mirrors and differs from Boyle's own in its description. Like Boyle, Panici offered his readers many details about the set-up and operation of the air-pump. He described how it was composed of two main parts and how to make them:

> A stand is found so that from the upper part a vessel, either of glass or of another material (which is commonly called a Receiver) is fastened to the table, by its orifice with wax and with turpentine applied to the table, so that no access is accessible from the surrounding air; from the other part of the table is a great hole, in which the pump is placed so that the piston is almost made equal to the flat table. Between the glass vessel and the pump is a copper tube, slightly curved, through which air, which is in the Receiver, can be brought into the pump: and in the outside part of the same tube, which ends in the pump, there is the *verticillus*, where the entrance of air … is possible, whenever it pleases. Above the piston, a small metal iron piece, made of iron and compressed, is fastened … Finally in the lowest part of the pump there is a small aperture, which with the finger having been applied, or something similar, can be closed or opened up conveniently.[37]

Panici provided students with the names of the parts, their relation, the materials used to assemble the pump (wax and turpentine), how it operates, and the role of the experimenter who needed to use his finger and evacuate the pump multiple times. Yet, lest we imagine that Panici provided all the specifics of the procedure for his readers, consider very briefly the beginning of Boyle's description:

> The shape of the Glass, you will find express'd in the first Figure of the annexed Scheme. And for the size of it, it contain'd about 30 Wine Quarts, each of them containing near two pound (of 16 Ounces to the pound) of water: We should have been better pleas'd with a more capacious Vessel, but the Glass-men professed themselves unable to blow a larger, of such a thickness and shape as was requisite to our purpose.
>
> At the very top of the Vessel, (A) you may observe a round hole, whose Diameter (BC) is of about four inches; and whereof, the Orifice is incircled with a lip of Glass, almost an inch high: For the making of which lip …[38]

In just this brief excerpt, Boyle included, among other details, the size of the Receiver, the measure of its hole or opening, and the fact that the opening is encircled with a lip of glass almost an inch high. He went on to explain how to make that lip. While much detail was communicated via text, Boyle's reference to the accompanying figure indicates that he judged some information—such as the shape of the glass—to be more efficiently communicated visually. While Panici included a substantial amount of detail in his rendition, both text and image, he did not duplicate all of Boyle's information. While Boyle may have intended for readers to be able to

[37] Panici n.d., 410r: "Ita aperitur mensa, ut ex parte superiore vas vel vitreum, vel alterius materiae, (quod vulgo Recipiens appellatur) mensae agglutinetur, orificio eius cum cera, et terebintina ad mensam applicato, ut nullus aere circumfuso pateat aditus; ex altera parte mensæ sit magnum forman, in quo anthlia sic disponatur, ut embolus planae mensae paene commensuretur. Inter vas vitreum, et anthliam sit tubus cupreus paululum curvatus, per quem aer, qui est in Recipiente, traduci possit in anthlia: et in tubi eiusdem extrema parte, quae in anthliam desinit, sit verticillus, quo aeri aditus .. possit, quandocumque libuerit. Super embolum lamella ferrea, atque densata affigatur …. Sit denique in infima anthliae parte exiguum foramen, quod apposito digito, vel alio simili, occludi commode possit, aut recludi."

[38] Boyle 1662, 49.

reproduce (or imagine reproducing) his experiments for themselves, Panici's decision to include much less information suggests that he did not have this same goal in mind.

Another account of the air-pump was presented at the Collegio Romano by Panici's colleague Antonio Baldigiani (1647–1711).[39] In his 1696 *Exercitationes Physicomathematicae* (*Physicomathematical exercises*), Baldigiani discussed more generally the behavior of tubes filled with water and other liquids before explaining the creation of the Torricellian void:

> Let there be an oblong and wide tube, penetrable from both ends, which the eighth figure exhibits. This tube should be perfectly closed at its end A by means of a skin or bladder placed around [it] and attached well. This extremity should be present, inserted in the little cloak or bladder XZ, and thereupon well glued. The cloak indeed should have in its thin upper part the opening B through which the whole cloak is filled with water to such an extent that nothing of air remains in it. Then the opening should be closed with glue and skin. Then the whole tube is filled with mercury. But with the usual immersion made, a Torricellian void must be noted as occurring if the bladder appearing in the extremity of the tube is not broken or dilated.[40]

Baldigiani's figure was included on a separate sheet in his manuscript, in the same ink as the manuscript; no information is provided to indicate its creator, and it seems likely that the image was executed by Baldigiani himself (Fig. 9.4). Unlike many other professors, Baldigiani labeled parts of the diagram and referred to them in his description. We can note that Baldigiani's Torricellian void differed from the typical set up (and described by Panici) in that Baldigiani described using a skin or bladder to close the upper end of the tube. Panici had described the more common set-up in which the tube itself was closed at the upper end. This detail and the labels suggest that Baldigiani may have intended his students to carry out such an experiment themselves or at least to understand the details involved in doing so.

After describing the behavior of the tube under various conditions, Baldigiani posed the question, "Whether it is possible to make a Torricellian void in another way without mercury or another liquid descending and evacuating tubes and the apparatus placed above."[41] He answered as follows:

> The Magdeburgenses and the English Royal Society devised certain clever machines, through which huge glass vessels which are accustomed to be called a Receiver, are purged of all air not however without tedious toil and time. They [accomplish] it with the work of a pump purging the enclosed air successively and by parts. The way of using such machines

[39] On Baldigiani, see Findlen 2009, 211–54.

[40] Baldigiani 1696, 26^{v}: "Sit tubus oblongus et amplus utrimque peruius, qualem exibet figura octaua. Hic tubus sit perfecte clausus in extremitate A mediate pellicula siue uesica circum posita et bene alligata. Haec extremitas adsit inserta lacernula seu uesicula XZ. et ibi bene aglutinata. Lacerna uerò habeat in parte superiori tenue aperturam B per illam impleatur aqua tota lacerna ita ut nihil aeris intra illud remaneat deinde claudatur apertura glutine et pellicula tunc tubus totus impleatur mercurio, at facta solita immersione fiat uacuum torricelliam num notandum occurret quod uesica existens in a̲ extremitata tubi non frangetur nec dilatabitur."

[41] "Quaeres 6.° in Uacuum Torricellianum alio modo fieri possit sine mercurio vel alio liquore descendente et evacuante fistulas et vasa superius imposita," Baldigiani 1696, 37^{v}–38^{r}.

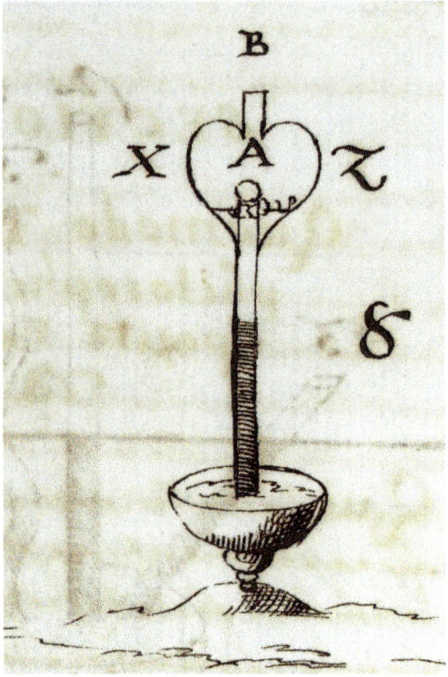

Fig. 9.4 Baldigiani's diagram of the mercury tube (Ink on paper, Biblioteca Casanatense, Ms. 1203, 64, detail, 1696)

> is very laborious and dangerous for by cracks often entering air slips by those observing and also many times a crack of the glass occurs.[42]

Baldigiani provided no other information about or depictions of the device or the experiments carried out with it. This lack of detail makes Baldigiani's attention to the process by which the recipient was evacuated all the more striking: without specific instructions on how to use or make a pump, why would students need to know about the possibility of the vessel cracking? According to Shapin and Schaffer, Boyle's inclusion of this and similar details communicated to the reader that the authors had done the experiment. It is less clear what purpose they serve in Baldigiani's account.

9.2 Text Versus Experiment

These examples offer a variety of descriptions, level of detail, and use of images. What stands out is the apparently random choice of key particulars to include in relating the procedures. Drawing on Shapin and Schaffer's analysis, historians have often interpreted the inclusion of details as a means of persuading readers of the

[42] Baldigiani 1696, 38^{r}: "Magdeburgenses, et societas regia anglicana excogitauit ingeniosas quasdam machinas, per quas ingentia uasa uitrea quae ab illis uocari solent recipientia omni aere euacuantur non tamen sine longo labore, et tempore. Id praestant ope Antliae evacuantis successive et per partes inclusum aerem. Modus utendi talibus machinis est ualde laboriosus, et periculosus nam per rimulas saepe ingrediens aer fallet obseruantes saepe etiam interuenit fractio vitri."

veracity of the text, which transformed readers into virtual witnesses. Could the inclusion of these details in these pedagogical texts function similarly, as a means of communicating to students that their professors were reporting experiments carried out and witnessed, either privately (by the professors themselves) or publicly, in front of the students?

By the latter part of the seventeenth century, experiments were becoming a standard part of philosophical courses at some universities. In 1675, the University of Leiden became the first university to offer courses in experimental physics and to create a facility dedicated solely to that purpose. These efforts were initiated by Burchard de Volder (1643–1709), a Cartesian professor of logic and physics, who saw experiments as pedagogical tools, in keeping both with Cartesian-inspired ideas of effective teaching and with Leiden's long tradition of teaching through observation. De Volder founded Leiden's Physical Theater and commissioned numerous scientific instruments for his demonstrations. As evidenced by his student notes, de Volder's early lectures centered on the demonstration of many experiments taken from Boyle's publications on the air-pump.[43]

In the Jesuit context, however, the carrying out of experiments was typically either a private activity or one intended more for public consumption and entertainment than pedagogical instruction. In preparation for writing his 1650–51 *Almagestum Novum*, Giovanni Battista Riccioli (1598–1671) worked with a group of students to make his own astronomical observations and carry out mechanical and optical experiments.[44] More generally, Jesuit mathematicians used their bedrooms as multifunctional spaces, where they stored mathematical instruments and directed the studies of select students who comprised the private mathematical academy of the college. The carrying out of experiments extended to a wider audience than the chief mathematician and his chosen pupils. According to the Jesuit's *Ratio Studiorum* of 1591, once or twice a month, one of these private students was to recount an illustrious mathematical problem to a large gathering of philosophers and theologians. Often these mathematical problems dealt with issues in the realm of mixed mathematics and included demonstrations of actual experiments, in addition to texts and images.[45]

While experiments were being carried out at the Collegio Romano, it seems unlikely that the descriptions included in these philosophical courses were intended as reports of professors' own trials or suggestions for students to carry them out on their own. For one, with the possible exception of Panici's text, most professors did not offer sufficient details to make such experiments possible. Moreover, the Jesuits who did perform experiments and wrote about them tended to make their participation clear in their accounts by writing in the first-person or through other literary devices. In contrast, these professors made no attempt to associate themselves with the carrying out of the experiments and instead invoked textual authorities.

---

[43] Nyden 2014 207–22 (esp. 207, 210–11); eadem 2013, 227–49 (esp. 227, 236).

[44] Dinis 2003 198.

[45] Gorman 2003, 13, 15–17, 26.

Yet the amount of detail Jesuit professors included about the experiments in their teaching may have marked them as unusual in the wider European context. Their attention to detail may partly reflect the Order's commitment to high quality and up-to-date scholarship, the additional specifics functioning to signal to readers these Jesuits' familiarity with the latest publications. It also might indicate that Jesuit professors expected their students to be unfamiliar with these experiments and thus requiring more information about them. In Geneva, the Cartesian-influenced Jean-Robert Chouet (1642–1731) addressed the same experiments in his *Syntagma Physicum* (*Physical syntagm*) as part of a larger argument that air pressure explained many phenomena once attributed to nature's fear of the void. Like the Roman Jesuits, Chouet referenced commonplace experiences with few details, as if he assumed readers would be familiar with them. Consider, for instance, his passing reference to divers at sea, an example intended to bolster his argument that air was heavy:

> It ought not to hinder us that we don't feel the heaviness and pressure of the air. For [that lack of feeling] arises from those causes by which divers at the bottom of the sea feel no weight of water.[46]

Chouet's discussion of the mercury tubes was similarly brief. Using the example to reinforce the same argument about air's heaviness, he noted that:

> If a glass tube is obtained, closed at one end, open at the other, whose length does not surpass twenty-six digits, and that tube is filled with mercury and its opening is inserted in some mercury contained in a vessel somewhere, it is known, I say, that the mercury in the tube, which is possessed of some heaviness, descends very little, but sticks suspended within the tube itself.[47]

The principal detail Chouet referenced was the length of the tube (twenty-six digits). The inclusion of this information, however, is not an indication of his desire to inform readers of specificities of the experiment. Rather, it was a necessary detail because his subsequent discussion centers on the behavior of tubes of lengths greater and less than twenty-six digits and how this varied behavior reflects not fear of the void but the air's pressure.

While Jesuit professors, in contrast, took care to reproduce numerous details from their sources when describing the experiments, the relationship between their teaching texts and their textual sources is not always clear. Semery, for example, cited Magni's *Demonstratio ocularis* specifically, and told readers to take a "glass tube" closed at one end so that nothing can enter. The tube was to be filled with "mercury or quicksilver" and closed up "with a thick finger." It was then to be

---

[46] Chouet 2010, vol. 2, 69: "neque id nos morari debet quod aëris pondus ac pressionem non sentiamus. Id enim ex iisdem oritur causis quibus urinatores in fundo maris nullum aquae pondus sentiunt."

[47] Chouet 2010, vol. 2, 70: "si sumatur tubus vitreus altero extremo clausus altero apertus, qui longitudinem viginti sex digitorum non superet, tubusque ille hydrargyro compleatur et eius orificium in aliud hydrargyrum in vase quopiam contentum imittatur, notum est, inquam, hydrargyrum tubi qualicunque praeditum sit gravitate minime descendere, sed suspensum intra ipsum tubum haerere."

immersed in a shell-shaped vessel filled with mercury, at which point the finger was to be withdrawn. The empty space that resulted after the vibrations of the mercury settled would measure three or four finger-lengths.[48]

Though Semery cited Magni's account for his students, the details included in the two texts do not correspond. According to Magni,

> I obtained a glass tube, with a length of more than two cubits, whose funnel receives a pea comfortably: the thickness of the glass is as much as that of a grain of corn. I closed one of its openings with a Hermetic seal, namely with liquid glass. I filled this tube with quicksilver, whose open opening I stopped up with a finger applied vigorously, above that opening, so far stopped up, I raised the tube, which I immersed in quicksilver, put into a suitable vessel, similarly immersed in a vessel filled with waters, in such a way that the waters stood out above the quicksilver by four finger-lengths.[49]

Unlike Semery, Magni gave a length for the tube and measurements for the width of its opening and the thickness of its glass. He also described immersing the vessel filled with mercury within a larger one filled with water. Magni continued by narrating that he repeated the process three times, uncovering the lower part of the tube first immersed in the mercury, the second time in the waters, and the third time outside both the mercury and the waters. He described how the mercury "jumps up again and jumps down and after some fluctuations of this sort, the bottom part of the tube, with a length of one cubit, is seen full with mercury, the upper part indeed is believed empty of all bodies: for there is nothing of water discerned by the eyes in that part."[50]

Semery's narration of the account resembles more that of other Jesuits. Consider, for example, the following description by Zucchi:

> If a glass tube, impassable along one extremity, is filled through the other open [extremity] with quicksilver having been poured in, and, upright, is submerged in other quicksilver, contained in a dish placed under it and with water poured over, according to the open extremity but closed by an applied finger; at such a time when the finger is moved from the mouth of the tube, the quicksilver contained in the dish descends; after the descent it rises

[48] Semery 1688, part 2, 552: "Objicitur primo, experimentum P. Magni. Accipe fistulam vitream ita clausam ex una parte, ut nullum corpus vel tenuissimum introgredi possit. Imple fistulam mercurio, sive argento vivo, et qua parte patet, crasso digito occlude. Deinde in concham alio mercurio plenam immerge; nec retrahe digitum donec fistula secundum illam extremitatem immersa sit. Tunc vero retrahe digitum, et videbis argentum vivum ex parte deorsum tuere, et in spatio relicto, quod erit trium, vel quatuor digitorum, pro magnitudine, vel parvitate tubi, edere vibrationes quasdam, et fluctuare, donec postmodum plane conquiescat."

[49] Magni 1648, 5: "Procuravi fisultam vitream, longam ultra cubitos duos, cuius canalis commode excipiat pisum: crassities vitri est tanta, quanta est grani frumentacei. Unum eius orificium occlusi sigillo Hermetis, scilicet, vitreo liquenti. Hanc fistulam implevi argento vivo, cuius orificium patens obturavi digito vehementer applicato, super quo orificio, eatenus obturato, erexi fisultam, quam immersi argento vivo, indito vasi congruo, quod vas erat pariter immersum vasi repleto aquis, ita ut aquae eminerent supra argentum vivum digitis quatuor."

[50] Magni 1648, 6: "salitque iterum, ac desilit, et post nonnulas eiusmodi vibrationes, inferior pars fistulae, longitudine cubitu uno, visitur plena mercurio, superior vero creditur vacua ab omni corpore: nam oculis nil aquae, in illa parte cernitur."

> a little; then it descends less and rises less; until with the number of repeated fluctuations always fewer, at last it rests in the bottom part of the tube, with the upper part abandoned not only by itself, but by any other body.[51]

Like Semery's, Zucchi's account says nothing about immersing the tube in containers of mercury and water. Zucchi's simpler narration also resembles Semery's: the tube is filled, stopped with a finger, flipped over, and the mercury fluctuates. At the same time, Zucchi mentions nothing about the length of the space at the top of the tube, while Semery specifies that it measures three or four finger-lengths. Examination of other sources indicates that it was more common for authors to report the final height of the mercury, not the length of the empty space at the top of the tube.[52] It is possible that Semery used Magni's or another author's report of the height of his tube and the final height of the mercury to calculate for himself the length of the empty space above the mercury, reporting this value alone to his students. But Semery may have relied on other sources as well. A possible source for Semery's alternative names for mercury is Casati's *Vacuum Proscriptum*, which states that "argento vivo" could also be called "mercurius" or "hydrargyrus."[53]

There are similar issues with regards to the textual sources for Panici's and Baldigiani's descriptions of the air pump. On the one hand, the current literature might suggest that Panici and Baldigiani were regurgitating a "Jesuit" interpretation of these experiments presented in Schott's *Technica Curiosa*, where Schott endorsed Boyle's view that the air's pressure explains the behavior of the mercury tubes and air-pump. Schott as a source—and especially the only source—is unlikely, however. For one, despite the plethora of authors cited by Panici and Baldigiani, including Lines, Fabri, Boyle, Borelli, Torricelli, Daniello Bartoli (1608–85), and Casati, neither one names Schott. It is certainly possible that Panici and Baldigiani purposefully obscured their own sources, but other details make Schott unlikely, or unlikely as the only source for their accounts. In his account, Schott described attempts to quantitatively measure the heaviness of the air. While Schott cited Galileo as an individual who made such measurements, he did not provide for his readers the value Galileo reported in his final publication, the *Two New Sciences*. Panici, in contrast, also referred readers to these measurements, but he offered a specific value for Galileo's reported measurement, namely the heaviness of air to water is in the ratio of 1 to 500. (Note that Panici's value is actually incorrect, as Galileo had claimed the ratio was 1:400.) Instead, Schott refers readers to Daniel Lipstorp's (1631–84) 1653 *Specimina philosophiae Cartesianae* (*Ideas of Cartesian*

[51] Zucchi 1649, 101–2: "Si Tubus vitreus, secundum alteram extremitatem impervius, ingesto per alteram patentem argento vivo, impleatur, et erectus immergatur secundum patentem, sed apposito digito clausam, alteri argento vivo in vase subiecto contento, et aqua superfusa; tum amoveatur digitus ab orificio Tubi, argentum vivum in Tubo contentum descendit, post descensum aliquantum assurgit; tum descendit minus, et resurgit minus; donec iteratis aliquot vibrationibus semper minoribus, tandem quiescat in parte Tubi inferiori, relicta superiori non ipso solum destituta."

[52] Another Jesuit writer, Paolo Casati, did just that, reporting the height of the mercury measured by Magni and Roberval, as well as those of his own experiments, Casati 1649, 84–5.

[53] "Argento vivo (seu malis appellare Mercurium, aut Hydrargyrum)," Casati 1649, 84.

Philosophy), which also only described Galileo's procedure without providing any quantitative value.[54] A more likely source for Panici and Baldigiani may have been Boyle himself, who discussed Galileo's and other attempts to measure quantitatively the heaviness of the air. Boyle reported the results of, among others, Riccioli (whose estimate of the weight of air to water he reported as being 1:10,000), Galileo (1:400), and Mersenne (1:1356), as well as his own (1:938).[55] Panici, in contrast, had claimed Galileo's ratio to be 1:500, while Mersenne's was 1:1000. He did not quote Boyle's measurement of the heaviness of the air. It seems likely that Panici and Baldigiani relied on a variety of sources when compiling their accounts and may have modified or massaged the values reported in their own sources.

Our inability to locate a precise textual source for the descriptions in the Jesuits' courses should not lead us to doubt the central role that textual practices played in the crafting of these teaching texts. Following the convoluted path of visible and invisible attributions in questions of chronology in the writings of Scaliger, Bodin, and Copernicus has led Anthony Grafton to argue that even early modern texts that appear coherent and organic may not be the result of original composition but instead of what he terms "systematic bricolage," or unacknowledged borrowing from disparate sources.[56] Unacknowledged reliance on or deliberately obscure use of textual sources was completely normal, as quotation practices in the pre-modern period did not conform to today's standards. John Whittaker has argued that ancient writers' mistranslations of earlier sources are not always due to faulty memory or carelessness but arose instead from the tendency of these authors deliberately to introduce into their quotations alterations, including transformations in word order and addition, subtraction, and substitution of words.[57]

9.3 Replicating Literary Technology

Historical scholarship on the rise of the experimental philosophy has emphasized the interest of period actors and their difficulties in replicating the material technology of their peers. This exploration reveals that the literary technology accompanying this material technology has its own convoluted history. It is not obvious whether readers wanted to replicate the literary technology of the texts they encountered, even when they incorporated some elements of that technology, namely certain details and images. In most, moreover, the details and images do not appear to function in the same way as they did in the original texts, as demonstrations that the experiments actually took place. In many ways, these details seem to be witnessing the act of reading, more than the act of experimenting.

[54] See Schott 1664, vol. 1, 228; Lipstorp 1653, vol. 3, 114.

[55] Boyle 1662, 149–50.

[56] Grafton 1989, 183.

[57] Whittaker 1989, 71.

The literary technology of replication also asked something of readers. Whereas analyses of virtual witnessing assume the details were included for the purposes of persuading readers that the reported experiments had been made, readers of these secondary reports of experiments were persuaded by different means. These accounts of the experiments were persuasive not because of their level of detail but because of their genre and because of the citation apparatus that accompanies them. These accounts were authoritative not for their ability to create virtual witnesses of their readers but rather because of their existence as teaching texts, which cited the writings of other authors.

Bibliography

Baldigiani, Antonio. 1696. Exercitationes Physicomathematicae et Progressiones Mathematicae. *Biblioteca Casanatense Ms* 1203.

Baldini, Ugo. 1999. The Development of Jesuit 'Physics' in Italy, 1550-1700: A Structural Approach. In *Philosophy in the Sixteenth and Seventeenth Centuries: Conversations with Aristotle*, ed. Constance Blackwell and Sachiko Kusukawa, 248–279. Aldershot: Ashgate.

———. 2011. Tra due paradigmi? La 'Naturalis philosophia' di Carlo Rinaldini. In *Galileo e la scuola galileiana nelle università del Seicento*, ed. Luigi Pepe, 189–222. Bologna: Clueb.

Blair, Ann. 1997. *The Theater of Nature: Jean Bodin and Renaissance Science*. Princeton: Princeton University Press.

Boyle, Robert. 1662. *New Experiments Physico-Mechanical: Touching the Spring of the Air and Its Effects*. Oxford: Printed by H. Hall for T. Robinson.

Brockliss, Laurence. 1987. *French Higher Education in the Seventeenth and Eighteenth Centuries: A Cultural History*. Oxford: Clarendon Press.

———. 1996. Curricula. In *A History of the University in Europe. Volume 2: Universities in Early Modern Europe (1500–1800)*, ed. H. de Ridder-Symoens, 563–620. Cambridge: Cambridge University Press.

Casati, Paolo. 1649. *Vacuum Proscriptum*. Genuae: Ioannes Dominicus Peri.

Chouet, Jean-Robert. 2010. *Corsi di Filosofia*, ed. Mario Sina, Marco Ballardin, and Elena Rapetti, 2 vols. Florence: Leo S. Olschki.

Dear, Peter. 1985. *Discipline and Experience: The Mathematical Way in the Scientific Revolution*. Chicago: University of Chicago Press.

Dinis, A. 2003. Giovanni Battista Riccioli and the Science of His Time. In *Jesuit Science and the Republic of Letters*, ed. Mordechai Feingold, 195–224. Cambridge: Cambridge University Press.

Findlen, Paula. 2009. Living in the Shadow of Galileo: Antonio Baldigiani (1647-1711), a Jesuit Scientist in Late Seventeenth-Century Rome. In *Conflicting Duties: Science, Medicine and Religion in Rome, 1550–1750*, ed. M.P. Donato and J. Kraye, 211–254. London: The Warburg Institute.

García Villoslada, Ricardo. 1954. *Storia del Collegio Romano dal suo inizio (1551) alla soppressione della Compagnia di Gesù (1773)*. Rome: Apud aedes Universitatis Gregorianae.

Gavagna, Veronica. 2002. I gesuiti e la polemica sul vuoto: il contributo di Paolo Casati. In *Gesuiti e università in Europa*, ed. Gian Paolo Brizzi and Roberto Greci, 325–338. Bologna: Cooperativa Libraria Universitaria Editrice Bologna.

Gorman, Michael John. 1994. Jesuit explorations of the Torricellian space: carp-bladders and sulphurous fumes. *Mélanges de l'Ecole française de Rome. Italie et Méditerranée* 106 (1): 7–32.

———. 2003. Mathematics and Modesty in the Society of Jesus: The Problems of Christoph Grienberger. In *The New Science and Jesuit Science: Seventeenth Century Perspectives*, ed. Mordechai Feingold, 1–120. Dordrecht: Kluwer Academic Publishers.

Gorman, Michael John, Nick Wilding, and Gaspar Schott. 2000. *La "Technica Curiosa"*. Rome: Edizioni dell'Elefante.

Grafton, Anthony. 1989. Editing Technical Neo-Latin Texts: Two Cases and Their Implications. In *Editing Greek and Latin Texts*, ed. John N. Grant, 163–186. New York: AMS Press, Inc.

Hellyer, Marcus. 2005. *Catholic Physics: Jesuit Natural Philosophy in Early Modern Germany*. Notre Dame: University of Notre Dame Press.

Lipstorp, Daniel. 1653. *Specimina Philosophiae Cartesianae*. Leiden: apud Johannem et Danielem Elsevier.

Maddison, R.E.W. 1959. The Portraiture of the Honourable Robert Boyle, F.R.S. *Annals of Science* 15 3–4: 141–214.

Magni, Valeriano. 1648. *Demonstratio ocularis, loci sin locato, corporis successive moti in vacuo, luminis nulli corpori inhaerentis*. Bologna: typis haeredis V. Benatii.

Mauro, S. 1658. *Quaestionum Philosophicarum Sylvestri Mauri Soc. Iesu. in Collegio Romano Philosophiæ Professoris. Libri Tres Pro Laurea Philosophica Andreae Portner Collegij Germanici, & Hungarici Alumni*. Rome: Ignazio Lazeri.

Middleton, W.E.K. 1971. *The Experimenters: A Study of the Accademia del Cimento*. Baltimore: The Johns Hopkins University Press.

Montacutelli, Stefania. 2009. Air 'Particulae' and Mechanical Motions: From the Experiments of the Cimento Academy to Borelli's Hypotheses on the Nature of Air. In *The Accademia del Cimento and Its European Context*, ed. Marco Beretta, Antonio Clericuzio, and Lawrence M. Principe, 59–72. Sagamore Beach: Science History Publications.

Nyden, Tammy. 2013. De Volder's Cartesian Physics and Experimental Pedagogy. In *Cartesian Empiricisms*, ed. Mihnea Dobre and Tammy Nyden, 227–249. Dordrecht: Springer.

———. 2014. Living Force at Leiden: De Volder,'s Gravesande, and the Reception of Newtonianism. In *Newton and Empiricism*, ed. Zvi Biener and Eric Schliesser, 207–222. Oxford: Oxford University Press.

Panici, J.J. n.d. In Libros Aristotelis de Physico Auditu. *Archivio della Pontificia Università Gregoriana, Rome (APUG) FC 1093.*

Schmitt, Charles B. 1983. Eclectic Aristotelianism. In *Aristotle and the Renaissance*, 89–109. Cambridge, MA: Harvard University Press.

Schott, Gaspar. 1664. *Technica Curiosa, Sive Mirabilia Artis Libri XII Comprehensa*. Nuremberg: Sumptibus Johannis Andreae Endteri, & Wolfgangi Junioris Haeredum, excudebat Jobvs Hertz, Typographus Herbipol.

Semery, A. 1688. *Triennium Philosophicum, Quod P. Andreas Semery Remus, è Societate Jesu, In Collegio Romano Philosophiae Iterum Professor Dictabat. Tertia Hac Editione Quae Est Prima in Germania Correctum*. Cologne: Joannis Caspari Bencardi.

Shapin, Steven, and Simon Schaffer. 1985. *Leviathan and the Air-Pump: Hobbes, Boyle, and the Experimental Life*. Princeton: Princeton University Press.

Shea, William R. 2003. *Designing Experiments and Games of Chance*. Canton: Science History Publications.

Sommervogel, Carlos, ed. 1960. *Bibliothèque de la Compagnie de Jésus, nouvelle édition*, 12 vols. Louvain: Editions de la Bibliothèque S.J., Collège philosophique et théologique.

Whittaker, John. 1989. The Value of Indirect Tradition in the Establishment of Greek Philosophical Texts or the Art of Misquotation. In *Editing Greek and Latin Texts*, ed. John N. Grant, 63–95. New York: AMS Press, Inc.

Wilding, Nick. 2014. *Galileo's Idol: Gianfrancesco Sagredo and the Politics of Knowledge*. Chicago: University of Chicago Press.

Zanfredini, M. 2001. Mauro, Silvestro. In *Diccionario histórico de la Compañía de Jesús : biográfico-temático*, ed. C.E. O'Neill, and J.M. Dominguez, 4 vols, vol. 3, 2583. Madrid: Universidad Pontificia Comillas.

Zucchi, Niccolo. 1649. *Nova de Machinis Philosophia*. Rome: Typis haeredum Manelphii.

Chapter 10
A Mirror of Wisdom: Simon Vouet's *Satyrs Admiring the Anamorphosis of an Elephant* and Its Afterlives

Alexander Marr

Abstract This essay addresses the invention, meaning and reception of Simon Vouet's *Satyrs admiring the anamorphosis of an elephant* (ca. 1627) and the engraving after it by Johannes Tröschel. It identifies the subject as a thesis print with theological and philosophical content, likely made in the orbit of the Jesuits of the Collegio Romano and under the patronage of Prince Maurice of Savoy. The image's deployment of catoptrical technology and anamorphosis is discussed in relation to period conceptions of *imprese*, including those of Emanuele Tesauro. The image's creation and reception are situated in the context of Italy's learned academies and pan-European interest in mathematical optics and instrumentation.

Keywords Simon Vouet · Johannes Tröschel · Maurice of Savoy · Anamorphosis · *Impresa* · Thesis print · Satyrs · Elephant · Jesuits

Early versions of this essay were delivered at the Renaissance Society of America Annual Meeting, the Early Modern French Seminar at the University of Cambridge, and the conference *Art and Science in the Early Modern Low Countries* at the Rijksmuseum and Trippenhuis. I am grateful to Eric Jorink and Tim Chesters for the invitations to present my work and to the audiences for their helpful questions and observations. I am also grateful to Louis Rice, Susanna Berger, Dan Garber, Paul Taylor, Liz McGrath, John de Boulton-Holland, Judit Sebö, Denis Ribouillault, and Simon Turner for their helpful comments and suggestions. The writing of this paper has been partially funded by the European Research Council under the European Union's Seventh Framework Programme (fp7/2007–2013)/erc grant agreement no. 617391. *In memoriam* Robert Oresko (1947–2010).

A. Marr (✉)
University of Cambridge, Cambridge, UK
e-mail: ajm300@cam.ac.uk

S. Berger, D. Garber (eds.), *Teaching Philosophy in Early Modern Europe*, Archimedes 61, https://doi.org/10.1007/978-3-030-84621-3_10

Fig. 10.1 Simon Vouet, *Satyrs admiring the anamorphosis of an elephant* (before 1627), red chalk on paper

In the mid-1620s, Simon Vouet (1590–1649)—then the leading French artist in Rome—made an elegant red-chalk drawing now in the Hessisches Landesmuseum, Darmstadt (Fig. 10.1).[1] It depicts a group of eight satyrs standing on a low mound overlooking a formal garden, which has been rendered in single-point perspective. Gathered round a table on a pedestal base formed by a pair of winged sirens, the satyrs gawk and gesticulate at the spectacle before them: the anamorphosis of an elephant in a cylindrical mirror. Above them floats a banderol, on which the faint letters "D O M" have been roughly inscribed.[2] Astonished, amused and bemused, these ancient, semi-bestial creatures have been captivated by a modern marvel: the cunning rectification of a distorted image through mirror technology and mathematical optics. The deployment of an elephant—well established by the early

[1] Hessisches Landesmuseum, Darmstadt, Inv. H. 21.762. The attribution to Vouet has occasionally been questioned, but there is now general consensus that it is an autograph work. The drawing's provenance cannot be traced earlier than the late eighteenth/early nineteenth century: it was purchased in Paris from the collection of the Marquis de Lagoy (1764–1829) by the ambassador Joseph van Dahlberg (1773–1833). See Thuillier et al. 1990, 107–8; Cordellier et al. 2007, 162. An anonymous, seventeenth-century copy of the drawing is in the Kupferstich-Kabinett, Staatliche Kunstsammlungen Dresden, Inv. C 6840. On early modern anamorphosis, see Massey 2007.

[2] Inspection of the drawing in person has shown these letters to be integral to the image—they are not a later addition. On the potential significance of the satyrs' gestures as recognizable rhetorical signs, see Dimova 2015.

Fig. 10.2 Johannes Tröschel after Simon Vouet, *Satyrs admiring the anamorphosis of an elephant (format et illustrat)* (before 1628), copperplate engraving

seventeenth century as a symbol of virtue and true religion—for the anamorphosis, along with the letters "D O M" (*Deo Optimo Maximo*; to God, the best and greatest), suggest that the scene has religious overtones.[3] We might interpret it as a reference to Christian truth defeating Pagan superstition or, since satyrs are notoriously lascivious, to the importance of pursuing virtue rather than vice. The image plays engagingly with a variety of themes connected to vision and knowledge, choreographed as a series of *paragoni*: order overcoming disorder, art compared with nature, truth against fiction, and wisdom versus ignorance, to name but a few.

In an engraving made after the drawing by the emigré Nuremberg printmaker Johannes Tröschel (1585–1628), the banderol's lettering has been replaced by the motto "format et illustrat" (it shapes and illuminates)—apparently a hapax legomenon in early modern sources (Fig. 10.2).[4] Nothing definite is known about the commissions or purposes of the drawing and the print, the dates of which are uncertain. However, the drawing was presumably made before Vouet left Rome for Paris in the late summer of 1627, the engraving after Tröschel's arrival in Rome in 1624 and

[3] For an overview of the symbolism of the elephant in the early modern period, see Orsi 1994.

[4] The engraving bears the signatures "Simon Voüet In:[venit]" and "Joan:[nes] Tröschel Sc:[ulpsit]". The use of the diacritical mark in Vouet's surname reflects the way he signed himself in his correspondence in 1627. See Thuillier et al. 1990, 107.

before his death there in May 1628.[5] As we shall see, there are good reasons for dating the images to Vouet's last year in Rome: 1627.

The production of the engraving implies that the design was intended for circulation, while its motto suggests an allegory, the interpretation of which has long intrigued historians of art and science. Baltrušaitis read Vouet's drawing as a pioneering example of the depiction of catoptrical anamorphosis in Western art, purportedly informed by Chinese examples transmitted via Constantinople (which Vouet visited 1611–12), while Chastel described the engraving and its motto as an exercise in early modern aesthetic theory, capturing the "implicit doctrine that painting functions to counter the dispersion of objects by unifying them into an image."[6] Scholars of Vouet and his milieu have speculated that the design may have been intended as a eulogy of Prince Maurice of Savoy (one of Vouet's Roman patrons), an *impresa* of the Roman Accademia di San Luca (of which Vouet was elected *principe* in 1624), or an emblem of so-called "baroque science."[7] Several have commented on the images' putative connection to Maurice's academies, specifically the Torinese Solinghi's symbol of a conical mirror anamorphosis of the motto (taken from Virgil) "omnis in unum" (all in one), illustrated and discussed at length by Emanuele Tesauro (1592–1676) in his *Cannocchiale aristotelico* (first edition, 1654).[8] As we shall see, this book plays an important role in the story of Vouet's drawing (Fig. 10.3).

Despite these speculations, the images' meaning and significance have remained elusive. In what follows, I will argue that they should be read primarily as an allegory of Christian "virtue as knowledge," promoting the collaborative pursuit of both under the benevolent guidance of a noble patron. Interpreted thus, they represent a striking example of a particular branch of moral philosophy (ethics) signified by an aspect of natural philosophy (optics), the fulcrum of which was "ingenuity" or "wit" (*ingenium*; *ingegno*; *esprit*).[9] Notably, the kind of wit the images promote is the early modern ideal of "collective ingenuity"—as opposed to the singular "genius" of Romanticism—propagated by the Italian academies and passed on to the scientific societies of the later seventeenth century. For, as Thomas Sprat (1635–1713) said of the Royal Society, "In assemblies, the wits of most men are sharper, their apprehensions readier, their thoughts fuller, than in their closets."[10]

During his Roman period, Vouet was an active member of the city's intellectual circles. In addition to receiving patronage from learned princes such as Urban VIII

[5] On Tröschel, see Marr 2011, passim; Beaujean and Turner 2020 (this engraving, Hans Troschel, no. 7).

[6] Baltrušaitis 1977; Chastel 1996, 51–2.

[7] Wazbinski 1992, 158–9; Préaud 1992; Brejon de Lavergnée 2008; Gómez López 2016. For Vouet's association with the Accademia di San Luca, see La Blanchardière 1972; Bissell 2011.

[8] See, in particular, Préaud 1992 who seems to have been the first to suggest that Vouet's drawing should be interpreted via Tesauro's *Cannocchiale* and Bonfait 2011. The motto "omnis in unum" derives from Virgil, *Aeneid*, X. 410.

[9] On which see Gensini and Martone 2002; and Marr et al. 2018.

[10] Sprat 1722 (first edition 1677), 98.

Fig. 10.3 Georges Tasnière after Domenico Piola, frontispiece to Emanuele Tesauro, *Cannocchiale aristotelico* (1670), copperplate engraving

(1568–1644) and Maurice of Savoy (1593–1657), he was close to Cassiano dal Pozzo (1588–1657) and friendly with literati such as Giovanni Battista Marino (1569–1625, whose portrait he took ca. 1623–4) and Marcello Giovanetti (1598–1631), designing the frontispiece for the latter's *Poesie* (1626).[11] While concentrating his efforts on religious commissions, in Rome Vouet was energetically active as a portraitist and genre painter of Caravaggesque subjects, made numerous drawings for printed books and thesis prints, and produced a small number of allegorical paintings and drawings. Two of the latter bear upon his *Satyrs admiring the anamorphosis of an elephant*, for reasons intellectual and patronal.

The first is the *Allegory of the human soul*, painted for the banker Marcello Sachetti (1586–1629) probably in late 1624 and engraved by Vouet's pupil Claude

[11] For Vouet's Italian period (1613–27), see principally Loire 1992; Bonfait and Rousteau-Chambon 2011; Méséguer 2008. For his connections with Cassiano, see Solinas 1992; Schleier 2008. For Vouet's portrait of Marino, see Brejon de Lavergnée 1980.

Fig. 10.4 Claude Mellan after Simon Vouet, *Allegory of the human soul* (1625), copperplate engraving

Mellan (1598–1688) the following year (Fig. 10.4).[12] As the inscription on the print explains, the picture shows personifications of three parts of the human soul—Intellect (*Intellectus*), Memory (*Memoria*) and Will (*Voluntas*)—from which derive man's capacities for reason, choice, and virtuous action. The trichotomic division of the soul was a staple of patristic and scholastic theology and natural philosophy, and remained current in early modern Rome's philosophical milieu, particularly amongst the Jesuits.[13] It is striking, though, that in the print's inscription Intellect is described as *ingenium*, which "leaps like a flame, flashes like a flame, that is why it crowns the

[12] On which see Crelly 1962, 212 and *passim*; Thuillier et al. 1990, 189–90; Ficacci 1989, 158–61 and 170–71. For the Sachetti, see Ceccarelli 1946; Zirpolo 2005.

[13] Ficacci 1989, 160 suggests that the subject and iconography were likely informed by the Roman Jesuits. The trichotomic soul was first proposed by St Augustine in *De Trinitate* (X, 17–19). For early modern scholarly treatment of the soul, see Vidal 2011. This division appears also in the chariot of Martin Meurisse (1584–1644) and Léonard Gaultier's (1560/61–1635) *Clara totius physiologiae synopsis* ("Tres sunt facultates animae rationalis, intellectus, voluntas, et memoria…"). See Berger 2017, 182. This broadside is reproduced in Susanna Berger's chapter in this volume.

hair of the head with harmless fire."[14] Indeed, Intellect/*Ingenium* is depicted as a young man holding a *tabula rasa*, whose head sprouts a lively flame.[15] Thus, while Vouet's *Allegory* is in many ways a conventional representation of a traditional subject, it partakes of the growing fascination for ingenuity and its properties, especially the vogue for sparkling quick-wittedness, promulgated particularly by the "poet of the marvelous," Marino.[16] In the philosophical arena, just a few years after Vouet had painted his *Allegory*, Descartes composed his (unfinished) *Regulae ad directionem ingenii* (ca. 1628; *Rules for the direction of the* ingenium), in which he anatomized *ingenium* as "the site of embodied cognition [that] could be trained or exercised" to make sharp philosophical judgments.[17] Vouet might well have discussed the matter with Descartes, since he took the philosopher's portrait around the time Descartes was completing his treatise.[18]

The second allegory which bears upon the *Satyrs* is, like that image, a red chalk drawing: the *Allegory of the House of Savoy*, drawn by Vouet ca. 1625 and subsequently engraved by Mellan (Fig. 10.5). Vouet had sought to gain Maurice's patronage some years earlier, requesting an introduction through Cassiano dal Pozzo.[19] Vouet hoped the connection with Maurice—who had close political ties to the French contingent in Rome—would impress his patron, the king of France.[20] In this he was successful, for Maurice commissioned a number of works from the French painter.[21] The *Allegory of the House of Savoy* depicts past members of the noble

[14] "Ingenium ceu flamma salit, ceu flamma coruscat/Iccirco innocuo circuit igne comas." The verse appended to the engraving is not signed, but it was likely composed by a poet in Vouet's circle. The Latin recalls a passage in Virgil's *Aeneid*, in which "innocuous fire" plays around the head of Iulis (Ascanius). I am grateful to Liz McGrath for bringing this reference to my attention.

[15] It has been speculated that the object held by Intellect/*Ingenium* is a mirror. See e.g. Rodari 2015, 50. However, while the object is rendered somewhat ambiguously in the painting, in the engraving the fall of a shadow from Intellect/*Ingenium*'s arm indicates that it is not a reflective surface, but a blank slate. In the *Synopsis* (see above, n. 13), Intellect also holds a tablet. As Crelly established (1962), the attributes of the personifications derive from the *Iconologia* of Cesare Ripa (1560–1622), although they have been adapted from several different figures. Will, for instance, has the double face of Prudence in Ripa's manual.

[16] On whom see e.g. Mirollo 1963. Notably, mirrors feature prominently as a theme in Marino's writings. See Rima 1991.

[17] Garrod 2021, 2.

[18] See Marr 2015; idem 2021.

[19] Thuillier et al. 1990, 97: "V.S., il quale per sua gratia mia semper procurato il mio bene come quomissco per la sua di qualo che mi dice avere trattato di me al sig[no]re Principe Card[ina]le di Scavoia, il quale averebe a caro di potere servir quoniosendo il jovamento che detto servitu mi potarebe aportar a presso il Re mio patrone." Simon Vouet to Cavaliere [Cassiano] dal Pozzo, from Genoa, 21 May 1621.

[20] Maurice was sent to Paris in 1618 to witness the marriage of his brother, Vittorio Amadeo (1587–1637), to the King's sister, Christine of France (1606–1663). Once in Rome, he became protector of France to the Holy See, having been charged by his father with the representation of French and Venetian interests. Made a cardinal in 1621, in the 1623 conclave that elected Urban VIII Maurice vigorously supported the French party. See Randi 1901; Osborne 2007.

[21] See Oberli 1999; Schleier 2008.

Fig. 10.5 Simon Vouet, *Allegory of the House of Savoy* (ca. 1625), red chalk on paper

dynasty standing on cylindrical pedestals and engaged in heroic actions or holding symbolic attributes, the significance of which are identified by inscriptions on the pedestals. At the center stands a youthful figure holding a bow in his left hand and, aloft in his right, an orb with the arms of the House of Savoy. He has been convincingly identified as Maurice (at that time a cardinal) in the guise of Apollo-Christ. The pedestal on which he stands bears the inscription "Sabaudica religionis simulacrum" (the likeness of Savoyard religion), indicating that Maurice—descendent of a noble line of virtuous defenders of the faith—is a shining embodiment of his *patria*'s commitment to true religion.[22] While the function of this image is not certain, it seems likely that it was devised as a thesis print. Indeed, it is probably the item described in the sale catalogue of the Crozat collection (1741) as a drawing "by Simon Vouet, which is the subject of a thesis dedicated to the House of Savoy, which has been engraved by Mellan."[23]

---

[22] See Préaud 1992; Di Macco 1995, 353. Tesauro associates Maurice with Apollo in his *Panegyrici* 1659, vol. 1, 155. For Maurice and Apollo-Christ symbolism, see Vagnoni 2008, 146–51.

[23] "de Simon Vouet, qui fait le sujet d'une thèse dédiée à la Maison de Savoye, lequel a été gravé par Mellan". See Bacou and Bean 1988; Brejon de Lavergnée 1990, 362.

There are good reasons for supposing that Vouet's *Satyrs* was also the model for a thesis print. First, its iconography is idiosyncratic but commensurate in both tone and theme with many other engravings securely identified as such.[24] Certainly, the format and size of the engraving after it, along with its emblematic mode (a group of figures clustered around a central motif, with a Latin inscription in a banderol above) are common to many *stemmata* (a word used in the period for thesis prints) produced in Rome at this time. Second, we know that Vouet designed models for a number of thesis prints, while the engraver of the *Satyrs*—Johannes Tröschel—specialized in this genre. Third, and especially compellingly, the banderol lettering "D O M" of Vouet's drawing (hitherto ignored in all studies of the image) appears on other, contemporaneous thesis prints made in Rome. Comparable lettering appears, for example, in an engraving of *Apollo and the Muses* (1614) by Luca Ciamberlano (1550–1641) after a design by Giorgio Ghisi (1520–1582) and in the *Emblematic garden of the Roman Seminary* (ca. 1610–20) by Matthaeus Greuter (1564–1638) (Fig. 10.6).[25] It seems likely that Vouet inserted the letters as a placeholder for the motto, which presumably had not been firmly settled upon at the time the drawing was executed.

Fig. 10.6 Luca Ciamberlano after Giorgio Ghisi, *Apollo and the Muses* (1614), copperplate engraving

[24] On thesis prints, see the indispensable work of Louise Rice, including: 2007 and 2009. See also Meyer 1990; Pampalone 2015; and Berger 2017.

[25] The latter is discussed in Rice 2007, 208 and 228.

Early modern thesis prints were gifts distributed to an audience during the academic examination of students, often those graduating from their studies. They were not infrequently accompanied by an engraved broadsheet bearing the student's dedicatory oration to their sponsor, along with the short "conclusions" to be defended, which were often Aristotelian in nature and regularly concerned matters of theology.[26] Together, the printed image (the *stemma* or *scudo*) and text were the main focus of the defence—an event sometimes punctuated by musical and poetic interludes.[27] Defences were often held in sumptuous surroundings (for example villas or palaces and their gardens, as depicted in the *Satyrs*) and were attended by cardinals, princes, diplomats, and churchmen, as well as scholars and students.

In early modern Rome, the most eager supporters of emblematic thesis prints were the Jesuits. Maurice of Savoy himself had paid for the apparatus and "bellissimo scudo" of the thesis defense of Count Giovanni Francesco Isnardi at the Collegio Romano in 1627.[28] Depicting an elaborate genealogy of the House of Savoy, its image was engraved by none other than Tröschel (Fig. 10.7). The Jesuits' particular interests in natural philosophy and mathematics—especially recondite topics in fields such as magnetism, astronomy, and optics—are represented in scores of thesis prints.[29] In many instances, such as a characteristic example by Johann Friedrich Greuter (1590–1662) after Antonio Pomarancio (1560–1620) from ca. 1625, the intellectual content is deployed on behalf of an allegory of illumination—of learning paving the way from darkness to enlightenment (Fig. 10.8). This is a theme obviously comparable to Vouet's drawing, given its mirror that "illuminates" (*illustrat*) and the satyrs, which were a common iconographic referent to ignorance contrasted with human intelligence. Satyrs are depicted in this manner by Mellan in a print which derives both thematically and compositionally from Vouet's drawing (Fig. 10.9). In *The mockers* (ca. 1636–45), a pair of ignorant satyrs laughs at a picture of an ass trampling the objects of the good arts and sciences, deriding the world on which one of them sits.[30] The "base ignorance" of monsters such as chimeras and satyrs was routinely contrasted to God-given human intelligence, a dialectical point stressed in Vouet's drawing, the protagonists of which marvel at what they see rather

[26] Sometimes, thesis defences were accompanied by illustrated pamphlets, on which see Collacciani and Roux in this volume.

[27] See Rice 1999.

[28] "Il Conte Gio. Francesco Isnardi … difesse quest'anno in Collegio Romano le conclusioni di Filosofia con belissimo apparato fatto dal Cardinale [Maurizio] di Savoia a cui erano dedicate le conclusione sotto un bellissimo scudo [.]" Quoted in Rice 1999, 160 and 167. See also Di Macco 1995.

[29] On early modern Jesuit science, see Feingold 2003.

[30] It is a scene not dissimilar to the iconoclastic asses that feature in the margins of Flemish "pictures of collections," on which see Marr 2010.

Fig. 10.7 Johannes Tröschel after Antonio Circignani, *Allegory of the House of Savoy*, for the thesis defence of Count Giovanni Francesco Isnardi (1627), copperplate engraving

than comprehending it.[31] In a delicious piece of irony, Vouet's satyrs—beasts which, as Tesauro noted, commonly stand for "deception"—have themselves been deceived by an optical illusion.[32]

The general themes of wisdom versus ignorance and truth against falsehood were staples of thesis prints. For example, in Camillo Cungi (1597–1649) after Giovanni Antonio Lelli's (1591–1640) *Wisdom confounding a magician* (*fraudes sapienta solvit*) (1637), a sage banishes the monstrous creatures of deception (notably similar to the sirens supporting the table in Vouet's image) near the motto

[31] Satyrs were, of course, ambivalent figures that—as half beast and half man—could display certain human attributes while being bound by their bestial natures. Strikingly, a treatise on this very topic was published in France in 1627, the year when Vouet's drawing was probably made: Hédelin 1627. On satyrs in the early modern period, see Kaufmann 1984; Roling 2010, 289–392.

[32] Tesauro 1670, 621: "L'ultima metafore è la decettione … Tal'è la chimera … le sirene, le harpíe, i satiri[.]"

Fig. 10.8 Johann Friedrich Greuter after Antonio Pomerancio, *Allegory of illumination (non tibi divisum)* (ca. 1625), copperplate engraving

"fraudes sapienta solvit" (wisdom undoes falsehood) (Fig. 10.10).[33] Like many Jesuit thesis prints, the action is set against the backdrop of a symbolic formal garden that served as the emblem of the Roman seminary, the motto of which ("pomis sua nomina servunt"; they preserve their fame with your fruit) is inscribed on the arched entrance portal behind the magician's head.[34] This metaphorical garden, generically synonymous with the groves of academe and with the theme of nursing the intellect like a plant, may well have informed the setting of Vouet's image, in which untamed nature—represented by both the satyrs and an untrained vine entwined about the tree to the right—is contrasted with nature ordered.[35] Coincidentally, the Minim Jean François Niceron (1613–46) connected the cylindrical mirror with the formal garden in the section of his treatise *La perspective curieuse* (1638) that deals with this device:

[33] Similar "monstrous" figures of ignorance appear in the thesis prints of Meurisse, Jean Chéron (1596–1673) and Léonard Gaultier, on which see Berger 2017; eadem 2014, 343–66.

[34] See Rice 2007.

[35] We may note here the similarity with Mellan's *The mockers* (Fig. 10.9) which features a similar vine in the foreground.

Fig. 10.9 Claude Mellan, *The mockers* (ca. 1636-45), copperplate engraving

> We call a cylindrical mirror that which is in its form similar to a cylinder, which is a long stone equal all the way round, which is sometimes used to bring together and flatten places where grain is threshed, and the avenues of gardens, as Virgil reports in the second book of his *Georgics*: "the threshing floor must be levelled with a heavy roller".[36]

[36] "Nous appellons un miroir cylindrique, qui est en sa forme semblable à un cylindre, qui est une pierre longue ronde également par tout, dont on se servoit autrefois pour unir & applanir les lieux où on batoit le grain, & les allées de promenades és jardins, au rapport de Virgile au 2. Des Georgiques. *Area cum primis ingenti aequanda cylindro*." Niceron 1638, 83; the quote is Virgil, *Georgics* I (not II), 178. On Niceron's anamorphoses, see De Rosa 2013; De Rosa and Bortot 2019; Collacciani and Roux in this volume. The frontispiece to Niceron's treatise (engraved by Pierre Danet, sometimes attributed to Vouet) features a cylindrical mirror anamorphosis and he may have based one of the anamorphosis illustrated in the book on Vouet's painting of Saint Francis of Paola, engraved by Jean Lenfant. See Gómez López 2016, 325.

Fig. 10.10 Camillo Cungi after Giovanni Antonio Lelli, *Wisdom confounding a magician (fraudes sapienta solvit)* (1637), copperplate engraving

Perhaps the prominent vine in the right-hand foreground of Vouet's image alludes to the setting of the thesis defence: the Roman villa of a prince or cardinal (such villas were colloquially called a "vigna"). Niceron, who had spent time in Rome, noted in his book that a wondrous optical machine could be seen in the "*vigne* of the Borghese" (i.e. the Villa Borghese). Indeed, by the early seventeenth century formal gardens had become a privileged setting for natural philosophical discussion and demonstration. Symptomatic of this trend is the frontispiece to the second edition of the Jesuit Mario Bettini's (1582–1647) *Apiaria universae philosophiae mathematicae* (1645; first edition 1642), in which a garden hosts a "philosophical fountain" (its base comparable to that of Vouet's table), surrounded by vases containing polygons that sprout topiary shaped in the form of mathematical instruments and other devices (Fig. 10.11). The vases are decorated with illustrations of mathematical topics, including a catoptrical anamorphosis, which by this time was certainly being taught in Rome by the Jesuits, as well as by Minims such as Niceron.[37]

[37] There is a further connection to Vouet's image, in that Bettini dedicated his book to Maurice of Savoy's nephew, Carlo Emmanuele II (1634–1675). On the iconography of Jesuit frontispieces, see Remmert 2011. On Roman gardens as sites of philosophy (which included a rich material culture), see Fischer et al. 2016.

Fig. 10.11 Francesco Curti after Giuliano Dinarelli, frontispiece to Maria Bettini, *Apiaria universae philosophiae mathematicae* (1645), copperplate engraving

Recently, it has been argued that such anamorphoses are exemplary of "Baroque science," since they hold both disorder *and* order in a tense relationship. According to Ofer Gal and Raz Chen-Morris, the sharp charge of this type of picture results from a "cluster of paradoxes":

> In Bettini's drawing (Fig. 10.12) mathematics is hardly a failsafe route to true order. In his hands, it can produce *both* correct and distorted representation. Indeed, one may ask which is which.[38]

[38] Gal and Chen-Morris 2013a, 4–5. See also Gal and Chen-Morris 2013b.

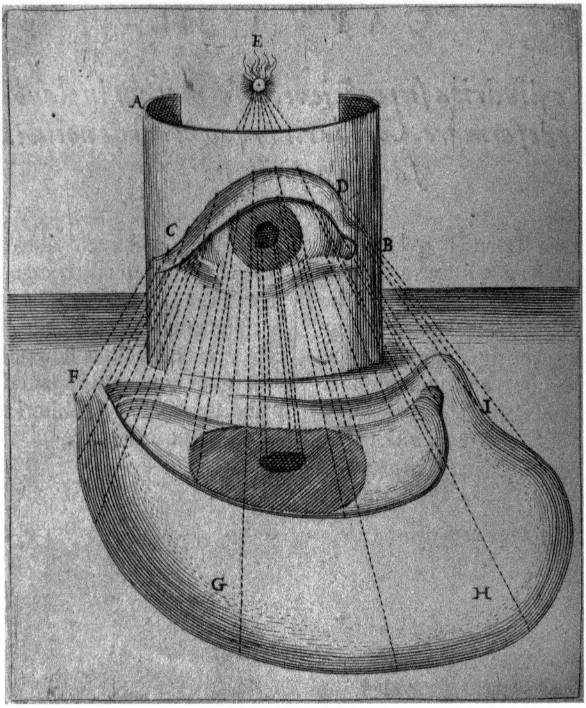

Fig. 10.12 Projected figure of an eye, from Maria Bettini, *Apiaria universae philosophiae mathematicae* (1645), copperplate engraving. Illustrated in Ofer Gal and Raz-Chen Morris, *Baroque Science* (2012)

Building on this claim, Susana Gómez López has interpreted Vouet's image as reflecting a world in which

> the dream of human reason conceived as the faithful mirror of nature, a passive perception that restricted itself to impinging on the mind with the simple information conveyed by the senses, had begun to fall apart.[39]

Yet this claim, like the paradigm of "Baroque science" on which it rests, is fallacious. As we have seen, Vouet's image was concerned with ordering and the capacity for comprehension (and faith) through right reason. It achieved this through the deployment of a novel optical technology, the mathematics of which were perfectly comprehensible to the highly educated class that comprised its audience. The deployment of anamorphosis for this thesis print is a good example of wonder as a spur to philosophizing (of which only humans are capable), as set out by Aristotle in the *Metaphysics* and thoroughly familiar to the Roman elite.[40] Indeed, the print's scholarly, Christian viewers of the image and auditors of the thesis were surely

[39] Gómez López 2016, 330.

[40] See Marr 2006.

intended to admire the effect sensorially *and* rationally, not remain mired in puzzlement like the semi-bestial, irreligious satyrs in the image. With this in mind, we would do well to recall Galileo's comments on anamorphosis. While Galileo looms large as a representative of "Baroque science," he specifically rejected the disordered elements of anamorphoses (in Vouet's *Satyrs*, the unresolved image of the elephant lying flat on the table) in a famous critique of allegory in *Gerusalemme liberata* by Tasso (1544–1595). Comparing that poem's mode to an anamorphosis, Galileo complained that such images, "show a human figure when looked at sideways but, when observed frontally (as we naturally and normally do), display nothing but a welter of lines, colors, and strange chimerical shapes." Likewise, Tasso's poetry "compels the straightforward narrative to adapt itself to an allegorical meaning, seen obliquely, as it were, and thus obstructs it by fantastic and superfluous figures."[41] Thus Galileo is no more a representative of "Baroque science" than Vouet's *Satyrs*: both advocate a highly classicising order and clarity.

Similarly, we would be wrong to interpret the elephant in Vouet's drawing as expressing enthusiasm for exoticism. There may be several reasons why this particular beast was chosen, not least because it was famed for being wise and for having a prodigious memory: both useful in a thesis defence. Moreover, elephants were considered to be noble animals and as such featured in the *imprese* of high-ranking Italians, while having various meanings in emblematics.[42] However, their most common meaning (and the most likely here, given the image's overall iconography and its motto) was as a symbol of religion (Fig. 10.13). This is the meaning assigned to the elephant in Cesare Ripa's popular guidebook to iconography, the *Iconologia* (first edition, 1593), where it is described as follows in the entry for *Religione*:

> The elephant is depicted behind her [the woman] because it is more religious than any other animal, it is said. Pliny narrates in chapter one of his eighth book, that this animal is rare in goodness, prudence, and love equal to humans. …But what strengthens our suggestion that this rare animal is the hieroglyph of religion, is that, as Pliny recounts in the place cited, he venerates the sun and the stars, and at the appearance of the full moon, goes promptly to wash himself in the waters of a flowing river and when sick, calls for help from heaven, throwing plants into the air by way of praying to be healed.[43]

---

[41] Galileo Galilei, "Considerazioni sul Tasso", quoted in Panofsky 1956, 4. See also Renn 2002.

[42] See Orsi 1994; Bouzy 2001. The elephant featured in the *imprese* of Emanuele Filiberto of Savoy and Gregorio Petrocchini. On the latter, see Ronzani 2013. On the elephant in the early modern visual imagination, see most recently Rice 2017.

[43] Ripa 1625, 554: "Gli si dipinge a lato l'elefante, per essere più d'ogn' altro animale religioso, come si dirà. Narra Plinio nel lib[ro] 8 al c[apitolo] 1 questo animale è raro in bontà, prudente, amator dell'equità e humano, percioché incontrando l'huomo. …Ma, quel che fa più a nostro proposito, è questo raro animale il hieroglifico della Religione, raccontando pur esso Plinio al luogo citato che egli ha in veneratione il sole et le stelle et, apparendo la nova luna, spontaneamente va a lavarsi in acqua di vivo fiume et amalandosi chiama aiuto dal Cielo, buttando verso il cielo dell'herbe come mezzi per inter-cedere gratia di sanità."

Fig. 10.13 "Religione", in Cesare Ripa, *Iconologia* (1613), woodcut

Furthermore, in the opening section of the entry, Ripa explains that religion

> Is a moral virtue, through which man gives honour and reverence internally in the soul and outwardly through the body to the true God. And thus religion has been placed naturally in men, because, as Aristotle says, through it we are able to reason, which is different to brute animals, seeing clearly from this that in sudden perils, without another thought, we turn to call on divine aid. The face is veiled in them, because religion in men looks on God, as St Paul says, *through a glass darkly*."[44]

Bearing in mind the fact that the imagery of thesis prints was oriented as much towards the defender's patron as the topic of his defence, we may assay an interpretation of Vouet's image.[45] The grammar of the motto indicates clearly that it refers to the cylindrical mirror which both "shapes and illuminates" the anamorphosis laid out on the table. Thus, in the elaborate exercise of epideictic rhetoric that

[44] Ripa 1625, 554: "è virtù morale per la quale l'huomo porta honore et riverenza interiormente nell'animo et esteriormente col. corpo al vero Dio. È anco ne gli huomini talmente inserta da natura la Religione che, come dice Aristotile, per quella, più che per essere ragionevole, sono differenti da brutti animali, vedendosi ciò chiaramente da questo, che ne' pericoli improvisi, senz'altra deliberatione, ci volgiamo a chiamare il divino aiuto. Si li fa velato il viso, perché la religione ne gli huomini riguarda Dio, come dice S. Paolo *per speculum in aenigmate*."

[45] For thesis prints as vehicles of epideictic rhetoric, see Rice 2009.

characterises thesis prints, the device may be explained as follows. The student presenting his thesis is like the distorted and unformed image of the elephant on the table, but now that he has completed his studies and is undertaking his defence, both his intellect and his character have been correctly shaped. Cultivated in the correct spiritual and academic setting, he has overcome base instinct and childish wonder, and arrived at proper, Christian understanding. While all men may look upon God "through a glass darkly" (a *topos* that resonates with the catoptrics and dioptrics favoured by the Jesuits and Minims), the student's mind and soul have been illuminated by learning undertaken through the munificence of his patron: a brightly shining mirror of virtue, who brings order through reason.[46]

In the absence of any surviving evidence for the precise subject or setting of the thesis, we cannot know definitely the exact meaning of the image, nor can we say for certain who the patron was. However, strong circumstantial evidence points to Maurice of Savoy.[47] We have already seen not only that Maurice was one of Vouet's Roman patrons and a supporter of theses, but that he was depicted in a thesis print by Vouet as the light-bringer: Apollo-Christ. More directly, we have the testimony of Emanuele Tesauro that at the very time Vouet's drawing was made, Maurice had seized upon the symbolic potential of catoptrical anamorphosis. In the *Cannocchiale*, Tesauro explained the circumstances leading to Maurice's choice of emblem for his Torinese Accademia de' Solinghi: a conical mirror anamorphosis of the motto "omnis in unum" (all in one; see Fig. 10.3). In the section titled "La perfettisima impresa voul'esser ingegnosa" (the most perfect *impresa* should be ingenious) he writes:

> But most ingenious is that [emblem] which Prince Maurice of Savoy … appointed for his heroic and famous academy of arms and letters, called *de' Solenghi*, which is the conical mirror, in which on the ground those things that appear blotted, reflected altogether above, become perfect and well assembled figures. Thus many desirable circumstances come together. Since, firstly, the body of the *impresa* could not be any more ingenious by this artifice, nor more noble, nor more marvelous, nor more strange: being a new part of speculative mathematics, the most miraculous of all the arts. …So, without any controversy, this may be called a most ingenious idea for *imprese*, since it brings together many facets, and chiefly novelty, which provokes wonder. It seems many other Italian academies have already (as I understand it) taken this mirror as the body of their *impresa*, with the same universal meaning, and it is therefore well known that in this His Highness was far ahead of all other intellects. Because, this miraculous mirror had been invented by a most subtle spirit in Paris, around the year 1627, and one of the first originals came quickly into the hands of their Royal Highnesses, before its fame spread quickly throughout Italy. This Prince [Maurice] who was at that time contemplating the *impresa* of his academy, seeing such a novel and wondrous invention, immediately applied it, and made it his symbol.[48]

---

[46] The mirror in Vouet's image would have been made not of glass but of brightly polished (and thus highly reflective) metal. As Niceron explained, it was "very difficult … not to say impossible" to make conical or spherical mirrors from glass or crystal, proposing instead a metal version, the manufacture of which he sets out in some detail. Niceron 1638, 81–3.

[47] This insight was first suggested in Préaud 1992 and has been repeated in subsequent studies.

[48] Tesauro 1670, 678–79: "Ma ingeniosissima è quella che il Principe Mauritio di Savoia, come perpetuo mecenate degli' ngegni, dirizzò per la sua heroica & famosa academia d'arme & di lettere

For Tesauro, the *impresa* of Maurice's Solinghi was the pinnacle of the quality he admired above all else: *ingegno*.[49] It seems safe to infer that Vouet's *Satyrs* would have been considered witty and original when it first circulated as a print. Yet since Tesauro's account is somewhat vague and has proven prone to misinterpretation, it requires some unpicking in order to establish properly its relation to Vouet's image.

First, we should note that Tesauro was writing almost thirty years after the events described, which he may well not have experienced firsthand.[50] Moreover, since in the 1650s he was in the service of Maurice of Savoy, to whom the *Cannocchiale* is dedicated, we can expect a degree of flattering embellishment in his account of the prince's brilliance.[51] Then there is the discrepancy between the conical mirror of the Solinghi and the cylindrical one of Vouet's drawing. Curiously, Tesauro seems to have used "conical" and "cylindrical" interchangeably when discussing Maurice's emblem. For instance, while he described the mirror as conical in the *Cannocchiale*, in his funeral oration for the prince, published in the *Panegyrici*, Tesauro dwelt on Maurice's

> heroic device of a mirror in the form of a cylinder, with those witty words "omnis in unum". …the generous Maurice used for his *impresa* a mirror shaped like a column, which the perspectivists call a cylinder; the centre of which brings together coloured images,

---

intitolata, de' Solinghi. Cioè; lo specchio conico; in cui quelle, che nella piana superficie paion macchie; unitamente riflettendo in alto, divengono perfette, & compostissime figure. Dove concorrono molte circonstanze considerabili. Peroche, primieramente il corpo della impresa non può esser più ingenioso per l'artificio: ne più nobile, ne più meraviglioso, ne più pellegrino: essendo un novello parto delle matematica speculare; la più miraculosa di tutte l'arti. …Talche senza controversia niuna, questa si può chiamare una idea delle imprese ingeniose: concorrendovi tante circonstanze quadranti; & principalmente la novità, che genera maraviglia. Che sebene alcun' altra nobile academia del'Italia, habbia dapoi (com' intendo) preso questo medesimo specchio per corpo di sua impresa, col. medesimo significato universale: egli è pertanto cosa troppo nota, che quest' Altezza in ciò prevene di molta lunga ogni altra intelletto. Peroch' essendo stato inventato questo specchio miracoloso, da un sottilissimo spirito in Parigi, d'intorno all'anno 1627 & venutone subito un degli primi originali à queste Regie Altezze, prima che ne passasse pur la fama più avanti nella Italia: questo Principe, che stava nel medesimo tempo meditando la impresa dell'Academia; veduto un sì pellegrino & ammirabil ritrovo, immediatamente l'applicò, & ne fabricò questo suo simbolo."

[49] There is a large literature on Tesauro's *Cannocchiale* and *ingegno*. See e.g. Hänsli 2008; Snyder 2017.

[50] In 1627 Tesauro was in Turin, since in that year he preached in the Duomo, before the princes of Savoy, a sermon on the Turin shroud (which, perhaps coincidentally, deploys the conceit of a mirror). See Maggi 2005. On Tesauro's employment as the iconographer for the court of Savoy, see Gauna 2014.

[51] Notably, he deploys the catoptrical metaphor in his dedication to Maurice, whom he calls "inspirer of the ingenious" (*fomentator degl'ingeniosi*). Tesauro 1654, a4^v-a4^r: "the selfsame imperfections of these pages, reflected in the high and most clear mirror of your incomparable judgment, received correction; just as in your heroic and sharp *impresa*, the conical mirror corrects the misshapen characters on a plane sheet, are reduced together in a right and perfect form." ("Ma le istese imperfettioni di queste pagine, riflettendo nell'alto e tersissimo specchio dell'incomparabile suo giudicio; riceveranno emendatione: sicome nella heroica & argutissima sua impresa, lo specchio conico emendo li caratteri disformati di un piano foglio; riducendogli in se stesso alla diritta, & perfettissima forma.")

> which outside the mirror look like malformed and contorted blots, in its crystalline breast they receive a right and most perfect form.[52]

Tesauro was a rhetorician rather than a mathematician, and it is clear from his elision of two quite different catoptrical devices that he was more interested in the symbolic valence of mirror anamorphosis than in its mathematical optics.

Evidently, though, Vouet had seen a mirror anamorphosis in action. While Baltrušaitis, in his influential account of the history of anamorphosis, argued that the *Satyrs* renders the cylindrical mirror's optical effect incorrectly, he based this claim only on the engraving, not the drawing. While the top edge of the elephant's reflection in Tröschel's engraving is more or less flat, in Vouet's drawing it is clearly ragged, accurately capturing the slight distortions that are visible in such a demonstration. This is an important point, since Baltrušaitis used Vouet's supposed error to assert that this type of catoptrical device was "not yet available in Rome."[53] He proceeded to engage in a series of scholarly contortions to argue that the print documents Vouet's familiarity with Chinese traditions of mirror anamorphosis, which he may have encountered on his visit to Constantinople in 1611—the Emperor Ahmed I (r. 1603–1617) having inherited a fine collection of Chinese *objets d'art*, assembled by Suleiman the Magnificent (1494–1566), that included catoptrical anamorphoses.

Baltrušaitis's arguments return us to Tesauro's testimony that the cylindrical mirror "had been invented by a most subtle spirit in Paris, around the year 1627" before coming to Italy, making its way swiftly into the hands of the princes of Savoy. Here again, we must clear up a misconception. Accepting Baltrušaitis's account, Gómez López has claimed that the "most subtle spirit" was Vouet, and that the invention to which Tesauro refers is the drawing, not the mirror. Yet it is unambiguous that Tesauro refers to the mirror itself, not the representation. Moreover, his claim that the mirror was originally known in Paris is supported somewhat by the first published treatise on the mathematical principles of cylindrical and conical mirror anamorphosis: Jean-Louis Vaulezard's *Perspective cylindrique et conique* (Paris: Julian Jacquin, 1630) (Fig. 10.14). Much has been made of the fact that this treatise postdates by at least three years Vouet's drawing. It has even been claimed that "it was precisely Vouet's anamorphic representation that, on his return to Paris [in 1627] stimulated the curiosity of geometers and encouraged them to search for the rules underlying this new phenomenon."[54] While it cannot be established for certain how and precisely when catoptrical mirror anamorphosis arrived in Europe, we may note that in his book (written, he says, at the request of his students), Vaulezard implies

[52] Tesauro 1659, 151–52: "heroica sua divisa lo specchio in forma il cilindro, con quelle argute parole, omnis in unum. …il generoso Mauritio usò per impresa lo specchia tondeggiato in figura di colonna, da' perspettivi detta cilindro: nel cui centro accogliendosi le colorate imagini, che fuor dello specchio paiono informi e scontorte macchie; nel suo seno cristallino ricevono diritta e perfettissima forma[.]" On Tesauro's eulogy, see Kolrud 2015.

[53] Baltrušaitis 1977, 167.

[54] Gómez López 2016, 323.

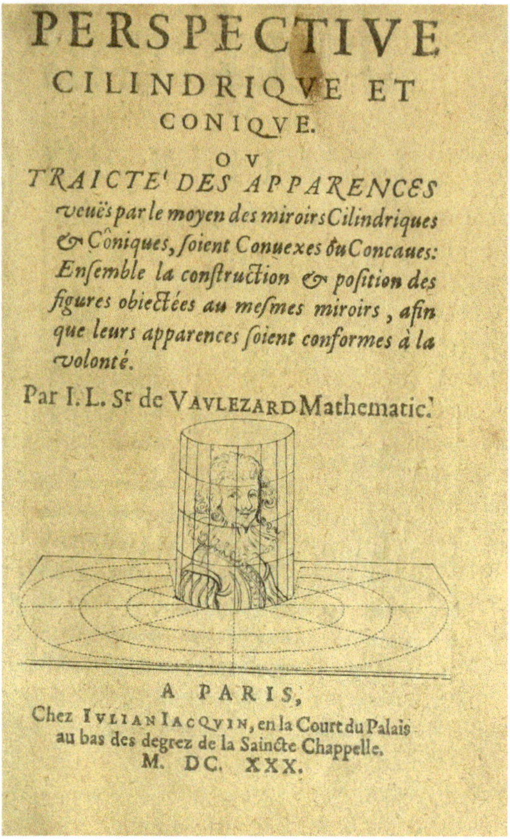
PERSPECTIVE
CILINDRIQVE ET
CONIQVE.
OV
TRAICTE' DES APPARENCES
veuës par le moyen des miroirs Cilindriques & Coniques, soient Conuexes ou Concaues: Ensemble la construction & position des figures obiectées au mesmes miroirs, afin que leurs apparences soient conformes à la volonté.
Par I. L. Sr de VAVLEZARD Mathematic.

A PARIS,
Chez IVLIAN IACQVIN, en la Court du Palais au bas des degrez de la Saincte Chappelle.
M. DC. XXX.

Fig. 10.14 Title page of Jean Louis Vaulezard, *Perspective cylindrique et conique* (1630)

that demonstrations of catoptrical anamorphosis were already well-known in the French capital:

> my friends advised me to bring this to light, persuading me that it could not be but agreeable to the public, since no author has so far written of it; equally because many who are little versed in mathematics, *who have already seen its effects in public*, will take in its causes with much pleasure[.]

It is much more likely that a mathematically inclined scholar or *reglioso* first demonstrated the technology of cylindrical mirror anamorphosis in Paris than Vouet, an artist who, other than in his *Satyrs*, showed little interest in this sort of material.[55] As

[55] Gómez López 2016, 329 claims that Niceron singled out Vouet as the "epitome of expertise in optics and, in particular, its application." The reality is rather more pedestrian, for he merely mentions Vouet, Laurent de la Hyre and "several other painters" as artists who, in all their works, follow optical principles in their designs and in the application of colour: "Monsieur Voüet, premier Peintre du Roy, Mo[n]sieur de la Hyre & quelques-autres: leurs ouvrages nous sont assex coignostre qu'ils suivent toutes les maxims de l'Optique dans la conduit de leurs desseins, & l'application de leur coloris." Niceron 1638, sig. [ëiv]^{r}.

early as 1631 Niceron was inventing examples of catoptrical anamorphosis, and while these may have been created in part in response to Vaulezard's book, they may equally have been a product of his education under another Minim, Marin Mersenne, some years previous.[56] It is to the intellectual circles of such figures in the 1620s—the Paris salons frequented by the likes of Mersenne, the frères Dupuy, and Descartes; the collections of *curieux* such as Louis Hesselin (1602–1662) and Pierre Trichet (1586/7-ca. 1644)—rather than to Vouet that we should look for the first stirrings of interest in this technology and the "subtle spirit" who "invented" it.[57] Indeed, by the time Niceron published *La perspective curieuse*, Parisian artisans could be counted on to fashion cylindrical mirrors on commission. Niceron employed one "Monsieur le Seigneur", of the faubourg Saint Germain, "to whom I have given models of one or the other, and whom I consider one of the best artisans that we have in Paris at the moment for making these mirrors in metal of all kinds."[58]

A further ambiguity of Tesauro's account concerns Maurice of Savoy's academies. Tesauro informs the reader that in 1627 Maurice was "contemplating the *impresa* of his academy [*De' Solenghi*]," which has led some to conclude that Vouet's drawing was an emblem for that very institution.[59] The situation is rendered confusing by the fact that in 1627 Maurice was in the process of relocating to Turin from Rome, in part to avoid his creditors following his lavish patronage of the arts. In Rome, he had established an academy—the Accademia dei Desiosi—about which we are well informed thanks to extensive manuscript evidence and the volume of discourses read at its gatherings, edited and published as *Saggi accademici* by Agostino Mascardi (1590–1640) in 1630.[60] Significantly, this volume includes a

---

[56] Niceron's earliest catoptrical mirror anamorphosis is an *Anamorphic portrait of Jacques d'Auzolles de Lapeyere* (1631) (Archives Départementales du Cantal, Aurillac Cedex). The attribution of roles in the image's production is notable: Niceron is named as having "invented" it (*invenit*), while Picard "drew and engraved" it (*delin[eavit] et incidit*). In his capacity as one of the theologians of the Order of Minims, Mersenne formally "approved" Niceron's treatise and his own *Optics* and *Catoptrics* were published with the second, 1652 edition of the book.

[57] In the section of *Perspective curieuse* that treats catoptrics, Niceron refers the reader to Mersenne's work on the burning mirror in *Traité de l'harmonie universelle* (1627) and mentions also the collection of Monsieur Hesselin (Louis Hesselin, Seigneur du Condé), whose *cabinet* featured wondrous mirrors, among other rarities. He refers also to examples of catoptrical anamorphosis in the library of the Place Royale (presumably that of the Minims' convent there). Niceron 1638, 76–7, 87. The large collection formed by Pierre Trichet in the first few decades of the seventeenth century similarly featured numerous optical instruments, including mirrors and prisms. See Schnapper 1988, 114 and (for Hesselin, who was close to Vouet), Schnapper 1994, 182–86. Given the literal meaning of *invenit* as "coming upon", the invention of catoptrical anamorphosis may well have been a discovery of foreign sources by Europeans, perhaps via Constantinople as Baltrušaitis suggested, but not necessarily by Vouet. There had been numerous European embassies to the Ottoman court before 1630. On early modern invention, see Marr and Keller 2014.

[58] "Mo[n]sieur le Seigneur au fauxbourg S. Germain, à qui j'ai donné les modelles de l'un & de l'autre, & que j'estime un des bons ouvriers, que nous ayons à Paris pour le present, pour faire de ces miroirs de metail de toutes sortes." Niceron 1638, 96.

[59] See e.g. Gómez López 2016.

[60] On Mascardi, see Bellini 2000.

Discorso dello specchio (*Discourse on the mirror*) by Vouet's friend, the poet Marcello Giovanetti. Indeed, if Vouet's drawing was not designed for a thesis print, it could quite plausibly have been made for the occasion of Giovanetti's lecture (the date of which is unspecified, but which was certainly delivered before 1627): the print would have prompted discussion and served as a memento for the academicians.[61]

The opening passages of Giovanetti's *Discorso* (addressed to Maurice) fit the themes of Vouet's drawing, albeit he refers to a clear glass rather than a metal one. Obliquely addressing his patron, Giovanetti writes:

> Under the light, which spreads so clearly from such a beautiful assembly, I could not imagine depicting a more proportionate thing than that other bright object, from which she herself is depicted with a beautiful reflection. …Since you cannot see yourself in this clear glass that I present to you; your image will not be lost only if I do not block your visible rays with the dense lead of my wit.[62]

Elsewhere, he describes a concave mirror in the armory of Venice, which the Cardinal Mondovì took for his *impresa*, with the motto "ulciscitur ultro".[63] This object, he says, "is not dissimilar to other types of mirror, which, located in the right place, show in the air things that cannot be painted with the brush, or that other one, which makes the mind immobile with its wonders, which is a mirror which shows to one the image of whatever one desires."[64] Yet the true mirror, Giovanetti concludes, is the mirror of the soul, composed of

> three luminous crystals: *Ex integritate charitatis*; through which the soul is cleaned of all stains; *Ex puritate humilitatis*; through which the rays of divine grace enter; *Ex subtilitate intellectus*; through which the sun of wisdom infuses his light, because in this guise also comes the beauty of God Himself[.] …But because the divine essence does not allow itself to be seen, even by means of a mirror *Videmus nunc per speculum in aenigmate*. We have no more than two ways to reach what we yearn for in that mirror so divine. The first is the way of learning[.] …The other is the contemplation of divine works[.][65]

The above passage indicates the scholastic context in which Vouet's image should be read, alongside the polymathic learning characteristic of Italian academies such

[61] See Bonfait 2011, 180–81 for this suggestion.

[62] Giovanetti 1630, 107: "Alla luce, che con tanta chiarezza si diffonde da si bella adunanza non ho saputo io immaginarmi di rappresentar cosa piu proportionata, ch'altro lucid oggetto, da cui essa medesima con bel riflesso venga rappresentata. …cosi non potete voi contemplare voi medesimi in questo terso cristallo che vi propongo; se io con l'oscuro piombo dell'ingegno mio non pongo il termine a vostre raggi visivi; per che l'immagine vostra non si transfonda."

[63] Giovanetti 1630, 110.

[64] On these sorts of "magic mirrors" and optical games, see Dupré 2008.

[65] Giovanetti 1630, 113: "composta da tre luminosi cristalli. *Ex integritate charitatis*; per cui l'anima si fa tersa d'ogni macchia; *Ex puritate humilitatis*; per cui entra il raggio della divina gratia. *Ex subtilitate intellectus*; per cui il sole della sapienza n'infonde il suo lume, perche in questa guida s'arriva anche a vagheggiare il medesimo Dio[.] Ma, perche la divina essenza non si lascia vedre, se non per mezzo dello specchio, Videmus nunc per speculum in aenigmate. Noi non habbiamo se non due strade per giunger a vagheggiarci in ispecchio cosi sublime. L'una è la strada della Sapienza[.] L'altra è la consideratione delle opere divine[.]"

as the Desiosi. We know from that institution's *Diario* that at its meetings, held at the Palazzo di Montegiordano, the academicians performed "diverse ingenious games" (*giuochi diversi d'ingegno*), studied mathematics, designed fortifications, discussed the motions of the heavens, engaged in drawing exercises, and—crucially—selected an *impresa* for the academy.[66] Each academician was invited to present a proposal, with Prince Maurice eventually selecting the figure of "five interwoven laurel crowns… with the motto "non uni insistere" taken from Sannazaro[.]"[67] True to Maurice's interest in ingenuity, the full lines of Jacopo Sannazaro's (1458–1530) epigram from which the motto is taken read "Non uni insistere palmae / Sed nomen factis quaere, et ingeniis" (Do not stop at one palm [laurel wreath], But seek to make a name through deeds and wit).[68]

Despite its connection to ingenuity, this *impresa* is dissimilar to Vouet's *Satyrs*, indicating that the latter could not have been intended as an emblem for Maurice's Roman academy. Moreover, Tesauro is quite clear that the symbol of the new, Torinese academy (the Solinghi) was the conical mirror with the motto "omnis in unum". Thus, Vouet's drawing seems to have been made in the interstices between the two institutions. As Merolla has suggested, there must have been a degree of continuity between these academies.[69] In relation to their mottoes, this can now be proven by way of the frontispiece to the physician Bartolomeo Torrino's *Parnassus triceps* (1657): a work of "physiatromathematics" dedicated to "Apollo, His Highness Maurice of Savoy" (Fig. 10.15). Despite having been published some thirty years after the Roman Accademia dei Desiosi supposedly closed, the frontispiece includes both the Solinghi's motto "omnis in unum" *and* the Desiosi's "non uni insistere," indicating that Maurice continued to be associated with (and perhaps even used) the motto of his earlier, Roman academy some years after it had closed.

In sum, then, we may conclude that while Vouet's *Satyrs* drawing was not designed as an *impresa* for either of Maurice's academies, the cardinal likely knew it in 1627 when, as Tesauro tells us, he was trying to choose a symbol for his new, Torinese academy. The drawing with its mirror anamorphosis was thus the inspiration for the *impresa* of the Solinghi, which was widely disseminated thanks to Tesauro's publications. Indeed, Tesauro, considered that *impresa* a perfect example of the genre, fitting not only the several virtues of Prince Maurice but also the collective endeavor of the teaching institutions he supported. For Maurice (who engaged in a long and bitter contest for the crown of Savoy during the minority of his nephew), the Latin motto signaled

[66] See Merolla 1995, 55–57 and, for the proposals for *imprese*, 57–68.

[67] Merolla 1995, 131: "Haverà l'impresa da levarsi nell'Accademia per Corpo, cinque Corone intrecciate insieme, cioè di Quercia, di Palma, di Hellera, di Gramigna, e d'Alloro … col. motto *Non uni insistere*, tolto dal Sannazaro[.]"

[68] Jacopo Sannazaro, "Epigram on Andrea Matteo Aquaviva", lines 9–10, in Sannazaro 2009, 305.

[69] Merolla 1995, 133–34. On the basis of evidence that Maurice welcomed academicians to his Torinese villa in 1622, Merolla argues that the Solinghi must have been founded in Turin prior to the Roman academy.

Fig. 10.15 Giovenale Boetto, frontispiece to Bartolomeo Torrino, *Parnassus triceps* (1657), copperplate engraving

> that his actions, however they might have been perceived or interpreted; all of them, however, within the sincere mirror of his mind, were rightly ordered to the one honorable end of virtue: *omnis in unum*.[70]

Since, as Tesauro explained (quoting Socrates), "all virtue is knowledge," Maurice's support of an academy of learning was yet further evidence of his virtue.[71] Moreover, the motto "omnis in unum" showed that in an academy the sum is greater than its parts:

> Ingenious also is the application [of the *impresa*] for whenever an academician is alone he is imperfect, but bringing his talent together with others in this academy he receives a perfect form.[72]

Yet the aspect of the *impresa* that most delighted Tesauro was a conceit that connects it even more firmly to Vouet's *Satyrs*: the formal garden.[73] As he explains,

> But here we enter into even more singular circumstances. The first is, that the seat of this heroic academy is the pleasure villa of that Prince [Maurice], just as the Academy of Athena had for its seat the Groves of Academe[.] ...But an even more remarkable circumstance is that the seat of the academy is a delightful garden in the form of a semicircular labyrinth. The body of the *impresa* represents this labyrinth, which, reflected in the mirror, forms in perfect characters the motto of the impresa: *omnis in unum.* Thus, through great force of wit and by a wonderful analogy, the figure forms the motto, the motto forms the figure, the soul serves as the body and the body serves as the soul.[74]

---

[70] Tesauro 1659, 152: "che le sue attioni, comunque dall'estimatione altrue venissero biecamente interpretate; tutte nondimeno dentro allo specchio sincero della sua mente, sarebbono diritamente ordinate al solo honoratissimo fine della virtù: *omnis in unum*". On the contest for the Savoyard crown, see Oresko 1997.

[71] Tesauro quotes Socrates via Aristotle, *Nicomachean Ethics*, 3.2: "Virtutum medium cognoscere non cuiuslibet est, sed sapientis." See Tesauro 1679, 152.

[72] Tesauro 1670, 677: "Dove concorrono molte circonstanze considerabile. Peroche, primieramente il corpo della impresa non può esser più ingenioso per l'artificio: ne più nobile, ne più meraviglioso, ne più pellegrino: essendo un novello parto della matematica speculare; la più miracoloso di tutte l'arti. Ingeniosa è dipoi l'applicatione. Volendo accennare; che quantunque ciascun'academico per se solo, sia quasi un'ente imperfetto: nondimeno, accomunando ciascuno il suo talento in questa erudita università; da questa ricevono perfetta forma."

[73] On formal gardens and academies, see Ribouillault 2018.

[74] Tesauro 1670, 677–78: "Ma vi entrano altre Circonstanze più singolari. L'una è, che questa heroica Academia de'Solenghi, ha per seggia principale la villa di piacere di questo principe: sicome l'Academia di Atena havea per seggia gli Horti delitiosi di Academo[.] ...Ma un' altra più ingeniosa circonstanza è; che sicome il luogo di quest'Academia, è un giardino amenissimo à modo di laberinto semicircolare in un verde teatro di colline: così nel corpo della impresa, quelle macchie, nel piano rappresentano il stessuoso laberinto: ma riverberate nello specchio, formano per vera regola di perspettiva perfetti caratteri, compneneti il motto della impresa; *Omnis in unum*: spiccato da quel di Virgilio, *Virtus coit omnis in unum.* Talche, per gran forza d'ingegno; & per maraviglioso riscontro, la figura forma il motto; il motto forma la figura: l'anima serve per corpo, & il corpo per anima."

Fig. 10.16 Anonymous oil painting of ca. 1670 showing Maurice's villa (known as *La Vigna*, later the *Villa della Regina*) with its semi-circular formal garden

As a painting of ca. 1670 shows, Maurice's villa (known as *La Vigna*, later the *Villa della Regina*) did indeed have a semi-circular formal garden (Fig. 10.16).[75] Yet the *impresa* is surely fantastical. The plantings of the villa's garden were not laid out to form the misshapen words *omnis in unum* and would anyway have required an enormous conical mirror to be rectified. However, the conceit was elaborated on in a little-known *stemma* for a thesis print made for the Solinghi academician Paolo Pasta, in which Vouet's initial design of a mirror anamorphosis has been greatly amplified (Fig. 10.17).[76] Invented by Tesauro and cut in 1654 by the leading Torinese engraver Giovenale Boetto (fl. 1642–1682), who also engraved the frontispiece to Torrino's *Parnassus triceps*, it depicts the Accademia de' Solinghi in action. Although the lower half of Pasta's thesis print—which would have contained the theses to be discussed, a dedicatory inscription, and information on when the disputation was to be held—seems not to have survived, Tesauro published the "supplementary adornments" to Pasta's defense in his *Inscriptiones* (1666).

Introducing the print, Tesauro explains that

> This far-sighted duke [Maurice] founded a most celebrated academy, which was twice as beneficial to the state as was that academy of Platonists in the gardens of Academus. For the latter, as Timon declared, educated its devotees in chatter, like crickets in the long grass, whilst ours, fulfilling the role of both mother and father, has proved to be the parent of great minds, and, by training them in both equestrian prowess and the arts, has initiated its off-

[75] On the Villa della Regina, see De Marchi and Garnero 1997; Mossetti 2007.

[76] On Boetto and the *Horti Academi* stemma, see Carboneri and Griseri 1966; di Macco and Romano 1989, 332–33; Failla and Goria 2003, 161; di Barme and Varallo 2009, 108–9.

Fig. 10.17 Giovenale Boetto, *The Accademia de' Solinghi in the formal garden of the Villa della Regina* (*Horti Academi*) (1654), copperplate engraving

> spring in civil and military matters so successfully that the most energetic military leaders, the most skilled administrators of justice, the most learned cardinals, and even the present pope, have proceeded from its bosom. Since this undertaking proved an entirely pleasurable and delightful diversion for our Duke from the affairs of state, he accordingly established a sequestered seat and home for his Academy in a most pleasant location just outside his city, near the river Po, and he gave the title "Solitarii" to the members of his Academy in remembrance of his own disposition and example[.][77]

In the foreground, the print's subject is announced by a topiary balustrade which forms the words "Horti Academi." This balustrade, Tesauro explains, "gives an entrance to scholars, but excludes outsiders, for whom the threshold is the limit of learning."[78] To its left, an artist paints the portrait of the thesis defender, Paolo Pasta,

[77]Tesauro 1671, 192: "Instituerat providus hic Princeps, celeberrimam Academiam, duplo utiliorem Reipublicae, quam illa fuerit Platonicorum in Hortis Academi. Illa enim, ut ajebat Timon, Sectatores suos ad garriendum, Cicadarum more inter arbusta, duntaxat educabat: haec Androgina Ingeniorum Parens, Literariis juxta, & Equestribus exercitamentis, Foro & Castris suos ita initiavit: ut strenuissimi Belli Duces, & consultissimi Justitiae Praesides: Doctissimi etiam Cardinales, & ipse, qui nunc regnat Pontifex, ab illius sinu prodierint. Et quoniam haec tota Principis voluptas, hoc delicium, hoc a negotius erat diverticulum; solivagam Academiae Sedem ac Domicilium amoenissimo in Suburbano suo prope Eridanum, constituerat: unde Academicis suis, suo de genio & exemplo, Solitariorum Nomen indidit[.]"

[78]Tesauro 1671, 194: "Academicis aditum pandit, occludit exteris, quibus sapientiae limen est limes."

whose coat of arms is being engraved in the left-hand corner. Pasta himself welcomes the viewer to the entrance of the *hortus*, holding his thesis print, while to his left the engraver Giovenale Boetto is shown drawing the *stemma*. Inside, the garden is elegantly laid out with formal parterres, in the centre of which is the labyrinth described by Tesauro, featuring misshapen words that, reflected in a conical mirror, are formed into the motto "omnis in unum."

Surrounding this central anamorphosis, academicians are engaged in scholarly and martial activities. From left to right, we see groups of figures using mathematical and optical instruments (including a telescope) and taking measurements with surveying devices; vaulting; playing music; gardening; reading and disputing with *religiosi*; riding; fencing; and examining fortifications designs. They represent precisely the activities recorded in the *Diario* of the Accademia dei Desiosi and all, Tesauro explains, "like ingenious bees, which harvest the flowery riches of the Muses from the delights of Academus' garden, are improving themselves so as to benefit others."[79] Behind the garden is an imposing colonnade: "shaped to resemble a theatre [it] curves round on both sides to embrace a galleried courtyard." Surmounting this feature—perhaps alluding to the architecture of Maurice's actual villa—is a small dome topped with a figure of Apollo, whose brilliant light illuminates the scene below.[80] In its central niche sits a bust of the academy's "Apollo": Maurice of Savoy, identified in a dedicatory inscription as the "patron of sacred wits" (*ingeniorum moecenati sacrum*). Below is his coat of arms, flanked by panels of interlocking laurel wreaths (akin to the emblem of the Desiosi) and crowns, presumably indicating his intellectual achievements and exalted rank.

The colonnade houses fourteen groups of figures, identified—in the arches and entablature above—as named worthies of the various arts and sciences and associated with the relevant muses. As Tesauro explains:

> The individual openings of the colonnades are fashioned with studies of the individual arts, like separate workshops. In these the individual originators of these arts are hard at work and all are directing their efforts to ensuring that the noble students of the academy exert themselves in this common Athenaeum. Furthermore, each individual arch bears the appropriate emblem of each art; the shape of this is never the same, but the title, *omnis in unum*,

[79] Tesauro 1671, 197: "Omnes denique, ingeniosarum instar Apum, ex hortensibus Academi deliciis, floreas Musarum divitias populantium, ita sibi proficiunt, ut aliis prosint." Tesauro, quoting Pasta's letter of dedication for the thesis print, explains: "You see how one contends with the sword, and another with the pen; this one practices the dance, and this one military campaigns; one thunders from the orator's chair, one hurls bolts of lightning on horseback; and the very leisure activities of your students strikes terror into the hearts of your enemies. You descry how the laurels of Mars vie in their endeavours with the laurel of Poesy, the priest's head-dress with the soldier's helmet, the lawyer's toga with the philosopher's cloak, all within one haunt of the Muses, and contribute the diverse fruits of their studies for the common good[.]" ("Cernis ut ille gladio, hic calamo certat: ille ex Cathedra tonat, hic ex Equo fulminat: alter Choris, alter Castris proludit; & Academicorum tuorum ipsa otia, hostibus terrorem incutiunt. Cernis ut Martia Laurus, cum Poetica Lauro: sacer Apex, cum Militari Galea; forensis Toga cum Philosophali Palliolo, uni Musaeo certatim allaborant; & diversa studia in commune conferunt[.]") Tesauro 1671, 194.

[80] Little of the original architecture or garden layout remains, the villa having been extensively remodeled in the eighteenth century. See Mossetti 2007.

> remains the same. Finally, trophies conspicuously decorate the entire top of the colonnades in front of the acroteria, and are made up of the tools of the individual arts and branches of learning.[81]

The worthies and their subjects are (from left to right) Sacrobosto for astrology; Euclid for geometry; Archimedes for mathematics; Orpheus for music; Zeuxis for painting; Homer for poetry; Aristotle for logic; Parmenides for physics; Hippocrates for medicine; Peter Lombard for theology; Socrates for ethics; Justinian for jurisprudence; and Carlo Emanuele, Duke of Savoy, for the military arts. Their symbols, whether the emblems accompanying the motto *omnis in unum* or the sculpted "trophies" surmounting the colonnade, are largely self-evident, although (somewhat unusually) for ethics Tesauro opted for the "Silenus of Socrates, misshapen on the outside, but full of jewels inside, which gleam forth from an open breathing air-hole in his breast[.]"[82] At the far corners of the colonnade, figures of fame—wearing (as per Ripa's *Iconologia*) garments decorated with eyes and ears—trumpet the triumph of the academy.

Inspired ultimately by Vouet's *Satyrs* and the Roman intellectual milieu that gave it birth, the *Horti Academi* is an immensely rich and representative allegory of the collective ingenuity culled from diverse natures ("tot ingenia, genio dissimilia") promoted in academies by Maurice and his kind. As Pasta eulogized:

> The lessons learned by individuals openly educate all. And so you showed the greatest understanding when you devised the cylindrical mirror as the heroic emblem of your academy, with the inscription *omnis in unum*. For just as characters, shapeless and scattered all over the ground, form a most perfect harmony when they have all come together in one mirror and have been integrated into one whole by what one might call the miraculous power of light, in like manner a similar number of wits, which are unlike one another in nature [*genio*], and once lay in a state of disconnection, achieve an appearance which is striking and perfect in all respects when enlightened by the learned intercourse of this one academy.[83]

---

[81] Tesauro 1671, 195: "Singulae igitur Porticuum apertiones, singularum Artium studiis, discretae quasi officinae, confinguntur: in quibus singuli Artium illarum authores, Auctoresque, operi intendunt suo: unaque omnes satagunt, ut communi velut in Athenaeo Nobiles, Academiae Alumni exerceantur. Singulis autem Fornicibus, congruum cuique Arti Symbolum praefixum est, figura sempervaria, manente semper Totulo eodem: Omnis in unum. Totam denique Porticuum summitatem, pro Acroteriis Trophaea eminenter ornant, ex singularum Artium Doctrinarumque Instrumentis composita."

[82] Tesauro 1671, 197: "Silenus Socratis, extime deformis, intime repletis Gemmis, quae de aperto pectoris spiraculo pellucent[.]"

[83] Tesauro 1671, 194: "omnesque palam erudit, quicquid singuli didicerunt. Argutissime igitur cylindricum speculum pro Heroico Academia Tuae symbolo commentus es, inscripto Titulo, Omnis in unum. Nam quemadmodum informes passimque solo dispersi Characteres, ubi omnes in unum Speculum coiverint, mira quadam luminis virtute integrati, perfectissime conformantur: ita tot Ingenia, genio dissimila, quae singulatim jacerent; eruditio unius Academiae illuminata commercio, absolutam undequaque speciem, conspicuamque consequentur."

Fig. 10.18 Leonard Heckenauer, illustration in Giovanni Palazzo, *Aquila Romana, overo, Monarchia occidentale* (1679), copperplate engraving

This notion of diverse wits of varying natures, brought together by a learned Maecenas to form a collective—and thus perfected—ingenuity is connected to the subject of a final variant of Vouet's *Satyrs*: an engraving by Leonhard Heckenauer (1655–1704), made to illustrate the life of Conrad IV of Germany (1228–1254) in Giovanni Palazzo's history of Western princes, *Aquila Romana, overo, Monarchia occidentale* (1679) (Fig. 10.18). In this modification of the Tröschel engraving, the satyrs (reduced in number from eight to five) gaze at an impossible anamorphosis. The distorted elephant has been replaced by the *disiecta membra* of a leopard, whose scattered parts are re-assembled in a cylindrical mirror surmounted by the bust of a man crowned with a laurel wreath. In this instance, however, the meaning is decidedly ambivalent. In the biography, Palazzo compares Conrad—renowned for his tyranny—to a "leopard, enclosed in a cylinder, whose nature it is to join together things that are scattered." Meanwhile, the satyrs, "who leap around it, are the peoples [of the Holy Roman Empire], who were transformed into satyrs by their leader's bad example."[84] Palazzo—who had likely read Tesauro—has inverted the virtuous sense of both Vouet's design and Maurice's emblem, using the mirror anamorphosis to symbolize a cruel individual as opposed to a heroic assembly.

[84] Palazzo 1679, 102: "un Pardo, e racchiusa in un Cilindro, natura del quale è, unire le cose disperse. Sotto figura de Satiri, che d'intorno faltellano, sono i popoli tramutati in Satiri di costumi dall'esempio del comandamte." I am grateful to Judit Sebö for having brought this example to my attention, and for having shared with me her unpublished work on the Heckenauer engraving.

In tracing the fortunes of Vouet's *Satyrs* and its afterlives, we have witnessed how—in the right patronal and pedagogical environment—an encounter with the material culture of philosophy could prompt a witty and exuberant flourishing of images. Thrust into the world of thesis prints and the discourse of the Italian academies, the cylindrical mirror and catoptrical anamorphosis "invented" in Paris sparked a whole range of ingenious conceits, whereby mathematical optics was deployed on behalf of learned sociability and the promotion of right religion, as well as being used—as in our final example—to characterize an individual nature. Although in many ways a singular iconography, its effect was anything but: the catoptrical anamorphosis both reflected and fostered the collective ingenuity that characterized philosophical pursuits and visual creativity in the courtly milieu of early modern Italy.

Bibliography

Bacou, Roseline, and Jacob Bean. 1988. *Le dessin à Rome au XVII siècle*. Paris: Editions de la Réunion des Musées Nationaux.

Baltrušaitis, Jurgis. 1977. *Anamorphic Art*. Cambridge: Chadwyck-Healy.

Beaujean, Dieter, and Simon Turner. 2020. *Hollstein's German Engravings, Etchings and Woodcuts 1400–1700, vol. XCVI: Hans and Peter Troschel*. Ouderkerk aan den IJssel: Sound & Vision Publishers.

Bellini, Eraldo. 2000. *Agostino Mascardi tra "ars poetica" e "ars historica"*. Milan: Vita e Pensiero.

Berger, Susanna. 2014. The Invention of Wisdom in Jean Chéron's Illustrated Thesis Print. *Intellectual History Review* 24-3: 343–366.

———. 2017. *The Art of Philosophy: Visual Thinking in Europe from the Late Renaissance to the Early Enlightenment*. Princeton: Princeton University Press.

Bissell, Raymond Ward. 2011. Simon Vouet, Raphael, and the Accademia di San Luca in Rome. *Artibus et Historiae* 63: 55–72.

Bonfait, Olivier. 2011. Vouet et le milieu des lettrés autour des Barberini. In *Simon Vouet en Italie*, ed. Olivier Bonfait and Hélène Rousteau-Chambon. Rennes: Presses Universitaires de Rennes.

Bonfait, Olivier, and Hélène Rousteau-Chambon, eds. 2011. *Simon Vouet en Italie*. Rennes: Presses Universitaires de Rennes.

Bouzy, Christian. 2001. El Tesoro de la Lengua: un cas d'éléphantiasis emblématique de la citation. *Cahiers d'études romanes* 5: 49–71.

Brejon de Lavergnée, Barbara. 1980. Portraits des poètes italiens par Simon Vouet et Claude Mellan à Rome. *Revue de l'art* 50: 51–57.

———. 1990. Simon Vouet et le dessin. In *Vouet*, ed. Jacques Thuillier, Barbara Brejon de Lavergnée, and Denis Lavalle. Paris: Réunion des musées nationaux.

———. 2008. Vouet dessinateur en Italie. In *Simon Vouet (les années italiennes 1613/1627)*, ed. Stéphanie Méséguer. Paris: Hazan.

Carboneri, Nino, and Andreina Griseri. 1966. *Giovenale Boetto*. Fossano: Risparmio.

Ceccarelli, Giuseppe. 1946. *I Sacchetti*. Rome: Istituto di Studi Romani.

Chastel, André. 1996. *French Art: The Ancien Régime, 1620–1775*. Paris: Flammarion.

Cordellier, Dominique, Pierre Rosenberg, and Peter Märker, eds. 2007. *Dessins Français du Musée de Darmstadt, XVIe, XVIIe et XVIIIe Siècles: Hessisches Landesmuseum Darmstadt Graphische Sammlung*. Montreuil: Gourcuff Gradenigo.

Crelly, William R. 1962. *The Paintings of Simon Vouet*. New Haven: Yale University Press.

De Marchi, Rebecca, and Stéphane Garnero. 1997. *La Vigna del Cardinal Maurizio: Il Racconto di Villa della Regina*. Turin: Beppe Grande.
De Rosa, Agostino, ed. 2013. *Jean François Niceron. Prospettiva, cattotrica e magia artificiale*. Rome: Aracne.
De Rosa, Agostino, and Alessio Bortot. 2019. Anamorphosis: Between Perspective and Catoptrics. In *Handbook of the Mathematics of the Arts and Sciences*, ed. Bharath Sriraman, 1–47. Dordrecht: Springer.
di Barme, Celia Arnaldi, and Franca Varallo. 2009. *Feste barocche: Cerimonie e spettacoli alla corte dei Savoia tra Cinque e Settecento*. Cinisello Balsamo: Silvana.
di Macco, Michela. 1995. 'L'ornamento del Principe'. Cultura figurativo di Muarizio di Savoia (1619-1627). In *Le Collezioni di Carlo Emmanuele I di Savoia*, ed. Giovanni Romano, 349–374. Turin: Fondazione CRT.
di Macco, Michela, and Giovanni Romano, eds. 1989. *Diana trionfatrice: Arte di corte nel Piemonte del Seicento*. Turin: Umberto Allemandi.
Dimova, Temenuzhka. 2015. Une anamorphose révélée par le langage des doigts chez Simon Vouet. *Interstudia* 17: 215–224.
Dupré, Sven. 2008. Images in the Air: Optical Games, Magic, and Imagination. In *Spirits Unseen: The Representation of Subtle Bodies in Early Modern European Culture*, ed. Christine Göttler and Wolgang Neuber, 71–90. Leiden/Boston: Brill.
Failla, Maria Beatrice, and Clara Goria. 2003. *Committenti d'età barocca: le collezioni del principe Emanuele Filiberto di Savoia a Palermo e la decorazione di Palazzo Taffini d'Acceglio a Savigliano*. Turin: Umberto Allemandi.
Feingold, Mordechai, ed. 2003. *The New Science and Jesuit Science: Seventeenth Century Perspectives*. Dordrecht: Springer.
Ficacci, Luigi, ed. 1989. *Claude Mellan, gli anni romani: un incisore tra Vouet et Bernini*. Rome: Multigrafica.
Fischer, Hubertus, Volker R. Remmert, and Joachim Wolschke-Buhlman, eds. 2016. *Gardens, Knowledge, and the Sciences in the Early Modern Period*. Dordrecht: Springer.
Gal, Ofer, and Raz Chen-Morris. 2013a. *Baroque Science*. Chicago: University of Chicago Press.
———, eds. 2013b. *Science in the Age of Baroque*. Dordrecht: Springer.
Garrod, Raphaële. 2021. Introduction: Descartes Re-imagined: Ingenuity Before and Beyond Dualism. In *Descartes and the Ingenium: The Embodied Soul in Cartesianism*, ed. Raphaële Garrod with Alexander Marr. Leiden/Boston: Brill.
Gauna, Chiara. 2014. L' 'iconomantia' d'Emanuele Tesauro: Paroles et images à la coeur de Savoie. *Dix-septiéme siècle* 262-1: 125–138.
Gensini, Stefano, and Arturo Martone, eds. 2002. *Ingenium propria hominis natura*. Naples: Liguori.
Giovanetti, Marcello. 1630. Discorso dello specchio. In *Saggi accademici*, ed. Agostino Mascardi, 107–113. Venice: Bartolomeo Fontania.
Gómez López, Susana. 2016. The Encounter of the Emblematic Tradition with Optics: The Anamorphic Elephant of Simon Vouet. *Nuncius* 31-2: 288–331.
Hänsli, Thomas. 2008. 'Omnis in unum': Inganno, Argutezza, und Ingegno als kunsttheoretische Kategorien bei Emanuele Tesauro und Andrea Pozzo. In *Wissenformen*, ed. Werner Oechslin. Zurich: Stiftung Bibliothek Wener Oechslin.
Hédelin, François. 1627. *Des satyres brutes, monstres et demons: De leur nature et adoration contre l'opinion de ceux qui ont estimé les satyres estre une espece d'hommes distincts & separez des adamicques*. Paris: Nicolas Buon.
Kaufmann, Lynn F. 1984. *The Noble Savage: Satyrs and Satyr Families in Renaissance Art*. Ann Arbor: UMI Research Press.
Kolrud, Kristine. 2015. The Gem and the Mirror of Heroic Virtue: Emanuele Tesauro and the Heroic at the Court of Savoy. In *Shaping Heroic Virtue: Studies in the Art and Politics of Supereminence in Europe and Scandinavia*, ed. Stefano Fogelberg Rota and Andreas Hellerstedt. Leiden: Brill.

La Blanchardière, Noëlle. 1972. Simon Vouet: Prince de l'Académie de Saint-Luc. *Bulletin de la Société de l'histoire de l'art français* 1972: 79–93.

Loire, Stéphane, ed. 1992. *Simon Vouet: actes du colloque international, Galeries nationales du Grand Palais 5-6-7 février 1991*. Paris: La Documentation Française.

Maggi, Armando. 2005. The World's Self-Portrait in Blood. The Shroud of Turin as Ecstatic Mirror in Emanuele Tesauro's Baroque Sacred Panegyrics. *The Journal of Religion* 85-4: 582–608.

Marr, Alexander. 2006. Introduction. In *Curiosity and Wonder from the Renaissance to the Enlightenment*, ed. R.J.W. Evans and Alexander Marr, 1–20. Aldershot: Ashgate.

———. 2010. The Flemish 'Pictures of Collections' Genre: An Overview. *Intellectual History Review* 20–1: 5–25.

———. 2011. *Between Raphael and Galileo: Mutio Oddi and the Mathematical Culture of Late Renaissance Italy*. Chicago: University of Chicago Press.

———. 2015. Crowned with Harmless Fire. A New Look at Descartes. *The Times Literary Supplement*, March 13, 14–15.

———. 2021. The Face of *Ingenium*: Simon Vouet's Portrait of Descartes. In *Descartes and the Ingenium: The Embodied Soul in Cartesianism*, ed. Raphaële Garrod with Alexander Marr, 204–216. Leiden/Boston: Brill.

Marr, Alexander and Vera Keller ed. 2014. *Early Modern Invention*. Special Issue of *Intellectual History Review*, 24–3.

Marr, Alexander, Raphaële Garrod, José Ramòn Marcaída, and Richard J. Oosterhoff. 2018. *Logodaedalus: Word Histories of Ingenuity in Early Modern Europe*. Pittsburgh: Pittsburg University Press.

Massey, Lyle. 2007. *Picturing Space, Displacing Bodies. Anamorphosis in Early Modern Theories of Perspective*. University Park: Pennsylvania State University Press.

Merolla, Riccardo. 1995. L'Accademia dei Desiosi. *Roma moderna e contemporanea* 3-1: 121–155.

Mersenne, Marin. 1627. *Traité de l'harmonie universelle*. Paris: Guillaume Baudry.

Méséguer, Stéphanie, ed. 2008. *Simon Vouet: les années italiennes 1613/1627*. Paris: Hazan.

Meyer, Véronique. 1990. Les frontispices de thèses: Un exemple de collaboration entre peintres italiens et graveurs français. In *Seicento. La peinture italienne du xvii*[e] *siècle et la France*, 105–123. Paris: La Documentation française.

Mirollo, James V. 1963. *The Poet of the Marvelous: Giambattista Marino*. New York: Columbia University Press.

Mossetti, Cristina. 2007. *La Villa della Regina*. Turin: Umberto Allemandi.

Niceron, Jean François. 1638. *La perspective curieuse*. Paris: Pierre Billaine.

Oberli, Matthias. 1999. *"Magnificentia principis": Das Mäzenatentum des Prinzen und Kardinals Maurizio von Savoyen (1593–1657)*. Weimar: VDG.

Oresko, Robert. 1997. The House of Savoy in Search of a Royal Crown in the Seventeenth Century. In Royal and Republican Sovereignty in Early Modern Europe: Essays in Memory of Ragnhild Hatton, ed. R. Oresko, G. C. Gibbs and H. M. Scott, 272–350. Cambridge: Cambridge University Press.

Orsi, Laura. 1994. The Emblematic Elephant: A Preliminary Approach to the Elephant in Renaissance Thought and Art. *Anthropozoologica* 20: 69–86.

Osborne, Toby. 2007. *Dynasty and Diplomacy in the Court of Savoy: Political Culture and the Thirty Years' War*. Cambridge: Cambridge University Press.

Palazzo, Giovanni. 1679. *Aquila Romana, ouero, Monarchia occidentale*. Venice: The author.

Pampalone, Antonella. 2015. *Cerimonie di laurea nella Roma barocca. Pietro da Cortona e i frontespizi ermetici di tesi*. Rome: Gangemi Editore.

Panofsky, Erwin. 1956. Galileo as a Critic of the Arts: Aesthetic Attitude and Scientific Thought. *Isis* 47-1: 3–15.

Préaud, Maxime. 1992. Simon Vouet, Claude Mellan et le cardinal Maurice de Savoie. In *Simon Vouet: actes du colloque international, Galeries nationales du Grand Palais 5-6-7 février 1991*, ed. Stéphane Loire, 557–562. Paris: La Documentation française.

Randi, Luigi. 1901. *Il Principe Cardinale Maurizio di Savoia*. Florence: Salesiana.

Remmert, Volker R. 2011. *Picturing the Scientific Revolution: Title Engravings in Early Modern Scientific Publications*. Philadelphia: Saint Joseph's University Press.

Renn, Jürgen, ed. 2002. *Galileo in Context*. Cambridge: Cambridge University Press.

Ribouillault, Denis. 2018. Hortus Academicus: Les academies de la renaissance et les jardins. In *Des jardins et des livres*, ed. Michael Jakob, 23–34. Geneva: Mètis Press.

Rice, Louise. 1999. Jesuit Thesis Prints and the Festive Academic Defense at the Collegio Romano. In *The Jesuits: Cultures, Sciences, and the Arts, 1540–1773*, ed. John W. O'Malley, Gauvin Alexander Bailey, Steven J. Harris, and T. Frank Kennedy, 148–169. Toronto: University of Toronto Press.

———. 2007. *Pomis sua nomina servant*: The Emblematic Thesis Prints of the Roman Seminary. *Journal of the Warburg and Courtauld Institutes* 70: 195–246.

———. 2009. Simon Vouet's Hesperus and the Mythopoetics of Praise. In *Dialogues in Art History, from Mesopotamian to Modern: Readings for a New Century*, ed. Elizabeth Cropper, 236–251. New Haven/London: National Gallery of Art.

———. 2017. Poussin's Elephant. *Renaissance Quarterly* 70–2: 548–593.

Rima, Beatrice. 1991. *Lo specchio e il suo enigma: vita di un tema intorno di Tasso e Marino*. Padua: Antenore.

Ripa, Cesare. 1625. *Iconologia*. Padua: Pietro Paolo Tozzi.

Rodari, Florian. 2015. *Claude Mellan, l'écriture de la méthode*. Milan: 5 Continents.

Roling, Bernd. 2010. *Drachen und Sirenen: Die Rationalisierung und Abwicklung der Mythologie an den europäischen Universitäten*. Leiden/Boston: Brill.

Ronzani, Rocco. 2013. Dell'impresa dell'elefante. L'arme del cardinale Petrocchini nell'erudizione dei secoli XVI-XVII. In *Il Cardinale Montelpare. Atti del convegno*, ed. AAVV, 65–97. Teramo: Archivio Diocesano di San Benedetto del Tronto.

Sannazaro, Jacopo. 2009. *Latin Poetry*. Trans. Michael C.J. Putnam. Cambridge, MA: Harvard University Press.

Schleier, Eric. 2008. Les commanditaires de Vouet à Rome. In *Simon Vouet (les années italiennes 1613/1627)*, ed. Stéphanie Méséguer, 67–80. Paris: Hazan.

Schnapper, Antoine. 1988. *Le géant, la licorne, la tulipe: Collections françaises au XVIIe siècle*. Paris: Flammarion.

———. 1994. *Curieux du grand siècle. Collections et collectionneurs dans la France du XVIIe siècle*. Paris: Flammarion.

Snyder, Jon R. 2017. Strokes of Wit: Theorizing Beauty in Baroque Italy. In *Aesthetic Philosophy After Early Modernity*, ed. Paul A. Kottman, 194–226. New York: Fordham University.

Solinas, Francesco. 1992. Ferrante Carlo, Simon Vouet et Cassiano dal Pozzo. Notes et documents inédits su la période romaine. In *Simon Vouet: actes du colloque international, Galeries nationales du Grand Palais 5–6-7 février 1991*, ed. Stéphane Loire, 135–147. Paris: La Documentation française.

Sprat, Thomas. 1722 [1677]. *The History of the Royal Society of London*. 3rd ed. London: Samuel Chapman.

Tesauro, Emanuele. 1654. *Il Cannocchiale aristotelico*. Turin: Giovanni Sinibaldo.

———. 1659. *Panegyrici*. Turin: Bartolomeo Zavatta.

———. 1670. *Il Cannocchiale aristotelico*. Turin: Bartolomeo Zavatta.

———. 1671. *Inscriptiones*. Cologne: Georg Schultz.

Thuillier, Jacques, Barbara Brejon de Lavergnée, and Denis Lavalle, eds. 1990. *Vouet*. Paris: Réunion des musées nationaux.

Vagnoni, Deborah. 2008. Motivi iconologici e immagini simboliche nel Diario dell'Accademia dei Desiosi di Maurizio di Savoia. In *L'Accademia dei Desiosi. Storia e testo*, ed. Riccardo Merolla, 130–151. Rome: Carocci editore.

Vaulezard, Jean-Louis. 1630. *Perspective cylindrique et conique*. Paris: Julian Jacquin.

Vidal, Fernando. 2011. *The Sciences of the Soul: The Early Modern Origins of Psychology*. Chicago: University of Chicago Press.

Wazbinski, Zygmunt. 1992. Simon Vouet et le cardinal Francesco del Monte: une hypthèse sur le mécénat d'Henri IV à Rome. In *Simon Vouet: actes du colloque international, Galeries nationales du Grand Palais 5-6-7 février 1991*, ed. Stéphane Loire, 159–167. Paris: La Documentation française.

Zirpolo, Lilian H., ed. 2005. *Ave Papa/Ave Papabile: The Sacchetti Family, Their Art Patronage, and Political Aspirations*. Toronto: Center for Reformation and Renaissance Studies.

Index

S. Berger, D. Garber (eds.), *Teaching Philosophy in Early Modern Europe*, Archimedes 61, https://doi.org/10.1007/978-3-030-84621-3

D

G

H